GCSE

Shaun Procter-Green
Paul Winters

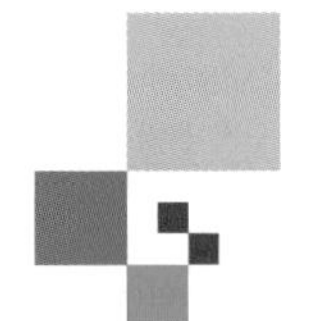

Nelson Thornes

Published in 2009 by:
Nelson Thornes Ltd
Delta Place
27 Bath Road
CHELTENHAM
GL53 7TH
United Kingdom

09 10 11 12 13 / 10 9 8 7 6 5 4 3 2 1

A catalogue record for this book is available from the British Library

ISBN 978 1 4085 0261 7

Cover photograph by Getty/Steve Murez
Illustrations by Roger Penwill
Page make-up by Hart McLeod

Printed and bound in Spain by GraphyCems

Photo acknowledgements
Alamy: eStock Photo / p.11; Ian Francis / p.17; JUPITERIMAGES/ Thinkstock / p.20 (2); Adrian Sherratt / p.37; Jeremy Hoare / p.48; Alex Segre / p.50; Chris Howes/Wild Places Photography / p.53; charcrit boonsom / p 55; Sally and Richard Greenhill / p.56; Colin Underhill / p.66; DBURKE / p.80; Sally and Richard Greenhill / p.212; David Kilpatrick / p.244; Doug Houghton / p.247; **Fotolia**: Bojan Pavlukovic / Chapter 3 banner; V. Yakobchuk / Chapter 7 banner; **iStockphoto**: Chapter 1 banner; p.14; p.20 (1); Chapter 2 banner; p.35; p.51; Chapter 4 banner; Chapter 5 banner; p.100; p.108; Chapter 6 banner; p.116; p.122; p.142; Chapter 8 banner; p.163; Chapter 9 banner; Chapter 10 banner; p.194; Chapter 11 banner; Chapter 12 banner; p.226; **PA Photos**: PA Archive / p.77; Koji Sasahara / p. 159; PA Archive / p.207; PA Wire / Chapter 13 banner; **Science Photo Library**: Cordelia Molloy / p.19.

Text acknowledgements
Page 53: British National Parks Office for National Statistics www.statistics.gov.uk © Crown Copyright reprinted under Crown Copyright PSI License C2008002303; page 54: Information relating to fires www.statistics.gov.uk © Crown Copyright reprinted under Crown Copyright PSI License C2008002303; page 54, Table C: Social Trends 2007 Office for National Statistics www.statistics.gov.uk © Crown Copyright reprinted under Crown Copyright PSI License C2008002303; page 55: Tides at Gorleston reprinted with kind permission from the Broads Authority www.broads-authority.gov.uk/boating/navigating/tide-tables.html; pages 55-6: Students in Higher Education statistics reprinted with permission from the Higher Education Statistics Agency, www.hesa.ac.uk; page 72: UK Population by age, Office of National Statistics www.statistics.gov.uk © Crown Copyright reprinted under Crown Copyright PSI License C2008002303; page 82: Population pyramid for UK in year 2000 Population pyramid for UK in year 2000; page 83: Population pyramid for Isle of Man 2009 www.census.gov/cgibin/ipc/idbpyrs.pl?cty=IM&out=s&ymax=300&submit=Submit+Query U.S. Census Bureau, International Data Base; page 122: Table of US lightning fatalities NOAA's National Weather Service www.weather.gov; page 172: Table of UK females not in full time education National Statistics website: www.statistics.gov.uk © Crown Copyright reprinted under Crown Copyright PSI License C2008002303; page 176: North Sea fish stocks The Centre for Environment, Fisheries and Aquaculture Science (CEFAS), International Council for the Exploration of the Sea (ICES), Department for Environment, Food and Rural Affairs (Defra)© Crown Copyright reprinted under Crown Copyright PSI License C2008002303; page 176: Table of post 16 pupil numbers in Devon www.devon.gov.uk/index/learning/educationstatistics reprinted with kind permission from Devon County Council www.devon.gov.uk; page 180: Z-chart unit sales, with kind permission from http://syque.com; page 182: Statistics for number of suicides in UK © Crown Copyright reprinted free under Crown Copyright PSI License C2008002303; page 188: "Endangered Birds" as found at www.teachervision.fen.com/tv/printables/cc/ma6-pr1-2.pdf, as it appeared on © 2000-2009 Pearson Education, Inc. All rights reserved; page 191: Number of jobs in the US 1990-1992 www.teachervision.fen.com/tv/printables/cc/ma6-pr1-2.pdf, © 2000-2009 Pearson Education, Inc. All rights reserved; page 191: Shrinking Family Doctor cartoon (Los Angeles Times 5 August 1979) reprinted with permission from the Los Angeles Times; page 192: Title of OECD table, OECD Health Data 2008: Statistics and Indicators for 30 Countries, OECD 2008, www.oecd.org/health/healthdata; page 193: Choropleth maps: population density for same area of England, © Crown Copyright reprinted under Crown Copyright PSI License C2008002303; page 195: Choropleth map: United States Permission is granted to distribute this document under the terms of the GNU Free Documentation License; page 198: Choropleth map: population density for countries of the world, England Permission is granted to distribute this document under the terms of the GNU Free Documentation License; page 208: Monthly RPI data for 2007/2008 National Statistics website: www.statistics.gov.uk © Crown Copyright reprinted under Crown Copyright PSI License C2008002303; page 210: Cost of grocery items over last 100 years question © Crown Copyright reprinted under Crown Copyright PSI License C2008002303; page 211: Allocation of items to CPI divisions in 2005 question © Crown Copyright reprinted under Crown Copyright PSI License C2008002303.

The publishers have made every effort to contact copyright holders but apologise if any have been overlooked.

Contents

Nelson Thornes and AQA

Nelson Thornes has worked in partnership with AQA to ensure this book and the accompanying online resources offer you the best support for your GCSE course.

All resources have been approved by senior AQA examiners so you can feel assured that they closely match the specification for this subject and provide you with everything you need to prepare successfully for your exams.

These print and online resources together **unlock blended learning**; this means that the links between the assessment questions in the book and the revision activities online blend together to maximise your understanding of a topic and help you achieve your potential.

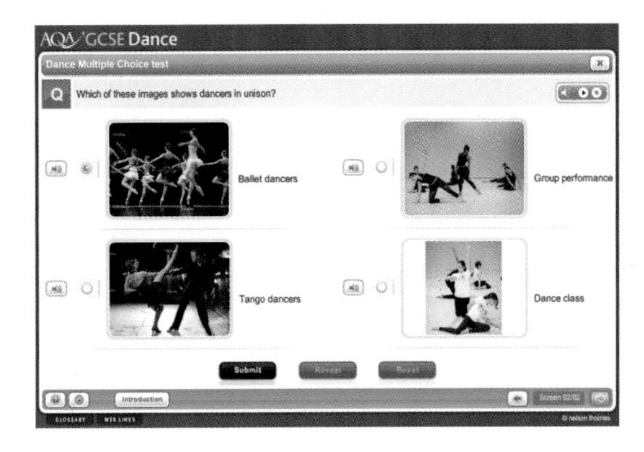

These online resources are available on kerboodle! which can be accessed via the internet at **www.kerboodle.com/live**, anytime, anywhere. If your school or college subscribes to kerboodle! you will be provided with your own personal login details. Once logged in, access your course and locate the required activity.

For more information and help on how to use kerboodle! visit **www.kerboodle.com**

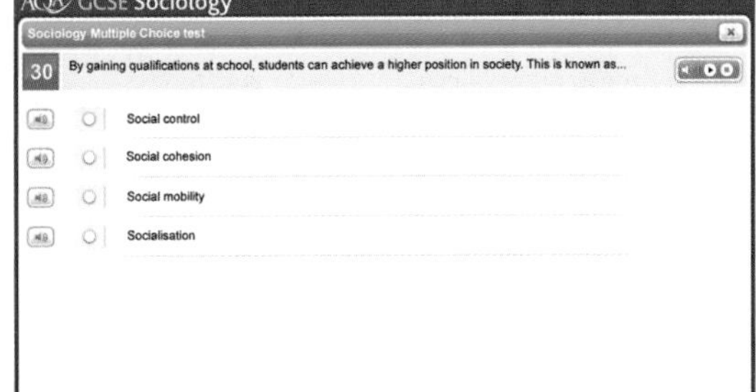

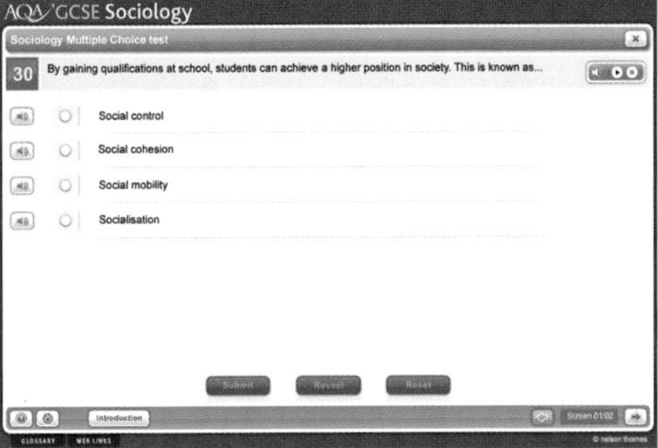

How to use this book

Objectives

Look for the list of **Learning Objectives** based on the requirements of this course so you can ensure you are covering everything you need to know for the exam.

AQA Examiner's tip

Don't forget to read the **AQA Examiner's Tips** throughout the book as well as practise answering **Examination-style Questions**.

Visit **www.nelsonthornes.com/aqagcse** for more information.

Introduction

For students who will use this book

This book is all about the wonderful world of statistics and how you can learn to understand and participate in this world.

If you look around, statistics are everywhere. Newspapers, the internet, on television and all around at school or college – you can't escape them. Can you trust every single fact and figure that is put forward as a statistical truth?

Doing well in GCSE statistics will show that you are well equipped to work your way through the labyrinth of numbers, statistics and statements that you will come across in everyday life.

In the 21st century it could easily be argued that after the need to be able to read, write and have basic numeracy there will be no skill that you will need more than the ability to understand the data that you will confront.

For teachers who will use this book

AQA provides the GCSE statistics course – the best available modern course for statistics. It has an updated specification to take account of current usage of statistics and of the need to understand and interpret (not just the ability to calculate measures and draw graphs).

Key to this modern approach to the subject is the **controlled assessment**. This is not just a rehash of coursework by another name. It provides students with a focussed short exercise in investigational statistics, with an associated written assessment to help students learn the underdeveloped skill of evaluation. AQA will provide both the theme for the controlled assessment and the required data, or places to find the data.

In this book there is a chapter devoted to the controlled assessment, as well as a chance to practise a full specimen version of one. Throughout the chapters there are many activities and features that will also benefit students in their preparation for this controlled assessment (see 'How to use this book'). The final examination for the new specification will gradually focus to a greater extent on interpretation rather than calculation and, wherever possible, there will be an increased use of real data.

The controlled assessment is worth 25 per cent of the final mark, with the investigation and written assessment equally weighted at 12.5 per cent each. Each year there will be two controlled assessment themes available, so students being taught over two years will have up to four opportunities to do as well as possible in the controlled assessment.

The final written exam is worth 75 per cent. It is two hours long at higher tier, and 1½ hours long at foundation tier.

In these pages you will have everything you need in order to run a successful course for GCSE statistics, so that your students can achieve grades that reflect their level of ability.

The book has been written by the senior examining team at AQA for this specification.

How to use this book

This book is designed to be as easy to use as possible, with the specification organised into a series of chapters that follow a possible teaching order through the handling data cycle and finishing with probability. Each chapter has been designed to introduce the material as clearly as possible with:

- learning objectives for the chapter, which are also listed as objectives for each topic
- prior knowledge that would be useful when starting the work for a chapter
- a clear division of work between the foundation and higher tiers
- a new feature that highlights the statistics-only content of the chapter.

Statistics-only content

This new feature has been added for the first time in a book of this type. In recognition that sometimes teaching time for the delivery of the course can be limited, a unique feature of this book, called 'focus on statistics' shows in each chapter's introduction the content required in GCSE statistics that is additional to that needed in GCSE mathematics. It has been possible to provide this feature because the authors are also senior examiners for GCSE mathematics. In the 'focus on statistics' feature, both the difference in material and the way the same material may be tested differently are featured.

Foundation and higher tier

In each chapter, the material is divided into topics based upon a particular theme and followed by a supporting exercise clearly divided into questions that are foundation or higher tier. Every question in this book has been written by the same people who will write the questions on the examination paper! Fully worked solutions to exercises are given in the answer section at the end of the book.

Higher tier material is highlighted on the page by a vertical rule that runs alongside the higher tier content to the left of the text. Higher tier learning objectives and higher tier questions are also highlighted in blue to distinguish them from foundation tier content.

Remember that higher tier students should know everything at foundation tier, as well as the material for higher tier, as anything from the foundation tier could appear as a common examination question.

Supporting features

Each topic is introduced clearly with full worked examples and a suite of other supporting features. **Key terms** are highlighted in the text, and a definition is given in the margin, as well as listing the key term in the glossary at the back of the book. The other supporting features fall under the following headings:

Be a statistician

This feature will enable you to put yourself into the shoes of a statistician. The feature will usually involve the need to collect or use some data in order to solve a practical problem, or to go through the steps required to address an issue with statistics. This feature can provide good practice at going through the stages required for the investigation part of the controlled assessment.

Spot the mistake

This is a very important aspect of understanding statistics. You will be presented with someone's work or ideas, and will be asked to spot the error(s) in their work. This is ideal practice for getting used to the often error-ridden presentation of statistics in the media.

Real world

One of the features of the new specification is the use of real-world statistics, where possible, in examination questions. In this feature you will be looking at real data or collecting real data of the type you might come across in your everyday life. Collection and use of real-world statistics could again be very useful in preparing students for the part of the written assessment for the controlled assessment, which tests understanding of the implications and issues surrounding the use of primary data.

links

There are many connections between the material found in different chapters of this book. Sometimes, where it would be especially beneficial to look at the related work in some material, the link to that related work will be given. Occasionally, this feature may link work in the book to a website, or other source of related data.

AQA Examiner's tip

It is important that you know how some of the issues that arise in statistics are actually examined in the examination papers. This feature will let you know how you should deal with issues in exam questions. It explains what you must do to get the marks and what you must avoid to prevent the frustrating loss of marks when you actually knew what to do.

Summary

All chapters include a bullet list summary of the key points covered in the chapter.

AQA Examination-style questions

Each chapter ends with a large set of past AQA examination questions, so that you can see how the issues within the chapter have been assessed in the past. Answers to these questions are provided online.

Controlled assessment

The controlled assessment tasks in this book are designed to help you prepare for the tasks your teacher will give you. The tasks in this book are not designed to test you formally and you cannot use them as your own controlled assessment tasks for AQA. Your teacher will not be able to give you as much help with your tasks for AQA as we have given with tasks in this book.

1 Data collection

In this chapter you will learn:

- to understand and use the data-handling cycle
- how to specify a hypothesis from a given situation
- how to make decisions about what data is needed to investigate a hypothesis
- to consider reasons for choosing a particular hypothesis
- to tell the difference between primary data and secondary data
- to identify different types of variables
- to classify data
- to identify a population
- to explain the difference between a census and a sample
- how to use pilot surveys and why they are used
- to write questionnaires
- to identify problems with questionnaires and how to deal with them
- to design opinion scales
- to explain how data is collected on sensitive issues
- how to design data collection sheets
- about different methods of data collection, including data logging
- to design and carry out a basic statistical experiment
- to identify the variables involved in a statistical experiment
- to collect data by observation and to understand that an observer can affect what is seen just by observing
- to appreciate some problems that occur when data is grouped
- about 'matched pairs'
- to identify and control extraneous variables.

You should already know:

- ✔ how to ask questions
- ✔ how to plan work carefully.

What's this chapter all about?

This chapter gets the ball rolling by making you think about what needs to be done when you start a statistical investigation. It's full of questions like, 'What data do you actually need?' and, 'Where should you get it from?' not to mention the questions you ask in a questionnaire when you collect your data. There is quite a bit about different types of data, which is important because the type of data you have determines what you can do with it.

This chapter also covers different ways to get hold of the data you want, such as surveys, observation and experiments – a bit like science really, but more interesting!

Focus on statistics

Some of the topics here are covered in maths but one of the most important bits is the data-handling cycle. You probably came across it in maths but without spending a lot of time on it. In statistics it is crucial!

- In 1.1 you will learn about hypotheses – these will be new to you.
- In 1.2 categorical data is often misunderstood, even by people who should know better – that's not taught in maths.
- In 1.4 you will meet opinion scales and the collection of sensitive data, which are not covered in maths.
- In 1.5 you will encounter the design of statistical experiments – very little of this is in maths, so we love to ask questions on this in exams!

1.1 Planning a strategy

The data-handling cycle

The data-handling cycle underpins almost all the work done in statistics. The diagram below shows the four stages of the data-handling cycle.

Stage 1

Specify the question

The starting point is the research question or area of interest.

This leads to the hypothesis.

Evaluate

Stage 4

Interpret and discuss

This stage involves interpreting the graphs and calculations you have generated, and deciding whether your starting question has been answered. This stage will also tell you whether more data needs to be collected, or whether more questions need to be answered. With evaluation this may then lead back to stage 1 and a refinement of the hypothesis.

Stage 2

Collect the data

To answer the question you need to decide what data to collect, how to collect it and where to collect it from.

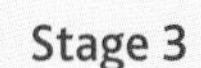

Stage 3

Process and represent the data

You will need to calculate some statistics, such as averages and measures of spread. You will find it useful to represent the data using diagrams.

The data-handling cycle

In any statistical project it is usual to go through this cycle at least once. The first time through could be in a 'pilot' survey, which is then used to decide what further information is needed to answer the research question. If you go through the data-handling cycle again the hypotheses could be refined or new questions could be posed.

Hypotheses

A **hypothesis** is a statement that might or might not be true. It is an unproven statement that can be tested by further investigation. Data can be collected to let you test the hypothesis and hopefully find out whether it is likely to be true. For example, if you are investigating reaction times you might make the assumption that girls are faster than boys or that students who play ball sports are quicker to react than students who do not play ball sports. Each of these is called a hypothesis. If there is more than one hypothesis then the statements are called **hypotheses**.

Objectives

In this topic you will learn:

- to understand and use the data-handling cycle
- how to specify a hypothesis from a given situation
- how to make decisions about what data is needed to investigate a hypothesis
- to consider reasons for choosing a particular hypothesis.

Key terms

Hypothesis: a statement you make that you think may be true.

Hypotheses: plural form of 'hypothesis'.

When collecting data to investigate a particular subject, you will usually start by writing down one or more hypotheses. The data you collect might provide evidence to support your hypothesis (that is, it might not contradict your hypothesis). However, it is also possible that your data might not support your hypothesis (that is, it might suggest your hypothesis is not correct). Data may disprove a hypothesis but it rarely proves that a hypothesis is correct. When data was collected to investigate whether smoking damages a person's health, it took many years of research before enough evidence was collected for this to be conclusive.

AQA Examiner's tip

Remember that a hypothesis is a statement – not a question.

Planning an investigation

When you have decided on an appropriate hypothesis the next step is to decide what data should be collected. Clearly, the data needs to be useful (this is not always as easy as it sounds) and you may need to collect more data than may first appear obvious.

For example, consider what data is needed to investigate whether girls have faster reaction times than boys. State the hypothesis to be tested:

Girls have a faster reaction time than boys.

You will need to collect some data about reaction times from both girls and boys. However, before you start collecting any data, you will need to consider which factors might be involved. If you just collect data about reaction times you might be missing something. For example, does age affect reaction times? What about intelligence? Perhaps musicians or people who play video games might have faster reaction times, or perhaps people are faster at different times of day. These, along with other possible issues, will need to be considered before you collect your data. But beware. It is possible to go into so much detail that it takes too long to collect the data and the data becomes difficult to analyse. A balance needs to be found. You also need to be careful that it is actually possible to collect the data you want. Of course, you may not need to collect your own data; someone may have already collected data that you can use.

AQA Examiner's tip

Candidates sometimes run out of space when answering 'wordy' questions in exams, so think about what you are going to write before you start and try to keep to the point when you answer the type of question below.

Example

In your exam you may be asked questions such as:

Rachel is researching, 'Who uses MSN more at home?' Is it her or her brother?

a State a hypothesis that could be used in this case.

b What data should she collect to investigate her hypothesis?

c What other factors could she consider?

d What might Rachel do next?

Solution

An appropriate answer could be:

a Rachel spends more time on MSN than her brother.

b Rachel needs to record how many minutes she spends on MSN each day and how many minutes her brother spends on MSN each day.

c Rachel could record when they each use MSN – the time of day as well as the time for each weekday and for the weekend. If she did this for her brother as well she would be able to see whether there are any patterns in when they use MSN.

d Rachel may decide to look at what else they do in their leisure time, such as watching television or films, playing computer games or even reading books.

Exercise 1.1

1. A company manager wishes to investigate whether her products are more popular than those of a competitor.
   a. Suggest a hypothesis that could be used in this case.
   b. What data would she need to collect to investigate her hypothesis?
   c. What other factors could she consider?

2. Sam wants to find out if his football team is improving.
   a. Suggest a hypothesis that could be used in this case.
   b. What data would he need to collect to investigate his hypothesis?
   c. What other factors could he consider?
   d. What problems might he have in collecting his data?

3. A local authority wishes to investigate recycling in its area.
   a. Suggest a hypothesis that could be used in this case.
   b. What data would they need to collect to investigate their hypothesis?
   c. What other factors could they consider?
   d. What problems may they have in collecting their data?
   e. What could the local authority do next?

4. Mary owns a hair salon in a town. She wishes to find out whether there is sufficient business for her to open a second branch in the same town. What data should she collect before making her decision, and who should she collect it from?

1.2 Data sources

When you obtain your data you need to decide where it will come from. Are you going to collect data yourself, or are you to use data that has already been collected by someone else?

Objectives

In this topic you will learn:

- to tell the difference between primary data and secondary data
- to identify different types of variables
- to classify data.

Key terms

Primary data: data that you collect yourself.

Secondary data: data that someone else has collected.

Primary data

Primary data is data that you collect yourself, or which you are responsible for having collected.

Secondary data

Secondary data is data that someone else has collected.

Many sources of secondary data are readily available, including *The Annual Abstract of Statistics, Social Trends and Regional Trends*. All of these are available for purchase or through libraries, or even for download from the internet. Most atlases contain useful information, as do some newspapers (such as the *Financial Times*) and journals. There are many data sources available on line, including the following websites: **www.statistics.gov.uk** (UK Statistics Authority website), **statistics.defra.gov.uk/esg/**, **www.stats4schools.gov.uk/**, **www.cia.gov** (which has a link to the *World Fact Book*). If you type 'statistics' into a search engine (such as **www.google.com**) there are lots of other interesting sources.

There are advantages and disadvantages to both types of data collection:

	Advantages	Disadvantages
Primary data collection	You know who the data has come from You know where and when the data was collected You know how the data was collected You should know whether the data is biased in any way	It can take a long time to collect data You can only collect data in your location It can be expensive It may not be possible for you to collect all the data you need
Secondary data collection	It is already collected so you can use it quickly There may be much more data than you could collect yourself You can use data from different areas	You do not know who the data has come from You do not know where the data was collected You do not know how the data was collected You do not know whether the data is biased in any way You do not know when the data was collected – it may be out of date

links

See Chapter 2 for more on bias.

You also need to think about the nature of any data that you collect because this will affect what you can do with it once you have collected it.

Types of data

Data is usually either **quantitative** or **qualitative**. Anything that varies and therefore can take values that vary is called a 'variable'. A variable is often denoted by a letter. Just as names of people begin with a capital letter, so the names of variables have a capital letter. For example, the number of advertisements in a commercial break on TV could be called N. The actual values that N can take would then be n_1, n_2, and so on, for as many commercial breaks you can bear to watch!

Quantitative data

Quantitative data is always numerical – it is always a number. The word 'quantitative' is similar to the word 'quantity' (an amount). There are two types of quantitative data: **continuous data** and **discrete data**.

Continuous data

Continuous data can vary on a continuous scale, so it is always a measurement of some kind, for example a distance or time. Continuous data is never exact and has to be rounded to be recorded.

Discrete data

Discrete data can vary only in steps – for example, shoe size, money or the attendance at a football match – and so it is usually counted.

Qualitative data

Qualitative data expresses a quality of something, such as taste, colour, smell, or touch.

Categorical data

Categorical data is data which has been put into groups in some way. For example, the size of an egg (small or large), the grade you get in your statistics GCSE (a letter), the level you got in your SATs in Year 6 (a number), or the class of an orange (class I, II or III). It may appear to be numerical sometimes, but must not be treated in the same way as a number.

Example 1

A shop's manager keeps a daily record of his customers.

State whether each of the following variables is qualitative, discrete or continuous.

- **a** The number of customers entering the shop.
- **b** The age of the customers served.

Solution

- **a** Discrete. (Remember that this is counted.)
- **b** Continuous. (Age is measured and it changes all the time. You are slightly older now than you were when you started reading this sentence!)

Key terms

Quantitative data: data that is a quantity and is therefore numerical.

Qualitative data: non-numerical data that describes a situation.

Continuous data: data that is measured on a continuous scale.

Discrete data: data that takes separate values.

Categorical data: data that is grouped into categories.

links

It is very important to know what type of data you are collecting as it will affect what you can do with it later – see Chapters 4, 5, 6 and 7.

That's the nth commercial break this evening…

AQA Examiner's tip

Every time questions like this are asked in an exam there are a number of candidates who give an answer such as 'three people entered the shop'. So remember to read what the question is actually asking!

Example 2

Larry collects stamps. He keeps records of his stamps.

The records he keeps are:

- the face value
- the area of each stamp
- the number of perforations.

Give an example of

a a discrete variable that Larry records

b a continuous variable that Larry records.

Solution

a Face value or the number of perforations (as either is counted).

b The area of each stamp (area is measured).

In the first example that we looked at in 1.1, Rachel was investigating the time she spent on MSN. She would need to collect primary data. If she were to compare how much time other people spent on MSN or the internet, it would be very time consuming for her to time her brother as well. If she wanted to extend this to other people it would not be possible to go and time them, so she would need to look for a secondary data source. This could be MSN or an internet company but, even then, the data may not be available.

links

See Chapter 4 for information on data presentation.

AQA *Examiner's tip*

A common error is to think that just because a variable is recorded as a whole number it is a discrete variable. Age is usually recorded as the nearest year below your current age but your actual age is continually changing!

Spot the mistake

Herbie said, 'Shoe size must be a continuous variable, because you can get half sizes.'

Why is Herbie not correct?

Exercise 1.2

1 In Question 1 of Exercise 1.1, should the manager use primary data or secondary data? Explain your answer.

2 In Question 2 of Exercise 1.1, should Sam use primary data or secondary data? Explain your answer.

3 In Question 3 of Exercise 1.1, should the local authority use primary data or secondary data? Explain your answer.

4 Jack has been given a learning activity by his teacher. He has cards with descriptions of variables on, which he has to put into a table. He has done the first one. Complete this activity for Jack.

Qualitative	Quantitative		Categorical
	Discrete	Continuous	
	Number of pips in an orange		

Number of pips in an orange | Speed of a car | Weight of an orange | Length of a pencil | Level in a SATs exam | Sweetness of a cup of tea

Colour of a car | Cost of a newspaper | Mark in a SATs exam | Amount of petrol in a car | Number of doors on a car | Size of an egg in an egg box

5 Lucy collects stamps. She records information about each stamp in her collection.
Write down a variable that she could record that is:

a quantitative and continuous
b quantitative and discrete
c qualitative
d categorical.

6 Here are some examples of different types of data connected with playing a game.

A Time spent playing
B Number of players
C Colour of the counters

Which one of these is:

a qualitative data
b discrete data?

1.3 Population, census and sample

Population and sampling

The first thing you need to do before you collect any data is to identify the **population**.

Think about the hypothesis given in 1.1 'Girls have faster reaction times than boys.' The population would be every girl and every boy. A population is 'everyone we are considering', so if your hypothesis was 'girls at school X have faster reaction times than boys at school X' then the population would be every boy and every girl at school X.

The owner of a newsagent's shop might wish to survey her customers about the arrangements of newspapers in her shop. In this case the population would be all the customers of her shop.

A **census** is when data is obtained about every member of the population.

The most well known census is the National Census, which is carried out every 10 years in the UK.

The National Census is a complete population count for the UK taken on a specific date. The 1841 census is considered to be the first modern UK census. The Census can provide the following information: full name, exact age, relationship to head of household, sex, occupation, parish and county of birth, medical disabilities and employment status.

The information provided by the Census is used by many different people for many different purposes. Amongst these are local authorities and the government, who use it for planning schools and hospitals, as the Census data allows changes in the population to be seen accurately.

Clearly it is not realistic to consider the reaction time of every girl and boy in the world for the hypothesis in 1.1, so you could only investigate the reaction times of part of the population. This is called a **sample**.

Objectives

In this topic you will learn:

- to identify a population
- to explain the difference between a census and a sample.

Key terms

Population: everyone that could possibly be considered in the hypothesis.

Census: the data is obtained from every member of the population.

Sample: part of a population from which information is taken.

There are advantages and disadvantages to be considered when deciding whether to use a sample or carry out a census.

	Advantages	Disadvantages
Census	Accurate (as far as is possible) Unbiased Everyone is represented	Takes a long time to do May not actually be possible to carry out May be difficult to ensure that every member of the population is included Very expensive Produces a lot of data
Sample	Quick to do Cheap to do Usually reliable if the sample chosen is representative Less data to analyse	Not guaranteed to be 100% accurate as not everyone is represented May be biased depending on the sample chosen

Real world

How has the real world changed since ...

This research is best done either individually or in small groups.

Type 'National Census' into your search engine when you are connected to the internet.

- Use one of the links to find out about people who have the same name as you in the 'National Archives' website. You will be able to access information from the 1901 census for free. How many people were there with the same name as you? How old were they? What jobs did they have?
- Use a different link from the search page to visit the 'National Statistics Online' website at **www.statistics.gov.uk/cci/nugget.asp?id=395** In 2001 an optional question collected information about ethnicity and religion. Do you see anything surprising about some of the people who have been classified as having no religion?
- Use the links from **www.statistics.gov** to find out other interesting facts about the UK and its population. Can you find out about whether the UK is meeting targets for carbon emissions? How much is the UK spending on health?
- Use the website **www.visionofbritain.org.uk** to find out about the place where you live. Find out how education, poverty, industry and 'life and death' has changed in your area by using the information from the National Census on this website.
- Use the information you have found to prepare a short presentation to report back to your class.

links

Later (in Chapter 2) you will be looking at different methods of taking samples from a population in order to obtain a representative sample.

Example 1

Lastlonger make batteries. They wish to carry out a study to see how long their batteries last.

Should they use a census or a sample? Explain your answer.

Solution

They should use a sample. If they used a census they would have to test every battery, then they would have none left to sell.

Example 2

Fatima is a student at a large comprehensive school. She is investigating whether girls at her school like chocolate more than the boys.

a Identify the population.

b Should Fatima use a census or a sample? Explain your answer.

Solution

a The population is all the boys and girls at Fatima's school.

b Fatima should use a sample. As the school is large it would take too long to collect data from every student at the school.

AQA *Examiner's tip*

A question may ask you to state what the difference is between a census and a sample. Remember to make it clear that a census uses the whole population, and a sample just uses part of the population.

Exercise 1.3

1 The owner of a garage believes that he sells cars which are coloured red quicker than cars of any other colour. What is the population in this case?

2 Brightlight is a company that makes light bulbs. They want to find whether the energy-save light bulbs last longer than their ordinary light bulbs.

a Suggest a hypothesis that could be investigated in this case.

b Identify the population here.

c Should a census or a sample be used by Brightlight? Explain your answer.

3 The school canteen staff in a small primary school wish to investigate whether students prefer healthy options for school dinners.

a Suggest a hypothesis that could be used in this case.

b Identify the population here.

c Should the staff in the canteen use a census or a sample? Explain your answer.

4 Mary says that the house prices in her neighbourhood are more expensive than other parts of England. Explain why she would be well advised to use sample data rather than a census of the house prices in England.

5 Georgia said that a census was the same as a sample survey but involved more people. Explain why Georgia is not correct.

1.4 Surveys

When you carry out a survey to collect data from someone, the person you are asking is called the **respondent**. There are several methods you can choose from in order to collect your data.

Interview

In an interview you will ask the respondent the questions directly. Care must be taken to ensure that questions are asked in the same way every time, otherwise, this could lead to **bias**. Interviews can take a long time to obtain sufficient information, and it can be difficult to get information from more than a small number of locations. One advantage of using interviews is that you can explain the questions. Clearly, if this is done, then it needs to be done in the same way every time.

Face-to-face interviews

These interviews are often carried out in shopping areas – this excludes people who do not go to those particular areas. A disadvantage of this method is that many people do not wish to stop and talk.

Interviews are often used by Mori and other companies that carry out **opinion polls**. By using this method they can ensure that they gain information from the particular sections of the population that they need. In this case the interviewers go to the houses of the respondents.

Telephone interviews

Interviews are sometimes carried out by telephone. For example, some local 'free' newspapers carry out surveys by telephone to ensure that their employees are covering all the areas that they are supposed to cover. The advantage of this method is that it is quick, easy and fairly cheap for a large area. This method has the disadvantage that not all numbers are listed in the telephone directory.

Objectives

In this topic you will learn:

- how to use pilot surveys and why they are used
- how to write questionnaires
- how to identify problems with questionnaires and know how to deal with them
- to design opinion scales
- to explain how data is collected on sensitive issues.

Key terms

Respondent: the person who replies to a questionnaire.

Bias: the name given to sample data that does not fairly represent the population it comes from.

Opinion polls: surveys that gather the opinions of the respondents.

links

See Chapter 2 for more on opinion polls.

Written questionnaire

A **questionnaire** is a list of questions that are given to the respondent.

Internet questionnaires

A questionnaire can be put on the internet for the respondent to complete. This is a very cheap way to ask questions and easily covers a large area, but questionnaires can be time consuming to complete and not everyone has easy access to the internet or likes to give information online, so they could lead to non-representative samples.

Postal questionnaires

Questionnaires are often posted to respondents. This method has the advantage that it covers a large area cheaply. The main disadvantage is that many are just not returned.

Key terms

Questionnaire: a set of questions that may be asked verbally or in writing.

Non-response

One of the main problems with gaining information from surveys is 'non-response'. The main method used to deal with this problem is to offer some kind of incentive, for example, entry into a prize draw, or discount vouchers of some kind. Another method that could be used in some cases would be to use a reminder in some way. This could be to send a follow-up letter, make a telephone call, or to call round and collect the questionnaire. However, this does require information about exactly where and to whom the questionnaires were sent, and is not practical if a very large number of questionnaires were sent out.

Questionnaire design

Questionnaires should always be written in a polite way and should explain their purpose. Respondents should be thanked for completing the questionnaire – even before they do so. It is important to assure respondents that the information they give will be treated confidentially and tell them who will have access to the information they are about to provide.

Questions themselves should:

- be concise (brief and to the point – do not ask long rambling questions)
- be unambiguous (be clear and easily understood by the respondent)
- be easy to answer
- not be 'leading questions' (avoid questions which use the word 'agree', for example, 'surely you agree that . . .')
- not be personal (questions asking about the age, or income of a person are often considered to be personal).

AQA Examiner's tip

Questions in exams often ask you to criticise an example of a question from a questionnaire. You may want to put 'it is biased' but to get the mark you must say why it is biased.

Types of questions

Questions may be either open or closed.

Open questions leave the respondent to write anything they wish as an answer to the question. For example:

'How do you get to school in the morning?'

Answer: ....

This gives the respondent no guidance about what type of response is required or expected and is likely to get a variety of responses, including a response such as 'Well, I go down Hazel Street, then Brow Ridge, then . . .' Responses to open questions are much more difficult to collect together, interpret and show in diagrams.

Closed questions will have a 'response section', which will guide the respondent and usually make the question both easier to understand and easier to answer. For example:

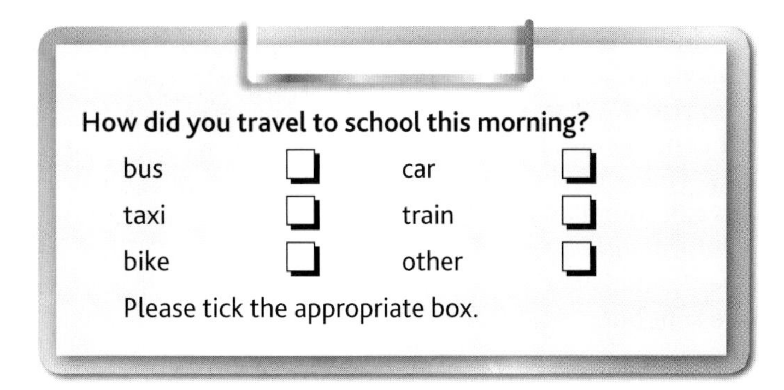

How did you travel to school this morning?

bus	☐	car	☐
taxi	☐	train	☐
bike	☐	other	☐

Please tick the appropriate box.

The response section here gives the respondent clear guidance about what is required to answer the question and covers all possibilities – anyone not using one of the first five responses can tick the 'other' box.

It is very important that response sections cover all possibilities and do not have overlaps.

The question 'How tall are you?' is another example of an open question that could get almost any response, including the following: 'quite tall', 'not very', '1.85 metres', '5 feet 10', 'not sure exactly'. The question itself has a lot of problems: no units are stated, no degree of accuracy is specified and not everyone can be expected to know how tall they actually are, although most people have a good idea. This question would be much better with 'option boxes', such as:

How tall are you?

Less than 1.5 m	☐
1.5 m to 1.9 m	☐
More than 1.9 m	☐
Not sure	☐

Please tick the appropriate box.

AQA *Examiner's tip*

There are no rules about the number of 'option boxes' but in your exam you should give at least three to indicate that you understand about 'overlaps' and 'gaps'.

AQA *Examiner's tip*

Always make sure that response sections do not have 'gaps' or 'overlaps'.

Example 1

Colin carried out a survey to find out how likely people would be to see his local football team play. One of the questions he asked was:

'How much would you be prepared to pay to see Grange United play?'

£5 to £10 ☐

£10 to £12 ☐

more than £13 ☐

a What is wrong with the question and response section?

b Write an improved question and response section that he could use.

Solution

a For the question – they may not like football. The question does not mention whether it is for home or away matches or whether it is for a particular seat (family area/terrace/directors box and so on). There is an overlap in the response section – if someone was prepared to pay £10 they would not know which box to tick. There is also a gap – what if someone was prepared to pay £12.50? Some might not like watching football and there is no box to tick for that. (There is also no instruction to 'tick a box' but this is a small fault compared to the others.)

b He could ask: 'How much would you be prepared to pay for a standard ticket on the terrace to see Grange United play?

Would not go to watch Grange United play ☐

Less than £5 ☐

£5 to £9.99 ☐

£10 to £11.99 ☐

£12 or more ☐

Please tick the appropriate box.

Note that here the classes are of different widths – this is not a problem and may produce more detailed information as a result.

links

See Chapter 3 for more on class widths.

AQA Examiner's tip

In this example, 'may not like football' occurs as a possible fault with both question and response section. In your exam this would only score for one mark. You must find two different faults for two marks.

AQA Examiner's tip

In your exam you might be asked to give criticisms of a question and a response section. If this is the case then the response section is the 'tick box' part and the question would be the part before the tick boxes. Make sure you get them the correct way around.

Pilot surveys

If a questionnaire does not obtain the information required in order to address the hypothesis, then a lot of time and money is wasted, so it is important to ensure that the questions work. This can be done using a **pilot survey**. A pilot survey is a small-scale survey carried out before the main survey. The three usual reasons for carrying out a pilot survey are:

- to decide whether the main survey is worth pursuing
- to test the questions
- to test the data collection method.

Although it is possible to eliminate many sources of bias when composing a questionnaire, it will never be possible to eliminate bias from the respondents totally.

Key terms

Pilot survey: a small-scale survey carried out before the main survey.

Opinion scales

When a survey includes questions to get an opinion on something there are two types of scale that can be used: continuous or discrete.

Continuous opinion scales

These use a continuous scale, for example:

Please indicate how much you agree or disagree with the following statement:

Cricket is exciting

Agree Disagree

Discrete opinion scales

These have separate options to choose from, for example:

Please indicate how much you agree or disagree with the following statement:

Cricket is exciting

- Strongly agree ☐
- Agree slightly ☐
- Neither agree nor disagree ☐
- Disagree slightly ☐
- Strongly disagree ☐

It is important that a balance between positive and negative responses is maintained in these.

Random response

Random response is a method that is sometimes used when collecting data on sensitive issues. Many young students at school might not wish to admit to (or might boast about) smoking cigarettes, or drinking alcohol, so they might not answer a direct question on these topics truthfully, making it impossible to collect reliable data.

Random response does not necessarily get 'the answer' correct, but it does allow you to get an idea of what a situation actually is.

Example 2

A teacher wishes to gain an idea of how many of the Year 10 students drink alcohol. A sample of 90 students are asked to toss a coin and if they get a Tail they tick the 'yes' box as their answer without looking at the question. If they get a Head when the coin is tossed, they answer the question 'Did you drink any alcohol last night? Yes ☐ No ☐.'

The results showed that 51 ticked the 'yes' box and 39 ticked the 'no' box. You would expect half the 90 (that's 45 altogether) to get a 'tail' and tick 'yes';

$51 - 45 = 6$, so about six students out of the 45 who answered the question drank alcohol last night. So that is about 12 out of the 90 in the sample, and

$\frac{6}{45} \times 100 = 13.3\%$

of the whole of Year 10.

Be a statistician

There are lots of 'local issues' in most areas, from the opening of a new car park or leisure centre to whether the one-way system is working. In schools issues may involve wearing jewellery, uniform, homework, school dinners, timing of the school day.

1 Find a local issue that affects you, your friends, your family or people in your area.

2 Write a questionnaire to find out what people think. Think carefully about the questions you ask and how you will ask them.

3 Collect some data using your questionnaire. If you use interviews you will need to make a data collection sheet – if you do, then see 1.5 for help.

4 Make a short presentation explaining what you have done and what you have found. Report your findings back to your class.

Exercise 1.4

1 Jay wants to find out whether girls are paid more for part time jobs than boys. Write a short questionnaire that she could use.

2 Hagred wanted to find out which was the most popular football team supported by people at his school. This is one of the questions he asked: 'Anyone who likes football supports either Manchester or Liverpool, which one do you support?'

a What are the main faults with Hagred's question?

b Rewrite his question in a way that could be used to find the information Hagred wants.

3 What is the main reason for doing a pilot survey?

4 George runs a hotel called 'The Dragon'. He wishes to find some information about the people who stay at his hotel. He wishes to find out whether their stay is for business or pleasure, how many nights they are staying for and how old they are. Write a small questionnaire that he could use.

5 In a questionnaire, Niall was finding out about pupils' taste in music. Why might he carry out a pilot survey first?

6 Beckie is considering opening a restaurant in the village where she lives. To find out the views of the people who live in the village she delivers a questionnaire to every house.

a Included in the questionnaire is a closed question asking for people's age.

i Explain what is meant by a 'closed question'.

ii Give one advantage of using a closed question for age.

b Only 15 per cent of the questionnaires are returned. What could Beckie do to improve the response rate?

c The returned questionnaires showed that some of her questions had been badly worded.
What should Beckie have done before she delivered her questionnaire to avoid this problem?

d One of Beckie's questions was 'How often do you eat out at a pub or restaurant?'
Give two criticisms of this question.

7 A travel agent decides to survey all existing customers using a postal questionnaire.
One question is shown below.

'How much do you spend each year on holidays abroad?'

Please tick one box.

£1000–£1500 ☐

£1500–£2000 ☐

£2000–£3500 ☐

£3500–£6000 ☐

£6000 and over ☐

a Give two criticisms of the response section of this question.

b Of 1000 questionnaires posted out, only 42 were returned. Give two ways in which the travel agent could increase the response rate.

8 Jane wants to find out how many of her Year 8 class smoke cigarettes. She knows that they would not answer honestly if she asked them directly. Describe a method she could use to get a rough idea of how many actually do smoke.

9 A survey of employees in a factory contains the following question: 'What is your opinion of the proposal to increase the number of hours worked each week to 39?' Describe two types of scale that could be used to measure the opinions of the employees.

1.5 Obtaining data

Observation

When data is observed you will need to record it in some way. Although you can just list the data as you see it, a table is usually much easier to use.

Ford wished to record the colour of vehicles going past his school to see which colour was most popular. He looked in the staff car park to give him some clues. The colours he found were red, blue, green, silver, one yellow car and one was grey and white. He decided to use the following table:

	Red	Blue	Green	Silver	Other
Cars					

He soon realised that there were more vehicles than just cars going past his school. Because he had organised his table in this way he was able to easily add new rows for buses, lorries and vans. He then decided not to include bikes or any other types of vehicles, as he thought that there would be very few of them.

Here is his completed sheet:

	Red	Blue	Green	Silver	Other	Totals
Cars	卌 |	卌 |||	|	卌 |	卌 卌 ||	33
Buses	||					2
Lorries	|				||||	5
Vans	||||	|||	|		||	10
Totals	13	11	2	6	18	50

Note the use of a 'five-bar gate' for tallying. This is used to make it easier to 'total up'. This type of table is called a two-way table.

Objectives

In this topic you will learn:

- how to design data-collection sheets
- about different methods of data collection, including data logging
- to design and carry out a basic statistical experiment
- to identify the variables involved in a statistical experiment
- to collect data by observation and to understand that an observer can affect what is seen just by observing
- to appreciate some problems that occur when data is grouped
- about 'matched pairs'
- to identify and control extraneous variables.

Sometimes you will need to record numerical data, such as age. Many people would not be happy giving their age as they consider it to be personal data. It is usual to get round this problem by using groups, or classes.

links

For more on two-way tables see Chapter 3.

Example 1

Jane wanted to find out whether younger or older people went to school events. For one event she recorded the age of each person entering using the following table.

Age, t	Tally	Frequency
$0 \le t < 10$	|||	3
$10 \le t < 20$	卌 卌 卌 ||	17
$20 \le t < 30$	|||	3
$30 \le t < 40$	卌 卌		11
40 and over	卌 卌	10
Total		44

You can see the age distribution quite clearly here. The 17 people aged between 10 and 20 are probably students at the school. Their parents are probably the people aged 30 and over. What you cannot see are the actual ages of individual people.

Although people are more likely to give you their actual age if you ask them which age group they are in, you will not get their precise ages. You will need to think about this when you generate your own data recording sheets using grouped data.

Effect of observing

Sometimes the very fact that you are observing something will affect what you see.

links

See Chapter 3 for an explanation of double inequality.

For example, PC Speed visited students in a school as part of a maths day. Having discussed road safety and some of the maths that the police use for calculating speeds, they went out to the edge of the school grounds with a 'speed gun'. The school was by a main road, which had a 30 mph speed limit. The students were very surprised to find that all the drivers were doing between 22 mph and 28 mph. PC Speed was wearing a bright yellow jacket and was clearly visible to people on the road.

The following day the same class again went out to the same place, this time they calculated the speeds of cars going past the school by measuring a fixed distance and dividing by the time it took cars to travel over the distance. This time they found the cars were travelling at between 28 mph and 35 mph.

They put the difference down to the bright yellow jacket worn by PC Speed. The drivers had seen the jacket from a distance and had slowed down.

Inter-observer bias

Sometimes you will observe things slightly differently from another person. This is called inter-observer bias. Bias between observers. For example, George and Mildred were investigating the colour of cars going past a school. They did not always agree on the colours of the cars they saw. George recorded one car as grey. Mildred recorded the same car as silver. This is an example of inter-observer bias. This can be major problem when you have to make subjective decisions.

Inter-observer bias can be overcome by careful training of observers. Direction and rules can be given or initially observers can work together and agree how to categorise what they observe.

Data logging

Data logging involves the mechanical logging of data and usually involves just counting. For example, if shop managers wish to record how many customers enter their shops, they can use a data logger, which would count customers as they enter. It would not distinguish between any characteristics of the customers.

Mechanical data loggers are used by organisations such as local authorities to count the number of vehicles that use a particular stretch of road.

You might have used data-logging equipment in science lessons.

AQA Examiner's tip

Exam questions often ask what variable could be found using data logging. Remember that the answer will always be something that is just counted, without noting any particular features.

Experimental design

You must always take care to design experiments carefully to avoid any bias. If the experiment you use is not well designed any data you collect may be biased.

Control group

Control groups are often used when designing an experiment to make comparisons. They are an 'untreated' group. The group that is 'treated' is sometimes called the 'experimental group'. It is important that the both control group and experimental group have the same characteristics; for example, if they are two groups of people then the groups should have the same average age and the same gender balance.

Example 2

Shirley grows tomatoes in her greenhouse. She wants to find out whether Bigga-crop fertiliser or Fasta-grow fertiliser produces larger tomatoes. Design an experiment that she could use.

Solution

She could buy nine identical tomato plants and plant them in the same brand of compost in identical pots. She would need to treat all nine plants in the same way.

Three plants are fed with Bigga-crop and grown in pots labelled B.

Three plants are fed with Fasta-grow and grown in pots labelled F.

Three plants are fed nothing and are grown in pots labelled C. (This is the control group.)

As she crops the tomatoes she records their weights. At the end of the summer she compares the total weights of cropped tomatoes to find which fertiliser is best.

The fertiliser is an example of an **explanatory variable**, and the weight of the crop is an example of a **response variable**. The fertiliser is the variable which Shirley introduces and wishes to see what effect it has. The weight of the crop is the outcome, the result of putting in the fertiliser, and so depends on the fertiliser which is put in. Explanatory and response variables are sometimes called independent and dependent variables.

Experiments in nature, the laboratory and in the field

A natural experiment is a naturally occurring event or situation that is observed. There is very little control over factors that may influence the response variable.

A laboratory experiment is carried out under controlled conditions. It is possible to control other factors that might have an effect on the outcome of an experiment. Sometimes laboratory experiments may be criticised as being artificial because of this.

Some experiments are carried out in the natural environment. Whether that is a street, a field, a school, or a hospital or somewhere else depends on the research question. Field experiments are more 'real world' than laboratory experiments; unfortunately, they are also more expensive, and often more time consuming. It may be easier to control most factors in the laboratory but this may not be possible in the 'field'.

In the previous example, Shirley needs to make sure that no other factors are allowed to affect the outcome of her experiment. For example, plants near the greenhouse door may be in a draft and experience cooler temperatures; plants in the middle of the greenhouse may receive less sunlight. Variables such as these, which may affect the outcome of an experiment, are called **extraneous variables**. In order to attempt to control these variables (reduce any effect they may have) Shirley could order the pots BCFBCFBCF in the greenhouse, so that no single type of fertiliser is in the middle or by the door.

Control groups are often used by the medical profession to test new drugs. Patients are offered the opportunity to 'opt in' to the trial. They are then either allocated a tablet with the drug in or a tablet without the drug (this is called a placebo). This allocation is done randomly. In a 'blind' trial patients do not know whether they receive the placebo or the drug. In a double-blind trial the doctor does not know either. The patients receiving the placebo form the control group. Blind trials and double blind trials ensure that psychological factors (people who think they are being treated often get better!) and effects from other extraneous variables are kept to a minimum.

Before-and-after experiments

An athlete wishes to know whether a new training method improves her performance. She records her times before using the new method and again after using the new method. She then compares her performance before and after the new method has been used. It is crucial that all the conditions before and after are the same. This enables her to make a genuine before-and-after comparison.

Key terms

Explanatory variable: the variable you control, whose effect you wish to investigate.

Response variable: the variable that depends on changing the explanatory variable; the outcome that you measure.

Extraneous variable: a variable that may affect the outcome of an experiment.

Matched pairs

A school wishes to find out whether a new method of revising improves exam performance. To test this it sorts students into pairs. Each pair consists of two students who are as similar as possible in every way. Half of each pair revises using the 'old' method and the half of each pair revises using the new method. These allocations are made using a random process. All students then sit the same test and their performances are then compared.

It is very important that each pair is made up of two members, as nearly identical as possible.

Exercise 1.5

1. Mrs G Rocer, a shop owner, wishes to find out about the people who enter her shop. She wants to know their ages and gender.
   a. Design a data-collection sheet that she could use.
   b. Later, she decides it would have been helpful to know how many of her customers are teenagers. What changes, if any, need to be made to the data-collection sheet you designed in (a)?
   c. Why would it not be possible to use a data logging machine here?

2. Mr T Rout, owner of a fish farm, wishes to find out whether his fish weigh more after being given a new type of food. He knows that weighing a fish causes distress to the fish and so it cannot be weighed a second time. Design a simple statistical experiment that he could use to discover whether the new type of food works.

3. A company produces compost for germinating seeds. It develops Supagro, a new compost, which it claims helps seeds germinate much quicker than its standard compost. Design a simple statistical experiment that could be carried out to test the company's claim.

4. John thinks that the more fertiliser he adds to his plants the more flowers they grow. What are the explanatory and response variables here?

5. Recall is a company that produces vitamin tablets. It claims that taking one tablet each day for two weeks can improve memory.
   a. Design a statistical experiment that could be used to test this claim.
   b. What extraneous variables would need to be considered?

Summary

You should:
know about writing hypotheses
know that you collect primary data yourself
know that secondary data is collected by someone else
know that a census uses all of the population; a sample uses part of the population
know that opinion scales can be discrete or continuous
know that different methods, such as surveys, observation and experiments, can be used to collect data
be able to identify which variables are needed when designing statistical experiments
know that data on sensitive issues is collected using random response
be aware that identification and control of extraneous variables is important.

AQA Examination-style questions

1. Peter receives a questionnaire in the post about a new local radio station.
   Three of the questions are shown below. Give **one** criticism of each question.
   Question 1
   How many hours have you listened to the radio during past six months?
   Question 2
   How much do you earn each year? Please tick one box.
   Less than £10 000 ☐ £10 000 up to £20 000 ☐ More than £20 000 ☐
   Question 3
   If you have already heard our new radio station, give one reason why you enjoyed listening to it. *(3 marks)*

   AQA, 2003

2. Staff at Tower Bridge wish to undertake a survey to find out how long visitors spend at the site. They decide to use face-to-face interviews to collect this information.
   Give **one** advantage and **one** disadvantage in using this method of data collection. *(2 marks)*

   AQA, 2003

3. State whether each of the following variables is qualitative, discrete or continuous.
   (a) The time taken for an estate agent to sell one of its three-bedroomed houses. *(1 mark)*
   (b) The sale price of a three-bedroomed house. *(1 mark)*
   (c) The gender of the estate agent. *(1 mark)*

   AQA, 2003

4. A new sports centre is to open on a housing estate. A questionnaire sent to all adults on the estate includes the following question:
   'How many hours each week would you plan to spend at the sports centre?'
   Design a response section for this question. *(2 marks)*

   AQA, 2004

5. A ticket machine records the number of vehicles entering a town centre car park each hour during the day. What type of data collection procedure is used in this case? *(1 mark)*

   AQA, 2004

6. Give **one** difference between a census and a sample. *(1 mark)*

   AQA, 2004

7. The following question was asked in a survey on the use of public transport.
   'Do you think the reliability of this transport is:
   A Very good; B Excellent; C Outstanding?'
   (a) What is wrong with this question? *(1 mark)*
   (b) Write a question which would be suitable for finding out how often a particular form of transport was used. *(2 marks)*

   AQA, 1999

8. A company proposes to build a large wind turbine close to a village. Ben designs a questionnaire to obtain opinions on the proposal from the villagers. One of his questions is:

Do you agree that the wind turbine will be a disaster for our village?

Yes, definitely ☐ No ☐

(a) Give **two** distinct criticisms of Ben's question. *(2 marks)*

(b) Rewrite Ben's question to make it more appropriate. *(2 marks)*

AQA, 2006

9. For her statistics coursework, Mina wanted to find out how popular her local clothes shop was. She wanted to know:

A the number of females going into the shop per minute

B the number teenagers going into the shop per minute

C the total number of people going into the shop per minute.

(a) Which one of the above could be found using a data logging machine? *(1 mark)*

(b) Write a suitable question that Mina could use to find out whether a person is a teenager or not. You **must** include a response section. *(2 marks)*

AQA, 2008

10. One of the questions used by a builder in a survey was:

'How much do you pay each month on your mortgage?

under £200 ☐ £200–£300 ☐ £300–£700 ☐

(a) Give one criticism of the response section. *(1 mark)*

(b) Give one criticism of the question asked. *(1 mark)*

AQA, 2008

2 Sampling

In this chapter you will learn:

- why it is usually necessary to take a sample rather than a census
- that different samples can produce different results and findings
- some of the ways that bias can occur in sampling and how to avoid this happening
- about random and stratified sampling and how to obtain such samples
- about stratified sampling with more than one category
- the dangers of convenience sampling
- about systematic, cluster and quota sampling
- about multistage sampling
- about sampling criteria used in opinion polls.

You should already know:

- ✓ how to work with fractions
- ✓ how to work with decimals.

What's this chapter all about?

You can be the most accurate calculator of answers in the world, you can be the designer of the most beautiful and effective graphs in the world, but if you are working on flawed data then all that is pointless. The best jockey in the world still needs a good horse!

This chapter looks at how you obtain the sample for your data collection. One of the biggest issues in good statistical work is getting a sample which fairly represents the population you are working on. Remember this little saying: GIGO – Garbage In, Garbage Out. In other words, if you are working on poor data, whatever you do will be worthless.

Focus on statistics

In this chapter you will meet some work that you will not have done in maths. This can be found in 2.2 and covers:

- convenience sampling
- systematic sampling
- cluster sampling
- quota sampling
- multistage sampling
- the sampling criteria used in opinion polls.

2.1 What is a sample and why are samples needed?

Most data collection is based on finding and using a **sample** of data from a **population**.

Remember that when data is collected from every member of a population, this is called a **census**.

Usually any member of the population should be able to appear in the sample although sometimes you might find that some of the population is not available. The actual list of population members available to appear in the sample is called the **sample frame**.

Some of the reasons why it is a good idea to take a sample rather than use the whole population include:

- saving time
- saving money
- the amount of data to be looked at is reduced.

In fact, the size of a sample can be quite small compared to the population, so a great deal of time and effort can be saved. Despite this, if the sample is collected carefully enough, the sample can give results that should be representative for the population as a whole.

If the population is also quite small, then 5 per cent to 10 per cent of the population can be thought of as a good sample size. Alternatively, if there are n items in the population a sample size of $\sqrt{n}$ is thought to be a good sample size. However, if items are 'used up' or destroyed by being sampled (for example, light bulbs or food items) a sample smaller than $\sqrt{n}$ can be used.

Sample size affects the variability and reliability of the findings from an experiment.

In simple terms, if you have one sample size four times bigger than another, the larger sample will produce results with only half the variability of the smaller sample size, therefore increasing the reliability of the results.

Objectives

In this topic you will learn:

- why it is usually necessary to take a sample rather than a census
- that different samples can produce different results and findings
- some of the ways that bias can occur in sampling and how to avoid this happening.

Key terms

Sample: part of a population from which information is taken.

Population: every possible item that could occur in a given situation.

Census: the data is obtained from every member of the population.

Sample frame: the list of all the members of a population available to appear in the sample.

AQA Examiner's tip

This is the only fact about variability and sample size that can be explicitly tested in the exam. Otherwise, you simply need an understanding of the fact that a larger sample means less variability and more reliability.

Example

A factory produces 5000 car tyres every day, each of which should last for 30 000 miles on a car.

The manager of the factory needs to check that the tyres produced do in fact last that long.

a Explain why the manager should check a sample and not test the population.

b What sample size should be tested?

Solution

a The checking process is expensive and time consuming as all tyres tested are 'used up' and cannot be sold.

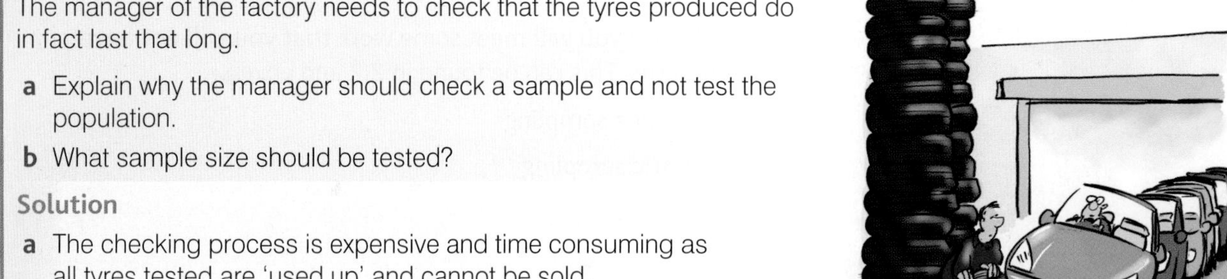

b A sample is needed that is big enough to show that the tyres last as long as stated but small enough to avoid using too many. Here 5 per cent is still 250 tyres, which is a lot of waste. Perhaps something in the region of 15–20 tyres could be tested.

It is important to know the population you are dealing with in order to be able to take a sample that is free of **bias**.

Some reasons that might lead to data from a sample being biased are:

- the choice of sample items is poor (the sample is unrepresentative)
- the population has not been correctly defined (so the sample is from the wrong set of items or people)
- the way the data is collected is faulty (for instance, the questionnaire is poorly designed or there are errors in recording answers)
- unexpected variables are present in the situation (**confounding variables**).

> **Key terms**
>
> **Bias**: the name given to sample data that does not fairly represent the population it comes from.
>
> **Confounding variable**: a variable that is unexpected and interferes with the variables under consideration.

Different samples in a given situation can lead to variations in results.

For example, if you take the lottery numbers for the last 100 Saturday draws and work out their mean, and then do the same for the last 100 Wednesday draws, you are actually very unlikely to get exactly the same result, even though they are both samples of numbers from 1 to 49.

> **Spot the mistake**
>
> Niles is collecting data about TV viewing habits amongst 15-year-olds. He asks a sample of 15-year-olds, all of whom are in his media studies class at school. What mistake has he made?

Exercise 2.1

1 Which of the following statements is true about a sample? Explain your answer.

- **a** A sample will always give the same results if the data is collected carefully.
- **b** A sample is quicker to do than a census.
- **c** A sample uses half of the population.

2 Natalie is a trained cake tester. She knows exactly how each cake produced by Bakebest should taste. Bakebest produces 200 cakes every day. How many should be sampled for taste? Give a reason for your answer.

3 Find three causes of bias in the situation below.

Wasim is interested in how many times adults go swimming in one year. He asks five parents of his friends to tell him how many times they went swimming in 2007. When Wasim got home he then recorded the data.

2.2 Sampling methods

The aim of taking a sample is to find data that is **representative** of the population, and so enables us to estimate population values. This means that is has roughly the same characteristics as the whole population – the same approximate values for its measures such as averages.

There are many different ways that samples can be taken. Here are the methods that you need to know for the examination.

Objectives

In this topic you will learn:

- about random and stratified sampling and how to obtain such samples
- about stratified sampling with more than one category
- the dangers of convenience sampling
- about systematic, cluster and quota sampling
- about multistage sampling
- about sampling criteria used in opinion polls.

Key terms

Representative: a representative sample will have the same characteristics as the population it is taken from and will be free from bias.

Simple random sample: every member of the population has an equal chance of being in the sample, and every possible sample has an equal chance of occurring.

Random sample: another name that implies the use of simple random sampling.

Random sampling

There are quite a few ways in which a sample can be chosen. The most popular is **simple random sampling** or **random sampling** as it is usually written. In a random sample, every member of the population has an equal chance of being in the sample.

Example 1

Sasha is surveying traffic in her street. She counts the type of vehicle for every tenth vehicle to go past her house. Explain why this is not a random sample.

Solution

Not every vehicle has an equal chance of being in the sample. After a particular vehicle passes her house, the next nine have no chance of being in the sample and the tenth is certain of being in her sample.

Obtaining a random sample

The best and most reliable way of obtaining a random sample is using random numbers. Random numbers are available in tables, on computers and on most calculators. Here is an extract from a small table of random numbers for you to use throughout the book where needed. The full table can be found at the back of the book.

51 38 42 50	40 27 59 68	04 27 38 22	91 04 73 82	85 40 82 84
09 56 76 51	04 73 94 30	16 74 69 59	04 38 83 98	30 20 87 85
55 99 98 60	01 33 06 93	85 13 23 17	25 51 92 04	52 31 38 70
72 82 45 44	09 53 04 83	03 83 98 41	67 41 01 38	66 83 11 99
04 21 28 72	73 25 02 74	35 81 78 49	52 67 71 40	60 50 47 50

Example 2

There are 740 houses in Jake's village. Jake wants to ask a sample of people from 50 houses about the possibility of having a village fete next summer. First he wants to do a pilot survey of five houses.

Use the random numbers 5138425040275968042738229104738285408284 as necessary to obtain the sample for the pilot survey. (This is the first row of numbers from the random number table on the previous page.)

Solution

First, Jake needs to number the 740 houses (or at least have them in an ordered list).

He now needs five three-digit random numbers. Starting from the left of the given numbers gives 513, 842, 504, 027, 596.

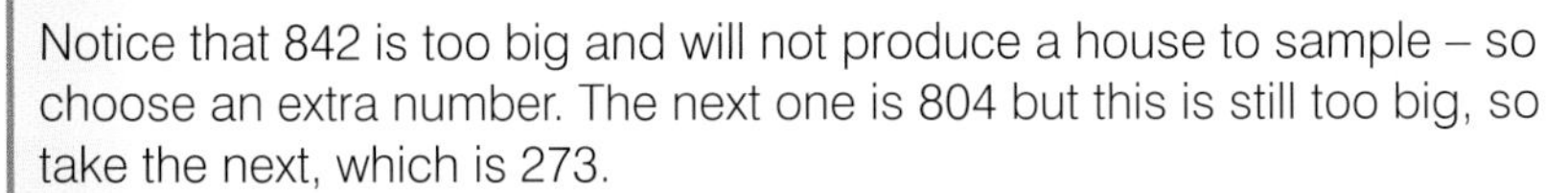

Notice that 842 is too big and will not produce a house to sample – so choose an extra number. The next one is 804 but this is still too big, so take the next, which is 273.

Jake also needed to check for repeats when choosing numbers as it would be pointless asking the same person twice.

Finally Jake can match the numbers to the houses in the list and he has his sample of five.

AQA Examiner's tip

Students often forget to mention the first and last points from Example 2. That is, they forget to number the list and they forget to match the outcomes to the list. This will lose marks.

The other way to obtain a random sample is far more difficult. Usually each item in a population is written on a piece of paper and all the pieces of paper are put into a container (often a hat!). The pieces of paper must then be well mixed up before the required number of pieces of paper is picked out of the hat.

Stratified random sampling

Most populations (especially human ones) have different well defined sections. For example, people can be divided into different genders, ages, social class, occupation types and so on.

These different sections are called **strata** and a good way of obtaining a sample which represents the population well is to get the *same proportion* of people from each stratum in the sample as exists in the whole population.

For example, a club has 100 members, 70 are male and 30 are female. A sample of 10 people from the club members, taking into account the strata proportions, would have seven males and three females. The seven males could be chosen using random sampling from all the males and the three females likewise from all the females. Then you would have a **stratified random sample**.

Key terms

Strata: different sections of a population such as gender or age groups.

Stratified random sample: a sample with consideration of the strata within the population – individuals from within each stratum are chosen using simple random sampling.

AQA Examiner's tip

Examiners will give credit for the 'hat' method of obtaining a random sample but you will need to be very careful. Make sure you say that the entire population is going into the hat on pieces of paper and state the exact number of pieces of paper you will take out.

It is possible to have more than one stratum taken into account at once.

Example 3

The number of girls and boys in each year of Scotter High School is given in the table.

	Year 7	Year 8	Year 9	Year 10	Year 11
Girls	106	121	98	142	111
Boys	91	106	125	135	130

A sample of 120, stratified according to gender and year group, is to be taken from the pupils at Scotter High School. Complete a new table showing the number in the sample for each gender and year.

Solution

First calculate the total number of pupils in the school, $106 + 91 + 121 + \ldots + 130 = 1165$

Now find the proportion of each type of pupil for the population (that is, the entire school).

So, for Year 7 girls the proportion is $\frac{106}{1165}$ and multiply this by the total sample size giving $\frac{106}{1165} \times 120$, which equals 10.92 (to two decimal places or d.p.).

It is important to keep the figures to 2 d.p. at this stage.

The table of all the two decimal place values is given below – check you can get the other figures.

	Year 7	Year 8	Year 9	Year 10	Year 11
Girls	10.92	12.46	10.09	14.63	11.43
Boys	9.37	10.92	12.88	13.91	13.39

Obviously, a sample of 10.92 Year 7 girls makes no sense and the figures need rounding but you must now check that when these are rounded to the nearest whole number the total sample size is the 120 you want $11 + 9 + 12 + 11 + 10 + 13 + 15 + 14 + 11 + 13 = 119$ As this is 119 one of the numbers that is rounded down must actually be rounded up. To do this choose the number whose decimal part is closest to 0.50 so that it wasn't far from being rounded up anyway. This will be the number of Year 8 girls which will round to 13 rather than 12. It is now possible to show the final table of sample sizes.

	Year 7	Year 8	Year 9	Year 10	Year 11
Girls	11	13	10	15	11
Boys	9	11	13	14	13

Convenience sampling

Convenience sampling is where the sample is chosen in a way that makes it as easy as possible for the person doing the sampling. For example, choosing the first 20 items from a production line to test.

This type of sampling is best avoided as it is likely to produce samples that are biased as they are not likely to be representative of the population.

Some people think that they can 'choose' randomly and avoid the work involved in selecting a genuine random sample. If you think you can do this – do the 'real world' feature and test yourself.

Real world

This is an example of real statistics that can be done as an individual or as a class. It does involve some preparation by the teacher involving a trip to a lake or the seaside!

Preparation: Collect 100 pebbles and stones. Most can be small but two or three need to be a little larger, say up to the size of a small apple. Number the unordered stones from 1 to 100 using a black marker. Also find the mean weight (mass) of this population of stones.

Experiment: Which is better at estimating the mean weight of the pebbles, a random sample of 10 pebbles using random numbers or a judgement sample of 10 pebbles chosen by a pupil to 'represent' the pebbles?

Systematic sampling

This is a method where items or people are chosen at regular intervals. This should produce a representative sample as long as there are no patterns in the items or people.

To obtain a systematic sample:

1 Number the population.
2 Decide on the sample size that you require.
3 Calculate the interval that matches your required sample size (see example).
4 Obtain a random starting point.
5 Count on to get the remaining sample members.

Example 4

Eighty people are running in a fun run. A sample of 10 is required to be interviewed about the run for the local newspaper.

Explain the steps required to obtain a systematic sample of 10.

Solution

1 Number the 80 runners.
2 Every eighth runner is needed to get a sample of 10 from 80 people ($80 \div 10 = 8$).
3 Use a random number list to get a random number between 1 and 8 (suppose it is 5).
4 The sample will now be the people numbered 5, 13, 21, 29, 37, 45, 53, 61, 69 and 77.

Cluster sampling

If a population is divided into **clusters** (another name for groups) then there needs to be a good mix of different types of people or items in each cluster. Then selecting a cluster at random and sampling everyone or every item within that cluster should produce a reasonably representative sample.

This can be used in situations such as mixed-ability tutor groups. Choosing one tutor group at random and sampling the whole group should be as good as randomly choosing 30 students from across the whole school year. If the tutor groups are organised using any criteria at all, though, this cannot work.

Key terms

Clusters: groups of items that have a distribution within them similar to the populations from which they come.

Quota sampling

Sometimes a particular number of items or people with a certain characteristic are needed. This number is known as the **quota**. The quota is usually filled by person obtaining the sample making a choice, rather than by using a more statistical method. For this reason quota sampling, often used in market research, can be quite unreliable.

Key terms

Quota: a specific number of people or items with a particular feature.

Be a statistician 

You are going to do some market research using quota sampling. You are going to find out whether pupils in your school prefer the taste of one brand of crisps to another (the make not the flavour). You must not reveal the identity of the brand of crisps at any time. Your quota of pupils is 15 per year group. Write up how you could do this research.

Now do the research. Did anything happen that you had not expected? Was quota sampling a good method to use? Why, or why not? Write a short report on your findings.

Opinion polls

These are normally large-scale attempts to discover what people think. They can be used to try to predict the outcome of events such as a General Election by taking a random or quota sample across the country. If it is a quota sample then the criteria used for selection of the sample would include geographical area, gender, age and social and economic background. In the UK, with a population of approximately 25 million voters, the sample size would typically be about 2000 people.

links

See the website **www.britishpollingcouncil.org** for more information on opinion polling and some of the issues surrounding the reliability and design of opinion polls.

Multistage sampling

This is a form of cluster sampling often used in opinion polls. A cluster or clusters chosen to represent the overall population may then be broken down into smaller sections, only some of which are then chosen to be part of the sample. Typically, when this approach is used in opinion polls, several towns will be chosen as the first stage of cluster sampling. In each town one or more wards will be the second stage of cluster sampling. (A ward is a small area of a town, often having a local councillor representing the people just in that small area.) It is important that at each possible stage of sampling the clusters fairly represent the larger area they are chosen from – a ward must be typical of the town.

The process is best illustrated using this series of diagrams for a country.

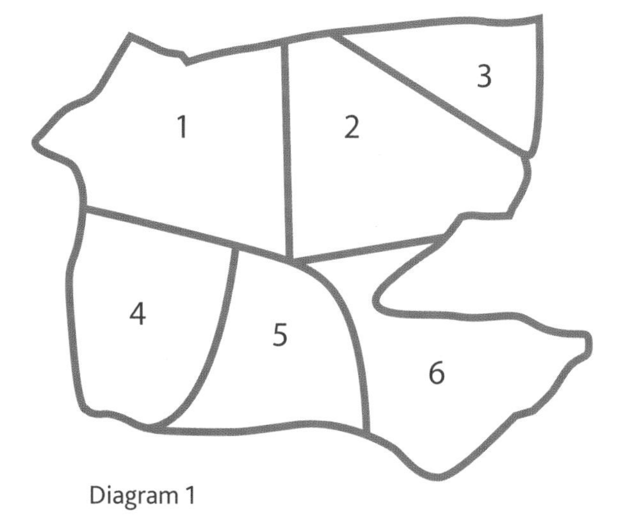

Diagram 1

Diagram 1 – the country is split into regions which are numbered. One region is selected usually using random sampling.

Diagram 2 – the chosen region (suppose it was region 4) is now split into constituencies. One of these is now selected using random sampling.

Diagram 3 – the chosen constituency (suppose it was constituency 8) can now be split into electoral wards. Quota sampling (for example for men and women) may now be used to get the sample from across the electoral wards.

Alternatively, further breaking down of a randomly selected ward into polling districts can be carried out before cluster sampling obtains the people to be asked.

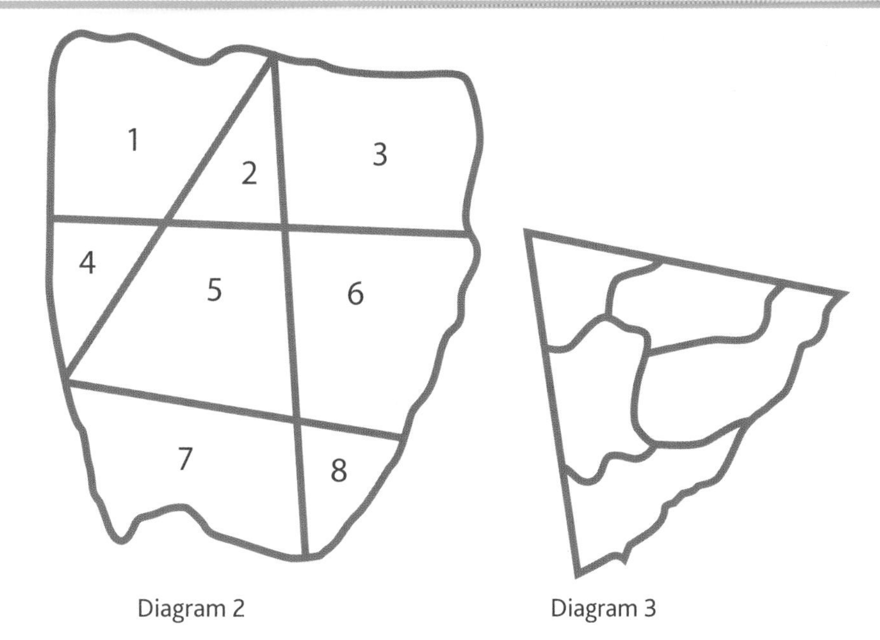

Diagram 2

Diagram 3

Exercise 2.2

1 Ash has 77 albums downloaded on his computer. He decides to play one of these albums at random. List the steps he should take to make a genuine random selection using random numbers.

2 Two thirds of the workers in a factory are part-time. The rest are full-time. The manager of the factory wants to ask a sample of 30 workers about changes to break times.

a Which sampling method would be the best to use in this situation? Explain your answer.

b How many part-time workers should be in the sample?

3 Quinlan is investigating the amount of sport played by Year 11 pupils. He uses a convenience sample of the other six members of the school band he plays in to fill in his questionnaire.

a Why is Quinlan's choice of sample poor?

b Given that there are 150 pupils in Year 11, explain how a simple random sample of six could be chosen.

c Comment on this sample size.

4 Ellie and Linda disagree about random and systematic sampling. Ellie says, 'Systematic sampling is random because every person has an equal chance of being in the sample.' Linda says that Ellie is wrong. Explain why Linda is correct to say that Ellie is wrong.

5 Martina runs a business looking after pets while people go on holiday. She looks after cats, dogs and rabbits for either one or two weeks. The table shows the number of each animal and the length of time they were looked after in 2008.

	One week	Two weeks
Cats	231	108
Dogs	406	323
Rabbits	103	66

Martina wants to send a questionnaire to 50 people who used her business in 2008, taking a stratified sample according to the type of pet looked after and the length of time they were looked after for.

a Complete this table showing the numbers in each category in the sample of 50.

	One week	Two weeks
Cats		
Dogs		
Rabbits		

b What assumption has been made in your calculations for part (a)?

6 Geoff and Gemma are collecting data about the lunchtime eating habits of 15-year-olds. They plan to ask 50 of the 15-year-old students at their school what they eat on a typical school day at lunch.

a Give one reason why they should consider obtaining a stratified sample.

b Forty-two per cent of the two hundred 15-year-olds in their school are boys. How many boys should be in a sample stratified by gender?

c How could these boys be chosen using a systematic sampling method?

7 Give an example of how multistage sampling could be used to obtain opinions about school uniform in a school with mixed-ability tutor groups.

Summary

You should:
know that samples are taken to save time or money or for convenience
know that the sampling frame lists all possible items that can appear in the sample
know that different samples can produce different findings
know about random and stratified sampling
know about systematic, cluster and quota sampling
know about multistage sampling
know that opinion polls use criteria to obtain representative samples.

AQA Examination-style questions

1. A survey of the eating habits of pupils at a school is to be carried out.

(a) Give one reason why it might be useful to undertake a pilot survey. *(1 mark)*

(b) Give one reason why each of the following methods would not give random results:

(i) Standing outside the school canteen on a Monday lunchtime questioning pupils as they arrive. *(1 mark)*

(ii) Sending a questionnaire to every pupil on the school register whose surname begins with S. *(1 mark)*

(c) Describe a method of choosing a random sample of pupils from the school. *(2 marks)*

AQA, 2004

2. A youth club has 72 members. The leader decides to select six members at random to go on a sailing course. He numbers the members 01 to 72 and uses the random number table below to make his selection. *(2 marks)*

29	44	76	56
44	51	38	00
07	21	92	17

(a) Starting with 29 and reading across each row, write down the number of each of the six members that he selects. *(3 marks)*

(b) The youth club has 48 boys and 24 girls as members. Calculate the number of boys and girls the leader should include in a stratified sample of members. *(3 marks)*

AQA, 2005

3. The table shows how many students are in a school.

Lower School	Upper School	Sixth Form
720	480	400

Chelsey wants to survey 100 students from the school using a stratified sample.

(a) Work out how many students Chelsey should include in her survey from the Lower School. *(3 marks)*

(b) Twenty-five students from the sixth form are to be chosen.
Describe a method of choosing a random sample of these sixth form students. *(2 marks)*

AQA, 2006

4. The table shows the number of GCSE passes for 30 year 11 students.
The gender of the students is also shown.

Student	Gender	Number of passes	Student	Gender	Number of passes
01	M	4	16	M	3
02	M	7	17	M	5
03	F	9	18	F	10
04	M	8	19	M	2
05	F	6	20	M	4
06	F	6	21	F	8
07	M	9	22	M	3
08	M	2	23	F	7
09	F	7	24	M	2
10	M	4	25	M	1
11	F	8	26	M	4
12	M	7	27	F	9
13	F	5	28	M	5
14	M	5	29	M	6
15	M	4	30	M	2

Here is a table of random numbers from 01 to 50.

Line 1	15	41	01	15	20	16
Line 2	32	22	33	30	19	08
Line 3	04	31	49	29	13	29
Line 4	14	23	37	11	24	29

(a) Starting with the first number on Line 1 of the random number table and reading across from left to right, select a random sample of size eight. *(3 marks)*

(b) Starting with the first number on Line 3 of the random number table and reading from left to right, select a random sample of size six stratified by gender. *(4 marks)*

(c) Give a reason why the sampling method used in part (b) is better than the sampling method used in part (a). *(1 mark)*

AQA, 2007

3 Tabulation

In this chapter you will learn:

- how to put data into frequency tables using a method of tallying
- about class intervals including open-ended classes
- about the effects of changing the number of class intervals
- how to choose class interval size and accuracy of values used
- how to design and use tables and two-way tables
- how to read and interpret data from media and national sources.

You should already know:

- ✔ inequality signs
- ✔ how to calculate with fractions and percentages.

What's this chapter all about?

There are two different reasons why this is an important chapter. First, you have just taken all this trouble to collect some good-quality data and now you need to consider what you are going to do with it. You can learn about different ways of summarising and tabulating data here. But you must also get used to the world of data that's out there. Whether you want a loan, a holiday, or just to see what the weather was like somewhere, you need to know how to deal with all the data in papers, shops, the internet – everywhere!

Focus on statistics

In this chapter you will meet material you have met before in maths, but you will need a deeper understanding for statistics. This is particularly true of the work in 3.1 on lower and upper class bounds and how these can differ for grouped data that is rounded to the nearest integer, compared with grouped continuous data.

In 3.3 you will also meet data tables in much greater detail than you have done in maths. Be aware that examination questions will frequently use real data in statistics, whereas in maths it is rare for examination questions to use real data in the same way.

3.1 Organising raw data – frequency tables

Data that have not been organised in any way are called **raw data**. It is difficult to use or draw conclusions from raw data. It is usually helpful for you to begin organising the data by putting them into a **frequency table**.

A method of tallying is often used. A single stroke | represents one item. Items are grouped in fives, and each fifth item is a cross stroke on the first four strokes ~~||||~~. Each set of five is sometimes known as a five-bar gate.

The tallies are then counted to give the total of each item called the **frequency**.

Example

Mrs Smith buys a bag of apples each week from the supermarket. The number of apples in each bag over six months is given below. Use a method of tallying to construct a frequency table of the data.

6 6 7 6 5 7 8 6 6 5 7 8 5
7 6 5 7 8 6 6 7 8 5 5 7 6

Solution

The numbers range from 5 to 8. This means the table needs a row for each number in this range.

This is how the table looks before you begin filling it in:

Number of apples	Tally	Frequency
5		
6		
7		
8		

The first value is a 6, so one strike | goes into the 6 row. Continue this procedure, but remember that every fifth strike goes across the previous four on that row. This is how the table looks after the tallies are completed:

Number of apples	Tally	Frequency
5	~~				~~		
6	~~||||~~ ||||	
7	~~||||~~ ||	
8						

Objectives

In this topic you will learn:

- how to put data into frequency tables using a method of tallying
- about class intervals including open-ended classes
- about the effects of changing the number of class intervals
- how to choose class interval size and accuracy of values used.

Key terms

Raw data: data that is unsorted in any way.

Frequency table: a table showing the number of times values or items that have occurred in a set of data.

Frequency: the frequency of an item is the number of times it has occurred.

links

See Chapter 1 for information relating to using tallies on data collection sheets.

AQA Examiner's tip

If you simply list tallies without five-bar gates it will lose marks in the examination.

Finally, you need to count the tallies for each row and fill in that number in the frequency column. The completed frequency table then looks like this:

Number of apples	Tally	Frequency
5	𝍸 I	6
6	𝍸 IIII	9
7	𝍸 II	7
8	IIII	4

You can see straight away that much more information can now be seen from the table than from the raw data.

Key terms

Groups: a range of values within which data lies.

Class intervals: a better statistical wording for groups.

Lower class bound: the lowest value that could be present in a particular class interval.

Upper class bound: the highest value that could be present in a particular class interval.

Class width: the upper class bound minus the lower class bound.

Open-ended class: a class interval where the upper class bound is not defined.

If the data are being given to you in a frequency table, the tally column is usually missed off, as it is the frequency column that is much more useful. Where data are continuous, or where there are lots of different discrete values, the frequency table rows will be for **groups** or **class intervals** of data.

For example, the weights of 100 people are shown in this frequency table. Everyone has a different exact weight but it would not be sensible to have a table with 100 rows – that is the same as just having a list of all the values – so the data is grouped.

Weight, w kg	Frequency
$40 \leq w < 60$	23
$60 \leq w < 80$	55
$80 \leq w < 100$	17
$100 \leq w < 120$	5

Look carefully at how the table is labelled. The expression $40 \leq w < 60$ is a class interval and means that all weights from and including 40 up to but not including 60. The lowest value that can go in the group is 40 – this is called the **lower class bound** (or lcb for short).

The **upper class bound** (or ucb) is harder to think about. Here it is the number only just below 60 (59.999 . . .), but to make things easier you can say 60 as long as you remember that the actual value 60 would go into the next class interval $60 \leq w < 80$.

The difference between the upper class bound and the lower class bound is the **class width**. In this table the class widths are all 20.

When you collect data you will not always know what the largest value will be. When this occurs you need to leave the final class interval **open-ended**.

For example, in the weights frequency table, if you wanted to leave the last class interval open ended you could write $w \geq 100$. This simply means all weights above and including 100 kg. You cannot work out the class width for an open-ended class interval.

Choosing class intervals

When drawing up your own frequency table, it is best to choose class intervals that have these features:

- make all class intervals the same width
- aim for at least four class intervals
- try to avoid having more than 10 class intervals
- make the lower and upper class bounds nice numbers such as multiples of 10.

Too many intervals make it difficult to see patterns in the data. Too few intervals make the class intervals wide, hiding the shape of the distribution, and much of the detail in the data is lost.

Real world

How warm was it yesterday?

Use a broadsheet newspaper or the internet to collect data about the maximum temperature around Britain yesterday.

Before you start to collect your data, look at the values so you can decide how many class intervals you need and how wide each should be. Collect data for at least 40 places.

Write a short summary of your findings.

Sometimes continuous variables are recorded to the nearest unit. This means that the lower and upper class bounds are half a unit either way of the recorded value.

For example, if weights in the example above had been rounded to the nearest kg, the value 50 would actually mean $49.5 \leq w < 50.5$, so, using the labelling above, it would not be possible to know which group to put 50 in. So in situations such as this we need a different labelling system, such as 40 – 49, 50 – 69, 70 – 109 and so on. It is then clear exactly which class interval a rounded value should go into.

Spot the mistake

Paul and Shelley use the raw data of the 100 weights between 40 kg and 120 kg to draw up frequency tables for the data. Here is Paul's table. What mistakes has he made?

Weight, w kg	Frequency
$40 \leq w < 50$	9
$50 \leq w < 70$	46
$70 \leq w < 110$	42
$110 \leq w < 120$	3

Here is Shelley's table. What mistakes has she made?

Weight, w kg	Frequency
$40 \leq w < 80$	78
$80 \leq w < 120$	22

Exercise 3.1

1 Paul collects data about the number of goals in 30 football matches over one weekend. Here are his results:

3 2 2 0 1 3 5 4 2 2 3 1 2 5 6 4 3 2 4 4 0 3 3 2 5 4 5 3 2 4

Complete this frequency table using tallying.

Number of goals	Tally	Frequency
0		
1		
2		
3		
4		
5		
6		

2 For each of these class intervals:

i write down the lower class bound

ii write down the upper class bound

iii work out the class width.

a $30 \leq x < 50$

b $100 \leq t < 200$

c $22 \leq l < 24$

d $1000 \leq w < 2500$

3 Here is a set of data about the time in seconds spent by 40 cars at a red traffic light.

30 22 45 67 12 35 60 55 16 44 48 37 33 28 65 49 61 30 44 43

43 38 65 71 23 48 11 26 47 34 27 21 36 57 40 32 27 38 71 57

a Using suitable class widths, construct a frequency table for this data using tallying.

b Give two reasons why you did not need to make your final interval open ended.

3.2 Two-way tables

A **two-way table** shows information about two features of data at the same time.

For example, here is a two-way table which gives information about the gender of visitors to a shop in a five-minute period *and* whether they were adults or children.

Visitors to shop	Adult	Child
Male	7	5
Female	23	16

Objectives

In this topic you will learn:

how to design and use tables and two-way tables.

Key terms

Two-way table: a table that shows information about two features of data at the same time.

A lot of information is shown by just those four frequency values. Can you find the nine completely different facts that you can get from this table?

1 There were 7 male adult visitors.
2 There were 5 male child visitors.
3 There were 23 female adult visitors.
4 There were 16 female child visitors.
5 There were 12 male visitors.
6 There were 39 female visitors.
7 There were 30 adult visitors.
8 There were 21 child visitors.
9 There were 51 visitors altogether.

Often row (across) and column (down) totals are given as part of the two-way table, so that this type of information is easy to obtain straight away. This is what the same two-way table looks like with these totals on:

Visitors to shop	Adult	Child	Total
Male	7	5	12
Female	23	16	39
Total	30	21	51

You need to be careful not to mix up the total with the values of the actual data. Other information such as comparisons can be made using fractions and percentages. Sometimes some of the values in a two-way table are missing and the totals or other information can be used to find them.

Example 1

The two-way table shows information about 80 sheep in a field. The sheep are either black or white and some of them have been sheared.

Sheep	Black	White	Total
Sheared	5		20
Unsheared		35	
Total			

a Complete the table.

b What percentage of the sheep has been sheared?

c What fraction of the sheep is black?

Solution

a White sheared sheep = 20 – 5 = 15

Total of white sheep = 15 + 35 = 50

Total of unsheared sheep = 80 – 20 = 60

Unsheared black sheep = 60 – 35 = 25

Total of black sheep = 5 + 25 = 30

The completed table looks like this:

Sheep	Black	White	Total
Sheared	5	15	20
Unsheared	25	35	60
Total	30	50	80

b $\frac{20}{80} \times 100 = 25\%$ **c** $\frac{30}{80} = \frac{3}{8}$

AQA Examiner's tip

Notice that in the question about sheep the total of 80 mentioned at the start was needed to find the answers – always read the question carefully as some information might only be given there.

Not all two-way tables only have two choices for each variable.

Example 2

Design a two-way table which could hold the following information:

Diners at a party can choose from soup, pate or melon for their starter course. They can choose from steak, chicken or vegetable lasagne for their main course.

Solution

There are three choices for starter and three choices for main. Do not forget that a row and a column are also needed to list the items, and a row and a column for the totals. Therefore a 5 by 5 table is needed (but it is still called a two-way table).

Starter/main	Steak	Chicken	Vegetable lasagne	Total
Soup				
Pate				
Melon				
Total				

This type of table can also be drawn to be used as a data-collection sheet or observation sheet.

Exercise 3.2

1 The two-way table shows information about the weather during the day and night during one month.

Night / Day	Clear	Cloudy but dry	Rain
Sunny	5	3	2
Cloudy but dry	3	4	1
Rain	3	6	3

a How many times during the month was there a sunny day but rain at night?

b Which combination of day and night weather was most common?

c How many times during the month were the day and night completely dry?

2 The two-way table has some of the information about the favourite healthy fruit snacks of 100 children.

a Complete the table.

	Raisins	Apple	Banana	Total
Years 2–4	15	1		21
Years 5–6		8		
Years 7–11	3		24	45
Total	28			

b What percentage of these children prefer bananas as their choice of healthy snack?

3 A cinema has three different screens that show films: Screen 1, Screen 2 and Screen 3. The cinema only shows either children's films, comedy films or adventure films. Barry has collected data about the type of film shown on each of the screens. Construct a two-way table that Barry could use to show this data.

3.3 Using published tables of data

The government, along with newspapers, magazines and websites, regularly publishes tables containing data on a wide range of topics and covering many issues. Sometimes the data may look simple, but interpreting the values may be difficult.

Look at this data for weekly number of deaths for one month in 2008. Each row has a footnote. This means that there is extra information related to the data in the row that might be important. Also, for no apparent reason one figure is missing. This type of issue can cause real difficulty when using real data.

Objectives

In this topic you will learn:

how to read and interpret data from media and national sources.

	17 Oct	24 Oct	31 Oct	7 Nov
Total deaths, all ages[1]	9501	9353	9280	9900
Deaths by underlying cause[2]	1174	1246	1159	-
Total deaths: average comparable week over last 5 years[3]	9395	9547	9678	9871

Footnotes

[1]Total deaths all ages.

[2]Deaths by underlying cause excludes deaths under 28 days.

[3]This average is based on the actual number of death registrations recorded for each week over each of the last five years.

Source:www.statistics.gov.uk/StatBase/ssdataset.asp?vlnk=6157&More=Y

It is important – and not just for this statistics course – that you understand how to read and interpret the data shown. Here are some more examples of real data tables. The source of each table (in other words where it came from) is given underneath each one.

A *Information relating to the place of birth of residents of British national parks*

	All people	People born in: England	People born in: Scotland	People born in: Wales	People born in: Northern Ireland	People born in: Republic of Ireland	People born in: other EU countries	People born elsewhere
Dartmoor National Park	33 552	30 652	516	702	111	152	412	1007
Exmoor National Park	10 873	10 114	127	182	29	38	93	290
Lake District National Park	41 831	38 538	1169	422	145	145	391	1021
Northumberland National Park	1935	1806	85	9	-	3	7	25
North York Moors National Park	23 939	22 667	384	146	57	48	223	414
Peak District National Park	37 937	36 024	449	408	87	113	246	610
The Broads National Park	5876	5473	88	61	21	20	46	167
Yorkshire Dales National Park	19 654	18 543	326	159	45	52	166	363
Brecon Beacons National Park	32 609	9159	380	21 760	116	113	342	739
Pembrokeshire Coast National Park	22 542	7526	254	13 901	71	127	224	439
Snowdonia National Park	25 482	8682	190	15 953	66	92	167	332

www.statistics.gov.uk/statbase/Expodata/Spreadsheets/D6559.xls

B *Information relating to fires in a Lincolnshire village/district/region and for England*

	Scotter	West Lindsey (Non-Metropolitan District)	East Midlands	England
Total Fires (Incidents)	15	276	11714	140 280
Total Fires; Accidental (Incidents)	~	146	5684	73 824
Total Fires; Deliberate (Incidents)	~	130	6030	66 456
Dwelling Fires (Incidents)	~	56	3244	46 606
Dwelling Fires; Accidental (Incidents)	~	52	2562	38 418
Dwelling Fires; Deliberate (Incidents)	~	4	682	8188
Other Building Fires (Incidents)	~	50	2486	29 375
Other Building Fires; Accidental (Incidents)	~	30	1386	17 715
Other Building Fires; Deliberate (Incidents)	~	20	1100	11 660
Road Vehicle Fires (Incidents)	~	113	4905	55 025
Road Vehicle Fires; Accidental (Incidents)	~	36	1347	14 690
Road Vehicle Fires; Deliberate (Incidents)	~	77	3558	40 335
Other Outdoor Location Fires (Incidents)	~	57	1079	9274
Other Outdoor Location Fires; Accidental (Incidents)	~	28	389	3001
Other Outdoor Location Fires; Deliberate (Incidents)	~	29	690	6273
Total Casualties (Incidents)	3	15	783	11 638
Fatal Casualties (Incidents)	~	0	22	387
Non Fatal Casualties (Incidents)	~	15	761	11 251

http://neighbourhood.statistics.gov.uk/dissemination/

C *Life expectancy, heathy life expectancy and disability-free life expectancy at birth: by sex*

	Males		Females	
	1981	2002	1981	2002
Life expectancy	70.9	76.0	76.8	80.5
Healthy life expectancy	64.4	67.2	66.7	69.9
Years spent in poor health	6.4	8.8	10.1	10.6
Disability-free life expectancy	58.1	60.9	60.8	63.0
Years spent with disability	12.8	15.0	16.0	17.5

Source: Government Actuary's Department, ONS

AQA Examiner's tip

Every examination paper is likely to have at least one question where you are given a table of real published data, on which you will be asked questions.

D *Information relating to tides at Gorleston on the Norfolk Broads*

	HIGH water				LOW water			
Date	**Time**	**Metres**	**Time**	**Metres**	**Time**	**Metres**	**Time**	**Metres**
August								
Fri 22nd	01:16	1.74	13:06	2.17	06:47	0.31	19:47	0.14
Sat 23rd	01:53	1.70	13:46	2.09	07:24	0.38	20:30	0.33
Sun 24th	02:38	1.66	14:39	1.95	08:12	0.49	21:27	0.53
Mon 25th	03:40	1.64	15:54	1.79	09:35	0.59	22:42	0.68
Tue 26th	05:05	1.66	17:51	1.71	11:22	0.57	*	*
Wed 27th	06:18	1.75	19:30	1.75	00:00	0.73	12:52	0.40
Thu 28th	07:16	1.86	20:47	1.84	01:12	0.70	14:07	0.18
Fri 29th	08:08	1.99	21:36	1.93	02:19	0.60	15:04	-0.03
Sat 30th	08:59	2.12	22:16	1.97	03:11	0.48	15:51	-0.20
Sun 31st	09:49	2.24	22:52	2.00	03:53	0.34	16:33	-0.30

E *A collection of tables about students in Higher Education – essential facts*

What graduates earn

		Starting salary
1	Medicine	£30 520
2	Dentistry	£28 030
3	Chemical Engineering	£25 136
4	Economics	£24 466
5	Veterinary medicine	£23 437
6	Physics and astronomy	£22 624
7	Social work	£22 566
8	General engineering	£22 488
9	Mechanical engineering	£22 364
10	Civil engineering	£22 364
11	Building	£21 914
12	Land and property management	£21 769
13	Mathematics	£21 760
14	Electrical and electronic engineering	£21 656
15	Aeronautical & manufacturing engineering	£21 527

Spending on facilities

		Per student per year
1	Imperial College	£3218
2	Oxford	£2884
3	Cambridge	£2299
4	Warwick	£1881
5	School of Oriental and African Studies	£1746
6	University College London	£1702
7	King's College London	£1696
8	Bedfordshire	£1685
9	Abertay	£1671
10	Aston	£1607
11	London School of Economics	£1562
12	Bristol	£1535
13	Newcastle	£1481
14	Southampton	£1479
15	Brunel	£1475

Staff/student ratio		
1	University College London	9.1
2	Imperial College	10.4
3	School of Oriental and African Studies	10.5
4	Oxford	11.6
5	King's College London	11.9
6	Cambridge	12.2
7	London School of Economics	12.6
8	St Andrews	12.6
9	Lancaster	12.7
10	Manchester	12.8
11	Queen Mary, London	12.8
12	York	13.1
13	Edinburgh	13.3
14	Liverpool	13.3
15	Glasgow	13.4

Top for living at home		
1	Wolverhampton	8900
2	Glasgow Caledonian	7845
3	Ulster	7210
4	London Metropolitan	6880
5	Westminster	6635
6	Manchester Metropolitan	6505
7	Glasgow	6170
8	Northumbria	6080
9	Kent	6040
10	Glamorgan	5905
11	Central Lancashire	5870
12	Kingston	5785
13	Hertfordshire	5635
14	Queen Mary, London	5630
15	Middlesex	5630

Top for ethnic minorities		
1	Open University	13 545
2	Middlesex	12 895
3	Greenwich	11 600
4	Westminster	11 285
5	Kingston	10 425
6	Hertfordshire	10 210
7	East London	10 190
8	City University	10 035
9	London South Bank	9905
10	Thames Valley	9075
11	Birmingham City	8925
12	King's College London	8475
13	Brunel	8205
14	Wolverhampton	8120
15	Manchester	8105

Top for disabled students		
1	Open University	11 770
2	Plymouth	2815
3	De Montfort	2450
4	Nottingham	2415
5	University of the Arts, London	2285
6	Birmingham City	2230
7	Manchester	2205
8	Manchester Metropolitan	2180
9	Southampton	2110
10	Ulster	2095
11	Leeds	2065
12	Sheffield Hallam	1995
13	Brighton	1975
14	West of England, Bristol	1970
15	Hull	1915

Source: HESA

F *Information about smoking rates in the USA*

	1965	1970	1974	1978	1980	1983	1985	1987	1990	1993	1995	1997	1999	2001	2002	2003	2004	2006
Smoking Status Total Population																		
Current	42.4	37.4	37.1	34.1	33.2	32.1	30.1	28.8	25.5	25.0	24.7	24.7	23.5	22.8	22.5	21.6	20.9	20.8
Former	13.6	18.5	19.5	20.8	21.3	21.8	24.2	22.8	24.6	24.6	23.3	22.8	23.1	22.2	22.6	21.8	21.4	21.0
Never	44.0	44.2	43.4	45.0	45.5	46.1	45.8	48.4	49.9	50.5	52.0	52.4	53.5	55.0	54.9	56.6	57.7	58.2
Sex																		
Male																		
Current	51.9	44.1	43.1	38.1	37.6	35.1	32.6	31.2	28.4	27.7	27.0	27.6	25.7	25.2	25.2	24.1	23.4	23.9
Former	19.8	26.3	27.7	28.3	28.1	28.3	30.9	28.9	30.3	29.9	27.5	27.0	27.3	26.4	26.4	25.2	24.8	24.5
Never	28.3	29.6	29.2	33.6	34.4	36.6	36.5	39.9	41.3	42.4	45.5	45.4	47.0	48.5	48.4	50.7	51.7	51.6
Female																		
Current	33.9	31.5	32.1	30.7	29.3	29.5	27.9	26.5	22.8	22.5	22.6	22.1	21.5	20.7	20.0	19.2	18.5	18.0
Former	8.0	11.6	12.7	14.2	15.1	15.9	18.1	17.4	19.5	19.7	19.5	19.0	19.2	18.3	19.1	18.7	18.3	17.8
Never	58.1	56.9	55.2	55.2	55.5	54.6	54.0	56.0	57.7	57.8	57.9	58.9	59.3	61.1	60.9	62.1	63.2	64.2

As you can see, tables come in varying shapes and sizes, and cover a multitude of areas, some more interesting than others.

Example

Use the tables to answer these questions.

a How many people living in the Lake District National Park were born in England?

b How many universities have more than 6000 of their students living at home?

c How much longer (in years) is a female born in 2002 expected to live compared with a male born in 2002?

d What times were low tide at Gorleston on 28 August?

e How many road vehicle fires were there in the East Midlands?

f Look at Table **F**. Explain why the total population column figures for 1970 do not add up to 100 per cent.

Solution

a Table **A** – 38 538

b Table **E** – 9

c Table **C** – 80.5 – 76.0 = 4.5 years

d Table **D** – 01:12 and 14:07

e Table **B** – 1347 + 3558 = 4905

f The figures add up to 100.1 per cent. This is quite common in tables of percentages (often if the percentages are whole numbers the total will be 99 per cent or 101 per cent). This will be due to rounding effects in the original figures when they are rounded, in this case to one decimal place. Other factors are a possibility in other situations, such as non-response.

Exercise 3.3

1. What percentage of males smoked in the USA in 2006?
2. What were the heights of the two high tides in Gorleston on Monday 25 August?
3. Which gender's life expectancy has gone up more between 1981 and 2002? Explain your answer.
4. Which National Park has the fewest people living in it?
5. True or false: Glasgow has the worst staff/student ratio in the UK. Explain your answer.
6. What was the total number of casualties from fires in West Lindsey district?
7. Look at this data from the Government Statistics website. Answer the questions that follow.

Total Fertility Rate (1.5 – 2.0)	Statistical Region	Total fertility rates 1986	2001	2006	Actual change in total fertility rates 1986–2001 (largest decrease ranked[1])	2001–2006 (largest increase ranked[1])
	North East	1.79	1.55	1.79	-0.24 (1)	0.24 (2)
	North West	1.86	1.65	1.89	-0.21 (3)	0.24 (3)
	Yorkshire and The Humber	1.80	1.66	1.85	-0.14 (4)	0.19 (10)
	East Midlands	1.74	1.61	1.82	-0.13 (6)	0.21 (7)
	West Midlands	1.85	1.74	1.96	-0.11 (7)	0.22 (4)
	East	1.75	1.67	1.87	-0.08 (10)	0.20 (8)
	London	1.72	1.62	1.86	-0.10 (8)	0.24 (1)
	South East	1.71	1.62	1.84	-0.09 (9)	0.22 (5)
	South West	1.71	1.58	1.79	-0.13 (5)	0.21 (6)
	Wales	1.86	1.65	1.85	-0.21 (2)	0.20 (9)
	England and Wales	1.77	1.63	1.86	-0.14	0.23

■ 1986 ■ 2001 ■ 2006

[1] The actual change in fertility rates represents the increase or decrease in the number of children per women over the period shown and has been calculated using the TFRs reported to two decimal places.

www.statistics.gov.uk/downloads/theme_population/PT133_part1.pdf

a What was the fertility rate for:
  i the North West in 2006
  ii the East in 2001
  iii the whole of England and Wales in 1986?

b Which region had the fourth largest decrease in rates from 1986–2001?

c Which region had the fifth largest increase in rates from 2001–2006?

d In what way might the bar graph from the left hand side be seen as misleading? (See Chapter 10.)

Be a statistician

You are a statistician who also works for a newspaper. You have been asked to write a short report using one of the above tables as your source of data. Write this report using no more than 200 words. Try to mention some of the most interesting facts from the table you choose to use. Give your report an amusing or appropriate title to grab the interest of potential readers.

Summary

You should:

- be able to use tallying with five-bar gates when compiling frequency tables
- know that class intervals are used for continuous data tables
- know that class intervals should be chosen to give 4–10 classes, where possible of equal width
- know how to use an open-ended final class if needed
- know how to use two-way tables to show data with two features
- be competent at interpreting data from published tables on the internet and in the media.

AQA Examination-style questions

1. Membership of Midton Golf Club is open to all women in the town. Of the members aged 40 and over, 58 are single and 30 are divorced. Altogether 52 of the women are married.

(a) Put these values into the table below: *(3 marks)*

Age	Single	Married	Divorced	Total
Under 40		25	12	45
40 and over				
Total				160

(b) Complete the remaining parts of the table. *(4 marks)*

AQA, 2003

2. The data show the number of televisions in 30 homes.

2	3	1	4	2	3	2	1	1	4
2	2	2	3	1	3	2	2	3	1
2	1	4	2	1	1	2	4	3	2

Complete the tally chart and the frequency column.

Number of televisions	Tally	Frequency
1		
2		
3		
4		

(4 marks)

AQA, 2007

3. The table below gives the results of a survey carried out on a housing estate.

		Number of adults in household				
		1	**2**	**3**	**4**	**Total**
Number of children in household	**0**	90	125	31	9	225
	1	9	40	15	6	70
	2	20	22	10	4	56
	3	12	7	4	0	23
	4	8	0	1	0	0
	Total	139	194	61	19	

For example, 40 households have 2 adults and 1 child.

(a) How many households:

(i) have three adults and two children *(1 mark)*

(ii) have two adults *(1 mark)*

(iii) took part in the survey? *(2 marks)*

(b) For this survey what is the greatest number of people in a household? *(2 marks)*

AQA, 2004

4. The two-way table shows the number of eggs and the number of slices of bacon bought by 35 boys at a school breakfast canteen one morning.

		Number of slices of bacon			
		0	1	2	3
Number of eggs	0	7	2	1	2
	1	1	5	4	3
	2	1	2	6	1

(a) How many boys bought three slices of bacon and one egg? *(1 mark)*

(b) How many boys bought exactly two eggs? *(1 mark)*

(c) How many boys bought at least two slices of bacon? *(2 marks)*

(d) Look at the 7 in the table. How does the breakfast of these seven boys differ from that of the other 28 boys in the table? *(1 mark)*

(e) Deborah said. 'Each boy bought more slices of bacon than eggs for their breakfast.' Is Deborah correct? Explain how you know. *(1 mark)*

(f) Show that the total number of eggs bought was 33. *(2 marks)*

AQA, 2006

5. The number of cars passing a school each minute is recorded for one hour. The frequency table shows some of the results.

Number of cars per minute	Frequency
1	1
2	3
3	2
4	4
5	4
6	3
7	8
8	11
9	
10	7
11	3
12	0
13	2
14	1
15	1

(a) Calculate the missing value in the frequency column. *(2 marks)*

(b) Complete the grouped frequency distribution for the same data.

Number of cars per minute	Frequency
1–4	10
5–8	
9–12	
13–16	

(3 marks)

(c) Give a reason why the original frequency table could be preferred to show the data. *(1 mark)*

(d) Give a reason why the grouped frequency table could be preferred to show the data. *(1 mark)*

AQA, 2006

6. The table shows the percentage by age for each ethnic group of the UK population 2001–2002.

	Age			
Ethnic Group	**Under 16**	**16–34**	**35–64**	**65 and over**
White	19	25	40	16
Mixed	55	27	16	2
Indian	22	34	38	6
Pakistani	35	36	25	4
Bangladeshi	38	38	21	3
Other Asian	22	36	38	4
Black Caribbean	24	25	42	9
Black African	33	35	30	2
Other Black	35	34	26	5
Chinese	20	40	35	5
Other	20	37	39	4

Source: Adapted from Office for National Statistics, Summer 2003

(a) Which ethnic group had the largest percentage of its population under 16 years of age? *(1 mark)*

(b) What was the difference between the percentages of Chinese ethnic group and Black African ethnic group aged 35–64 years? *(2 marks)*

(c) Give one similarity and one difference between the age profiles of the White ethnic group and the Indian ethnic group. *(2 marks)*

AQA, 2006

7. This table gives the size of households in Great Britain between 1971 and 2001. For example, in 1991, 34 per cent of households consisted of two people.

	1971	1981	1991	2001
One person	17	22	27	28
Two people	33	32	34	35
Three people	19	17	16	16
Four people	17	18	16	14
Five or more people	14	11	8	7

Source: Adapted from Social Trends 2005

(a) What percentage of households in 1981 consisted of four people? *(1 mark)*

(b) Throughout the period 1971–2001 what size household accounted for about a third of the households? *(1 mark)*

(c) The total of the percentages for 1991 is 101 per cent. Give a possible reason for this. *(1 mark)*

AQA, 2007

4 Basic forms of data representation

In this chapter you will learn how to draw and interpret:

- pictograms
- pie charts
- proportional pie charts
- bar charts
- multiple bar charts
- composite bar charts
- dot plots
- vertical line diagrams
- stem-and-leaf diagrams
- frequency polygons
- histograms with equal class intervals
- histograms without equal class intervals
- population pyramids.

You should already know:

- ✔ your multiplication tables
- ✔ that there are 360 degrees in a circle
- ✔ that the area of a circle is found using πr^2
- ✔ how to use a protractor to draw and measure angles.

What's this chapter all about?

This is the place to find out about all the simple and standard types of diagrams you might want to use to display the data that you have so carefully collected and tabulated. But be careful – the real skill here is to avoid using the wrong type of diagram for the wrong type of data. So you will find there is a separate topic for each of the possible types of data you could have. Remember, too, that it's not just about drawing things. Can you interpret what you see and write about what these diagrams show you? Making good choices about your diagrams and interpreting them well will help in your controlled assessment too!

Focus on statistics

You will not have met some of these diagrams before, at least not from doing your maths GCSE work. At the end of topic 4.1 there are proportional pie charts – you will definitely not have met these in maths. Similarly, the simple idea of dot plots in 4.2 is not covered in maths, although the dot plots are easy to draw. In 4.4 you will also find work on population pyramids, which you might have seen in other subjects such as geography but are very unlikely to have met in maths.

4.1 Representing qualitative data

Remember that **qualitative data** is not numerical but describes a situation. This could be qualities such as colour, grade or opinions.

This type of data is represented by pictograms, bar charts, pie charts, dot plots and comparative pie charts. Some of these diagrams can also be used to represent numerical information.

Objectives

In this topic you will learn how to draw and interpret:

- pictograms
- pie charts
- proportional pie charts.

Key terms

Qualitative data: non-numerical data that describes a situation.

Pictogram: a pictogram uses symbols to represent items of data.

Pictograms

A **pictogram** uses symbols to represent a number of items. You need a key to show the number of items that one symbol represents. Often the symbol has a connection with the item it is representing. For example, if you kept a record of how many DVDs you bought each month.

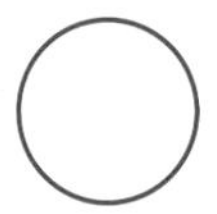

could be used to represent four DVDs bought one month.

You could then use part of a symbol to represent fewer than four DVDs:

would represent two DVDs

would represent one DVD.

Example 1

The table shows the number of people waiting for a shop to open each morning one week.

Day	Number of people
Monday	8
Tuesday	5
Wednesday	2
Thursday	3
Friday	7
Saturday	12

Draw a pictogram using the symbol to represent two people waiting. Remember to provide a key.

Solution

A pictogram looks tidy if you draw it within a table or lines. Each day needs its own row. All symbols should be lined up, and be drawn to about the same size.

Day	
Monday	
Tuesday	
Wednesday	
Thursday	
Friday	
Saturday	

Key: represents 2 people

Pictograms are good for comparison and it is easy to see patterns. A completed pictogram is visually similar to a simple bar chart. However, one disadvantage is that it can take time to draw lots of symbols.

AQA Examiner's tip

If you are not given a key in the question you must write one of your own. If you do not give a key you will lose a mark in the exam.

Pie charts

A **pie chart** shows information in a circle. The angle of each sector in the circle is in proportion to the amount of information it shows. The angles in a pie chart must add up to 360°. A key must be shown for a pie chart unless you label every sector in full.

Key terms

Pie chart: a circular diagram where the angle of a sector represents the frequency of an item.

Example 2

The table shows the sales of some items in a bakery in one hour.

Item	Meat pie	Sausage roll	Doughnut	Apple pie	Date square
Number Sold	12	19	8	11	10

Draw a pie chart to show the information.

Solution

Follow these steps to complete the pie chart.

- Find the total number of items to be represented by the pie chart.

$$12 + 19 + 8 + 11 + 10 = 60$$

- Find the angle that will represent one item.

$$360^\circ \div 60 = 6^\circ \text{ per item}$$

total degrees around centre of a circle — total number of items in this example

- Find the angles for each of the items.
  To do this multiply the answer to step 2 by each of the item totals.
  For example, meat pie = $12 \times 6^\circ = 72^\circ$
  Similarly we get the remaining angles as below.

Item	Meat pie	Sausage roll	Doughnut	Apple pie	Date square
Number sold	12	19	8	11	10
Angle (degrees)	72	19 × 6 = 114	8 × 6 = 48	11 × 6 = 66	10 × 6 = 60

Check that your angles total 360°. If they do not total 360°, then you know you have made a mistake.

- Draw a circle and mark the angles around the circle. They should fill the circle completely and have no overlap.

AQA Examiner's tip

In the exam there will be a circle on the page on which to draw your pie chart, which will look like this:

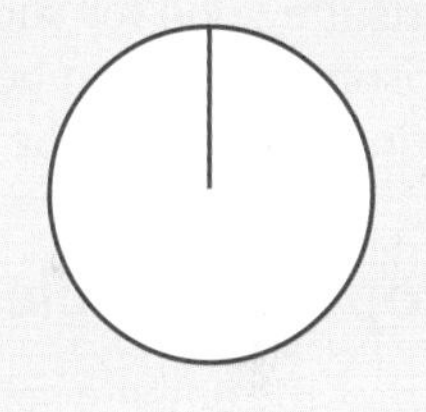

- Label each sector with the correct label, or it may be better to use shading and a key, especially if some of the sectors are small. Also give your pie chart a title.

Here is the finished pie chart for the information about the bakery. This example uses shading and a key.

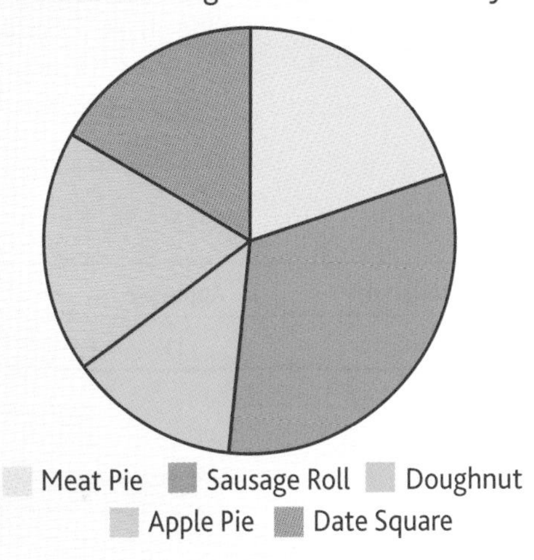

AQA Examiner's tip

It is important that you have the appropriate equipment with you for drawing a pie chart. If you do not have a protractor, the maximum marks you could earn are 2 out of 4 for calculating the angles you would draw.

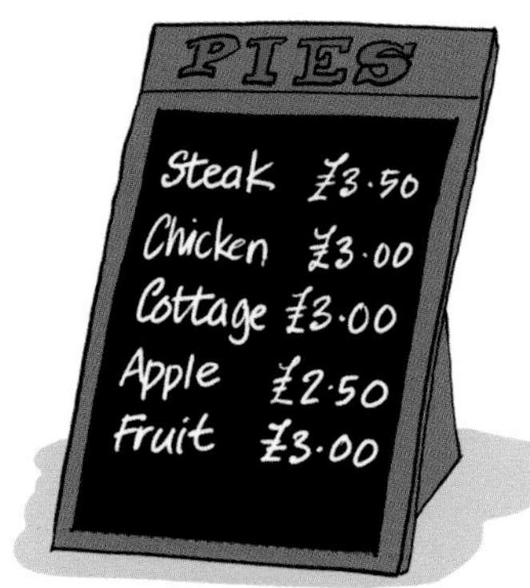

Pie charts are good for showing proportions and also good for comparisons. But it can be time consuming to do the necessary calculations and draw the charts.

Proportional pie charts

If you want to compare sets of related qualitative data that have different totals using pie charts, then **proportional pie charts** should be used.

The difference between these and ordinary pie charts is that the circles will have different areas according to the total frequency for each data set. This then gives a visual comparison of both the different items in a data set and the different total number of items in the two (or more) data sets.

Key terms

Proportional pie chart: the areas of proportional pie charts are used to represent and compare the frequency of items from two different samples.

Example 3

a Draw comparative pie charts to show the data for the bakery in two different hours. Use a radius of 5 cm for the 11 am–12 pm data.

11 am–12 pm

Item	Meat pie	Sausage roll	Doughnut	Apple pie	Date square
Number sold	12	19	8	11	10

12 pm–1 pm

Item	Meat pie	Sausage roll	Doughnut	Apple pie	Date square
Number sold	18	32	11	8	20

b Make two comparisons between the two times.

Solution

The first pie chart can be drawn as in the last example, making sure the radius is exactly 5 cm. (Please note the pie charts on this page are not shown at full size.) The angles for each of the items for the 12 pm–1 pm time can be worked out as before (work through these carefully – notice that they are not nice numbers, so round them to the nearest whole number).

The main work for the 12 pm–1 pm pie chart is to calculate the area that the circle needs to have, in order to be in proportion to the number of items that will be held within it. The first chart was for 60 items. The second chart is for 89 items, so the radius will be larger than 5 cm.

The ratio of the areas will be the same as the ratio of the total frequencies.

$\frac{\pi \times r^2}{\pi \times 5^2} = \frac{89}{60}$ where r is the radius of the new circle

So $r^2 = \frac{89 \times 25}{60}$ cancelling π and rearranging

$r = \sqrt{37.083\ldots}$

$= 6.0896086\ldots$

For drawing purposes the radius can be rounded to 6.1 cm.

AQA Examiner's tip

You will be given the radius for the first circle and will be expected to calculate the radius for the second circle.

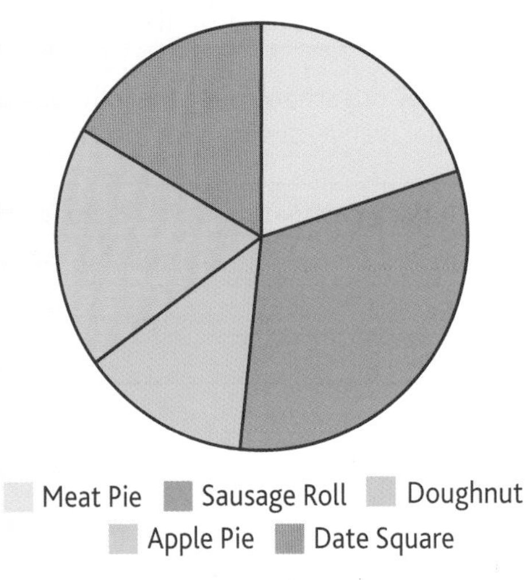

Pie chart showing items sold in a bakery 11 am–12 pm

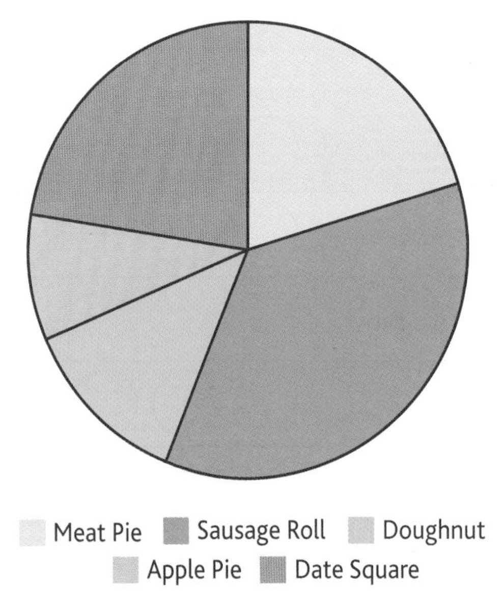

Pie chart showing items sold in a bakery 12 pm–1 pm

Exercise 4.1

1 Annette went fishing on each of five days. The table shows how many fish she caught on each of the days. Illustrate this data with a pictogram. Use this symbol to represent two fish: 🐟

Day	Monday	Tuesday	Wednesday	Thursday	Friday
Number of fish	8	4	7	3	11

2 A survey in a school found the following countries were the most popular holiday destinations for families.

Holiday destination	Frequency
UK	9
USA	6
Spain	12
France	7
Canary Islands	5
Other	6

Draw a fully labelled pie chart to illustrate this data.

3 The pictogram shows the number of DVDs hired from a small shop over five days.

Monday	○ ○ ○
Tuesday	○ ○ ○ ○ ◖
Wednesday	○ ○ ◖
Thursday	
Friday	○ ○ ○ ○ ◖

Key: ○ represents 20 DVDs

a How many DVDs were hired on Monday?

b How many DVDs were hired on Friday?

c On Thursday 70 DVDs were hired. Complete the pictogram.

d Draw a pie chart to show the number of DVDs hired throughout this week.

4 A pie chart is drawn with radius 6 cm to show information about the reading habits of 100 men. A proportional pie chart is to be drawn to show information about the reading habits of 150 women. Calculate the radius of the pie chart for the women.

4.2 Representing discrete data

Bar charts

You need to know about the three different types of bar chart. For a single set of data, an ordinary **bar chart** looks like this:

For two or more sets of data to be shown on the same chart you have two options. A **multiple bar chart** (or **dual bar chart** if there are only two sets of data) looks like this:

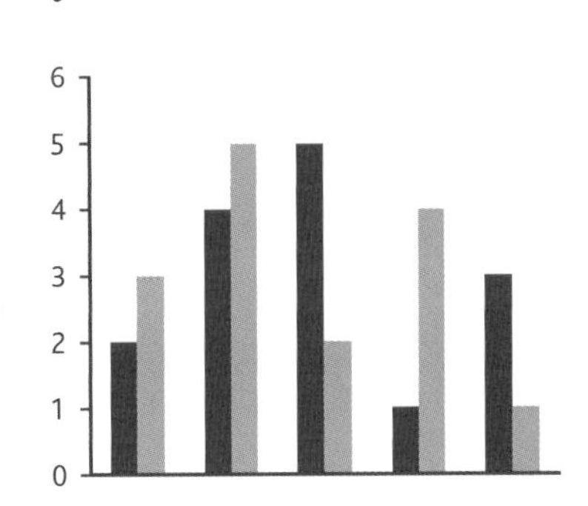

In a multiple bar chart it is easier to compare the different coloured bars in each pair. A **component** or **composite bar chart** (of the same data) looks like this:

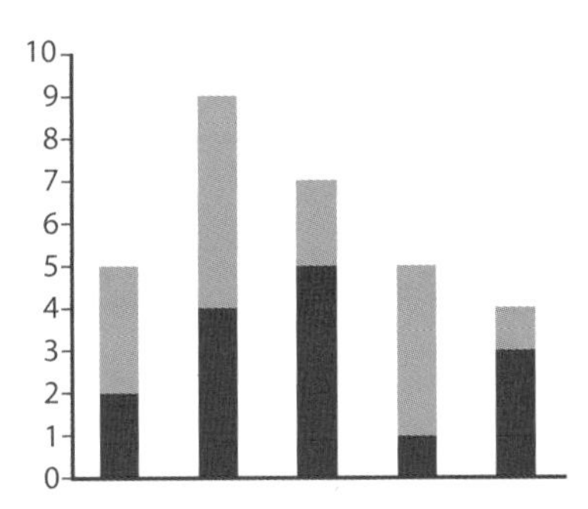

In a component bar chart is it easier to compare the total of the two bars across the chart.

For each type of chart it is possible, although unusual, for the bars to run horizontally.

All bar charts are types of **frequency diagrams**.

They should have

- a title
- labels on each axis
- values on the vertical axis evenly spaced starting at zero
- bars of equal width
- gaps between the bars (or sets of bars on a multiple bar chart).

Objectives

In this topic you will learn how to draw and interpret:

- bar charts
- multiple bar charts
- composite bar charts
- dot plots
- vertical line diagrams.

Key terms

Bar chart: a frequency diagram where the height of a bar represents the frequency of an item.

Multiple bar chart: a frequency diagram where two or more sets of data are presented in groups of bars for comparison.

Dual bar chart: a multiple bar chart with specifically two sets of data.

Component bar chart: a frequency diagram for two or more sets of data with each set of data wholly contained within one bar.

Composite bar chart: another name for a component bar chart.

Frequency diagrams: all charts or diagrams that compare the frequencies of objects.

Spot the mistake

What is missing from the bar charts above? There are at least three answers.

Example 1

The dual bar chart shows the numbers of boys and girls missing from each of Tacey's lessons one day.

a Which lesson had the most pupils missing?

b Why would part (a) be easier to answer using a component bar chart?

c Did more boys or girls miss the lessons Tacey was in on that particular day? Justify your answer.

AQA **Examiner's tip**

In questions such as part **(c)** Example 1, you will only get marks if you justify your answer.

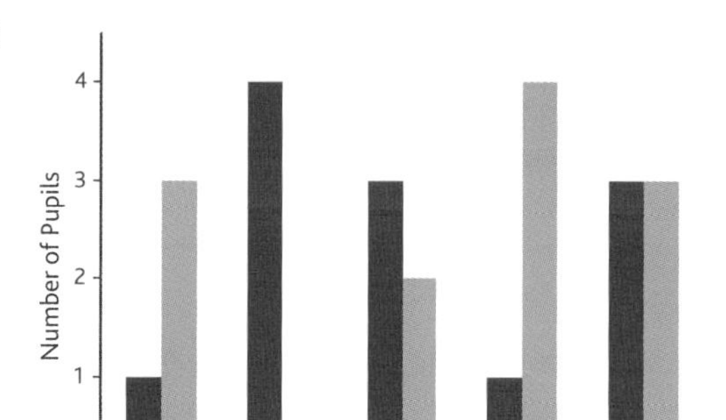

Number of pupils missing from lessons

Solution

a Add up the totals for each lesson.

Lesson 1 is 1 boy and 3 girls, which is $1 + 3 = 4$

Lesson 2 is 4 boys and no girls, which is $4 + 0 = 4$

Lesson 3 is $3 + 2 = 5$

Lesson 4 is $1 + 4 = 5$

Lesson 5 is $3 + 3 = 6$

So Lesson 5 had the most pupils missing.

b A component bar chart would show the total for each lesson, as the bars are stacked on top of each other, so you could simply read off the total at the top of the pair of bars.

c Boys $= 1 + 4 + 3 + 1 + 3 = 12$

Girls $= 3 + 0 + 2 + 4 + 3 = 12$

The same number of boys and girls missed lessons.

Real world

Ageing Population

16 per cent of the UK population is aged 65 or over

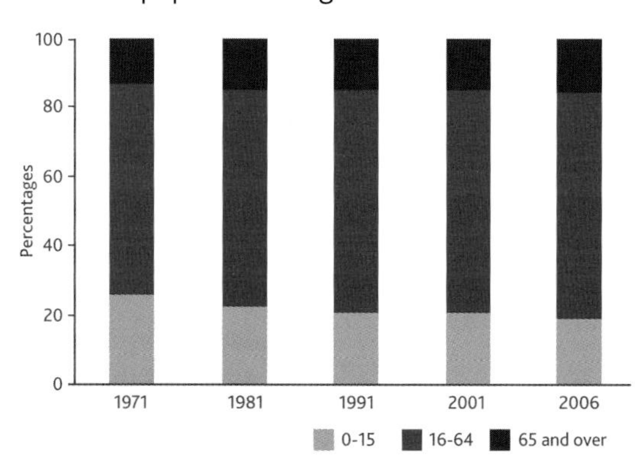

Population: by age, United Kingdom

This graph is another type of bar chart which is from the government website **www.statistics.gov.uk**

It is a percentage component bar chart and is drawn so that the total heights of all the bars represent 100 per cent.
This can be used to make comparisons between the different percentages of the UK population in each age group during the 35 years up to 2006.

With a friend, write down three facts that you can get from this chart. Design a poster to show examples of different types of bar charts used on the government's website.

Dot plots

A dot plot is very similar to a bar chart but is used for small data sets only. One dot represents one item or number.

Example 2

A class of 20 statistics pupils were each asked to choose a digit between 1 and 9 'at random'. The table shows the results.

Digit	Number of pupils choosing that digit
1	2
2	0
3	0
4	1
5	4
6	1
7	8
8	1
9	3

a Illustrate the data on a dot plot.

b Do you think that the pupils made random choices? Explain your answer.

Solution

a Notice the dot plot has no vertical axis. The dots are spread out evenly and are the same size.

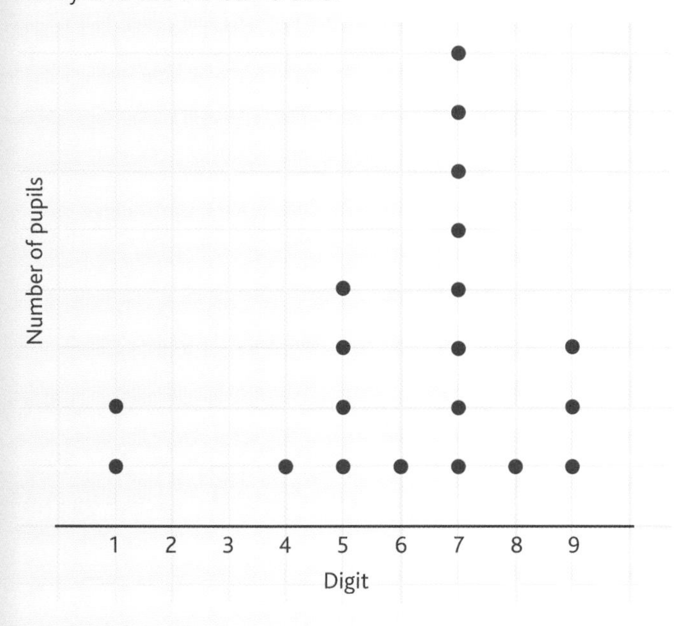

b If the pupils had made random choices, the dots should be fairly evenly spread across the chart. This has not happened as the choices seem biased towards 7.

Be a statistician 

Conduct an experiment like the one above. Ask a group of pupils to give you a number between 1 and 9. What precautions do you need to take to get fair data? Write a possible hypothesis about this situation. Show the results in a dot plot. Make a conclusion referring to your hypothesis.

Vertical line diagrams

A **vertical line diagram** is used to show quantitative discrete data. It looks very much like a bar chart with exceedingly thin bars. Another way of thinking of a vertical line diagram is that it is like a dot plot but with lines instead of dots! A vertical line diagram also has a vertical scale.

For example, here is the data for the dot plot above redrawn as a vertical line diagram.

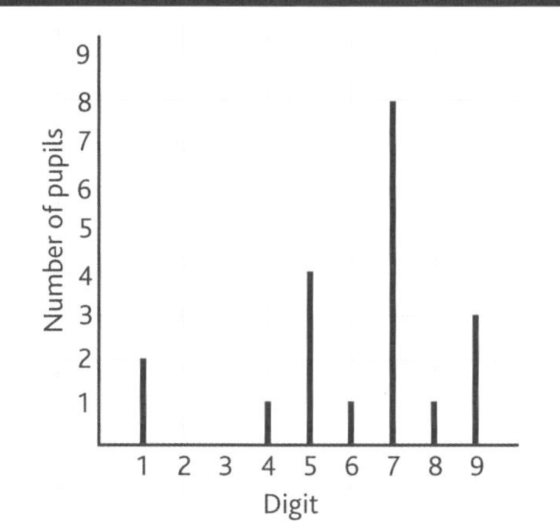

Key terms

Vertical line diagram: a diagram that is like a bar chart but with pencil thin bars and is more often used for discrete quantitative data.

Exercise 4.2

1 The bar chart shows the drinks sold by a machine one Monday.

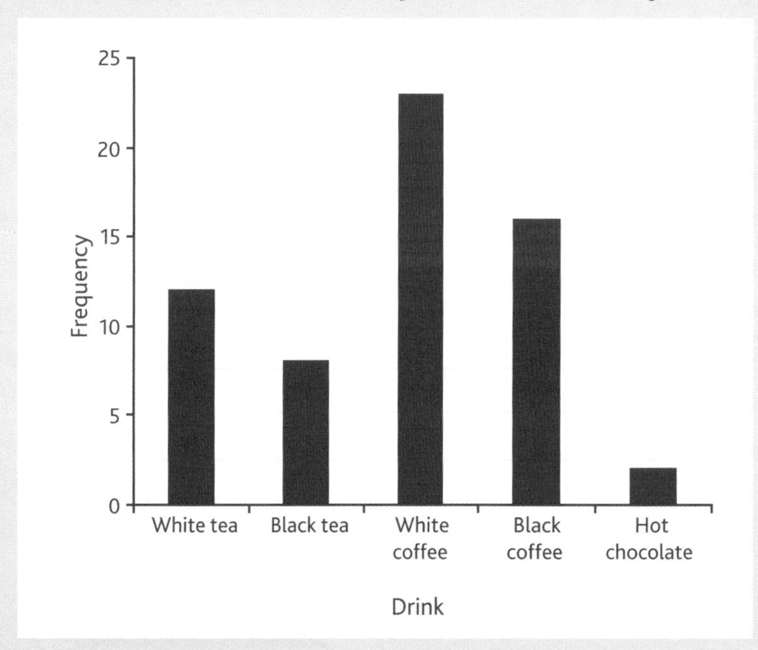

a Which drink was most popular?

b Which drink was sold on 16 occasions?

c How many coffees were sold altogether?

d On the Tuesday the following drinks were sold:

Drink	White tea	Black tea	White coffee	Black coffee	Hot chocolate
Frequency	15	11	22	23	6

i Use this information to draw a dual bar chart showing Monday's and Tuesday's information on the same chart.

ii Write down one similarity and one difference between the drinks sold on the two days.

2 Mr Grammar gave his class two spelling tests, each containing 15 words. The table shows the performance of four pupils in the two tests.

Pupil	Louise	Mutasem	Niles	Quinlan
Test 1	11	7	10	12
Test 2	8	10	12	11

a Draw a component bar chart to show this data.

b Explain why a component bar chart is better than a dual bar chart in this case.

3 A cinema records the number of tickets sold for each classification of film shown one day.

Classification	U	PG	12	15	18
Tickets sold	20	35	45	15	5

a Draw a bar chart to illustrate the data.

b Draw a pie chart to illustrate the data.

c Give one reason for drawing a pie chart rather than a bar chart.

d Give one reason for drawing a bar chart rather than a pie chart.

4 A restaurant has tables set up for between three and six diners at one table. The dot plot shows how many of each table there are.

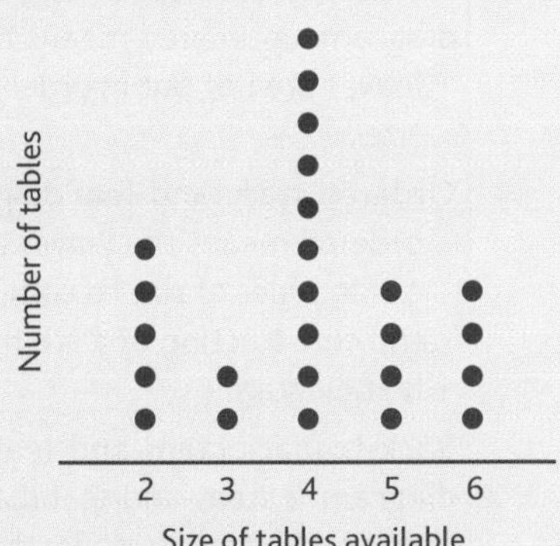

a How many tables for four does the restaurant have?

b How many tables for five or more does the restaurant have?

c If the restaurant were completely full, how many diners would there be?

d Draw a vertical line diagram to show these data.

e One night the restaurant had the following people wanting to eat there: four couples, four groups of three, eight groups of four and three groups of six.

i Draw a dot plot to represent the groups of people wanting to eat at the restaurant that night.

ii With this information, make one recommendation for the restaurant to change its available seating.

iii Give one reason why the restaurant should **not** change its available seating.

4.3 Stem-and-leaf diagrams

A **stem-and-leaf diagram** is used to show discrete data or rounded continuous data.

The stem-and-leaf diagram shows information about how the data is spread out, and also keeps all the original values visible within the diagram.

Example 1

A group of children keeps a record of how many portions of fruit and vegetables they eat in one week. The results are:

23 37 14 32 42 38 15 33 27 20 31 19 18 26 25 38 31 32 28 34 25 22 17 13 22

a Show this data in an ordered stem-and-leaf diagram.

b What is the range of the results?

c Fraser had the third highest number of fruit and vegetable portions that week. How many?

Solution

a The **stem** will, in this example, be the tens digits for the data. The data runs from 13 to 42, so there might be numbers in the 10s, 20s, 30s and 40s. The stem is 1, 2, 3 and 4, placed in a vertical line. You then go through the data in order, putting the **leaves** on one by one. For example, for the value 23, place a 3 alongside the stem of 2. For 37, place a 7 alongside the stem value of 3 and so on. You will now have an **unordered** diagram, like this:

1	4	5	9	8	7	3			
2	3	7	0	6	5	8	5	2	2
3	7	2	8	3	1	8	1	2	4
4	2								

Now draw a new version with the leaves placed in order – this gives you an ordered stem-and-leaf diagram. Notice in the ordered diagram that a key is essential to explain what the diagram means.

1	3	4	5	7	8	9			
2	0	2	2	3	5	5	6	7	8
3	1	1	2	2	3	4	7	8	8
4	2								

Key: 1 | 3 represents 13 portions

Note it is also possible to order the data first, and then you can draw the **ordered stem-and-leaf diagram** straight away.

b The range is the highest value minus the lowest value.

42 – 13 = 29

So the range is 29.

c Fraser had 38 portions of fruit and vegetables that week.

Back-to-back stem-and-leaf diagrams can show two sets of data at the same time. This is very helpful when comparisons need to be made.

Objectives

In this topic you will learn how to draw and interpret:

stem-and-leaf diagrams.

Key terms

Stem-and-leaf diagram: a frequency diagram which uses the actual values of the data split into a stem and leaves, with a key.

Stem: the left part of a stem-and-leaf diagram indicating the main digit or digits of the data.

Leaves: the figures to the right of a stem-and-leaf diagram indicating the final digit of the data.

Unordered stem-and-leaf diagram: unordered means the leaves are not put in order of size.

Ordered stem-and-leaf diagram: ordered means the leaves are put in order of size to complete the construction of a stem-and-leaf diagram.

Back-to-back stem-and-leaf diagram: a stem-and-leaf diagram where the stem is down the centre and the leaves from two distributions are either side for comparison.

AQA *Examiner's tip*

Lots of marks are lost in exams on stem-and-leaf diagrams, with keys forgotten and values misread or missed off diagrams. These need to be done with care and checked a second time.

Example 2

The back-to-back stem-and-leaf diagram shows the percentage turnout for males and females in 15 constituencies at the last general election.

Female turnout (%)		Male turnout (%)
9 9 8 7 3 2	3	7
9 7 4 2 2	4	2 3 3 4 6 7
5 4 0	5	0 1 5 7 7 9
6	6	1
	7	2

Key: 2 | 3 | 7 represents 32% for female and 37% for male (though not necessarily in the same constituency).

a What proportion of constituencies had over 50% turnout for males?

b What proportion of constituencies had over 50% turnout for females?

c Use your answers to (a) and (b) and one further piece of evidence to say which gender generally had a higher turnout.

Solution

a The proportion is $\frac{7}{15}$ – notice that there are seven values over 50%.

b The proportion is $\frac{3}{15}$ (or $\frac{1}{5}$).

c Male turnout is generally higher – the highest value is also much higher for males (72%) than for females (66%).

links

Measures of average such as mean, median and mode would be useful here in making comparisons – learn about these in Chapter 5.

Exercise 4.3

1 The waiting times (in minutes) for 24 patients seen one morning at a doctor's surgery are given below:

22 12 8 10 17 21 24 34 12 19 7 23 30 24 19 11 12 16 23 27 16 34 20 9

a Use a stem-and-leaf diagram to show the data. Remember to write a key.

b Find the range of the waiting times.

c The Health Trust running the doctor's surgery aims for waiting times of less than 15 minutes. Comment on the extent to which this target has or has not been achieved.

2 The number of men and women seated in each row of a theatre audience is given below:

Row	Men	Women	Row	Men	Women
A	17	22	H	9	30
B	21	20	I	22	15
C	13	29	J	11	27
D	18	19	K	17	19
E	22	16	L	7	17
F	8	32	M	10	15
G	16	26	N	9	9

a Draw a back-to-back stem-and-leaf diagram to show the data for men and women.

b Compare the distribution of men and women at the theatre.

c Do you think the theatre was full? Explain your answer.

3 The stem-and-leaf diagram shows the cost of 35 properties in one street, rounded to the nearest thousand.

Stem	Leaves								
18	4	6							
19	0	2	3	5					
20	3	3	7	9	9	9	9		
21	0	1	5	5	6	7	8	8	9
22	3	4	4	6	9				
23	2	5							
24	7	8							
25	3								
26	9	9							
27									
28									
29	5								

Key: 29 | 5 represents £295 000

a What was the value of the cheapest house?

b What was the most common cost of a house?

c Illustrate this data on another suitable diagram.

4 The weights of some parcels posted first class and some parcels posted second class are shown in the back-to-back stem-and-leaf diagram. The weights are recorded to the nearest 0.1 kg.

1st class								Stem	2nd class						
				7	6	6	4	0	9						
			9	9	7	2	0	1	2	4	6				
8	5	5	4	3	3	2	1	2	3	3	5	8			
				8	6	4	3	3	2	4	6	7	7	7	7
						8	5	4	3	3	5	8	9	9	9
							3	5	0	1	8				
						4	1	6	0	3	4	7	9		

Key: 4 | 0 | 9 represents a 1st class parcel weighing 0.4 kg and a 2nd class parcel weighing 0.9 kg.

For each of these statements say:

a whether the statement is true or false, or that you cannot tell

b how you came to your decision in part (a):

- Statement 1: 30 parcels were sent second class.
- Statement 2: the lightest 1st class parcel was 400 g, to the nearest 100 g.
- Statement 3: the heaviest 2nd class parcel was 6.0 kg.
- Statement 4: there were 20 1st class parcels lighter than the most common weight for a 2nd class parcel.
- Statement 5: the three heaviest parcels sent were sent 2nd class.

4.4 Frequency diagrams for continuous data

Frequency polygons and histograms with equal intervals

If data is continuous it will most probably be in a grouped frequency distribution. The best way to show this type of data in a diagram is to use a **frequency polygon** or a **histogram**. This is straightforward if the groups or classes are of equal width. For the histogram the diagram consists of bars like a bar chart, except the bars must be joined and the horizontal scale must be labelled as a continuous scale. The bars can then simply be drawn to the heights according to the frequencies. For example, the table of the weights of 100 people from Chapter 3 is reproduced below.

Weight, w kg	Frequency
$40 \leq w < 60$	23
$60 \leq w < 80$	55
$80 \leq w < 100$	17
$100 \leq w < 120$	5

Notice again that the classes are all of equal width. This enables the histogram to be drawn as shown in Graph **A**.

For the frequency polygon the diagram consists of points drawn to the height of the frequency and plotted at the midpoint of the group they represent. It is just like joining the middle of the top of each bar of the histogram. The frequency polygon for the weights data can be seen in Graph **B**.

Notice the use of the (-∧∨-) to avoid wasting unnecessary space on the horizontal axis.

There is usually no reason to make the polygon 'land' by extending the lines beyond the range of the data. The advantage of a frequency polygon over a histogram is that it is possible to draw more than one on the same axes, making comparisons easy to do.

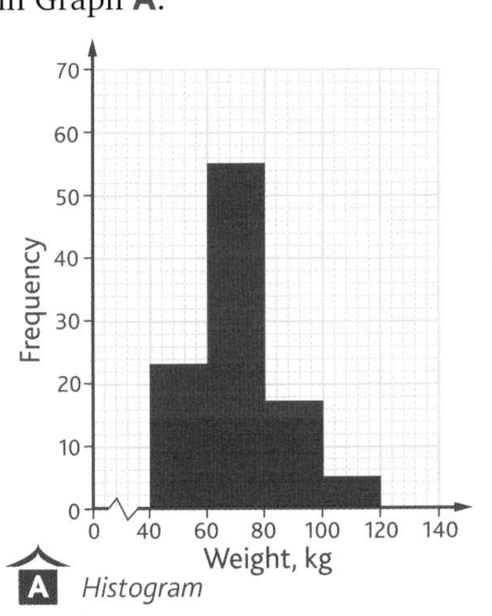

A *Histogram*

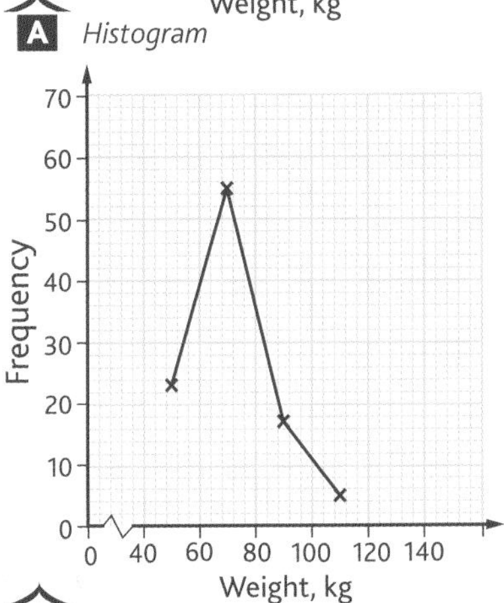

B *Frequency polygon*

Objectives

In this topic you will learn how to draw and interpret:

- frequency polygons
- histograms with equal class intervals
- histograms without equal class intervals
- population pyramids.

Key terms

Frequency polygon: a frequency diagram for continuous data with a line joining the midpoints of the class intervals using the appropriate frequencies.

Histogram: a diagram for continuous data with bars as rectangles whose areas represent the frequency.

Histograms with unequal class intervals

If you have continuous data grouped into classes where the widths of the intervals vary, it is not acceptable to draw a histogram like the one above with each bar drawn to the frequency of the class. This would overemphasise the amount of data in the wider class intervals.

Instead you should use a **frequency density** method, where the area of the bar is proportional to the frequency represented by the bar. This is given by the following formula:

$$\text{Frequency density} = \frac{\text{frequency of class}}{\text{class width}}$$

This ensures that the frequency for each class is in proportion to the area of the bar you draw to represent it.

Key terms

Frequency density: the frequency density is used to complete a histogram and is calculated using actual frequency divided by class interval width.

Example 1

A trading standards team tests petrol station pumps to ensure that they are giving the correct amount of petrol.

The team takes 1 litre of petrol from each of 200 pumps and measures the exact amount given in millilitres.

The results are given in the table.

a Comment on the methods used by the team.

b Draw a histogram to show the data.

c Estimate the percentage of pumps giving less than the expected one litre.

Actual Amount of petrol, x ml	Frequency
$940 \le x < 980$	8
$980 \le x < 995$	15
$995 \le x < 1010$	140
$1010 \le x < 1020$	30
$1020 \le x < 1060$	7

Solution

a The team is only taking a single litre from each pump, so effectively the sample size is one. A greater number of samples taken at intervals would be needed if there were any suspicions about the accuracy of a particular pump.

b You need to calculate frequency density for each class. To do this you need to work out the class widths. It is often a good idea to redraw the given table with two extra columns in which you can show your working.

Now the data can be drawn on the graph. Remember to have a continuous scale on the horizontal axis. Do **not** label with the classes $940 \le x < 980$, $980 \le x < 995$ and so on.

Actual Amount of petrol, x ml	Frequency	Class width	Frequency density (frequency ÷ class width)
$940 \le x < 980$	8	40	0.20
$980 \le x < 995$	15	15	1.00
$995 \le x < 1010$	140	15	9.33
$1010 \le x < 1020$	30	10	3.00
$1020 \le x < 1060$	7	40	0.18

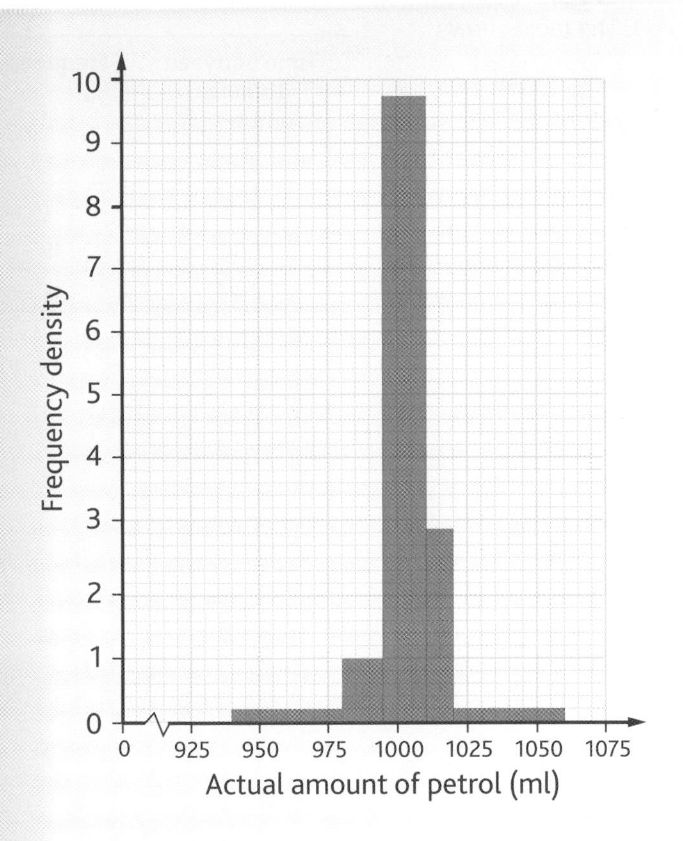

c This estimate can be made by counting the smallest size of square on the histogram, and finding the percentage of these to the left of an imaginary vertical line drawn at 1000 ml. Alternatively, use the original table.

$$\text{Number below } 1{,}000 \text{ ml} = 8 + 15 + \frac{1}{3}(140)$$

$$= 69\frac{2}{3} \text{ out of } 200$$

This is approximately 35%.

Some data sets have a different labelling system to consider. For example, data could be grouped as 4–9, 10–14 and so on. In this case the class widths are found by going half a unit either side of the given label.

AQA Examiner's tip

On examination papers, tables will be set to the left-hand side of the page to allow you room to draw on these extra columns.

links

See Chapter 5 to find out how a histogram can be used to estimate or calculate some measures of average.

Example 2

The time between 50 rumbles of thunder was timed. The table shows the frequency distribution for these times.

Calculate the frequency density for each class.

Time between thunder rumbles (s)	Frequency
1–3	4
4–6	12
7–10	19
11–20	15

Solution

Add two columns to the table. The first new column is for the class width. The data here is continuous but has been put into groups with discrete labels. So the first class width for 1–3 is for all data between 0.5 and 3.5 and so has a class width of 3. The second new column will again use frequency density = frequency ÷ class width.

Time between thunder rumbles (s)	Frequency	Class width	Frequency density
1–3	4	$3.5 - 0.5 = 3$	$4 \div 3 = 1.33$
4–6	12	$6.5 - 3.5 = 3$	$12 \div 3 = 4$
7–10	19	$10.5 - 6.5 = 4$	$19 \div 4 = 4.75$
11–20	15	$20.5 - 10.5 = 10$	$15 \div 10 = 1.5$

Population pyramids

These are a form of 'back-to-back' histograms, showing the age distribution for the population of a particular country split into the genders. Here is an example showing the UK for the year 2000.

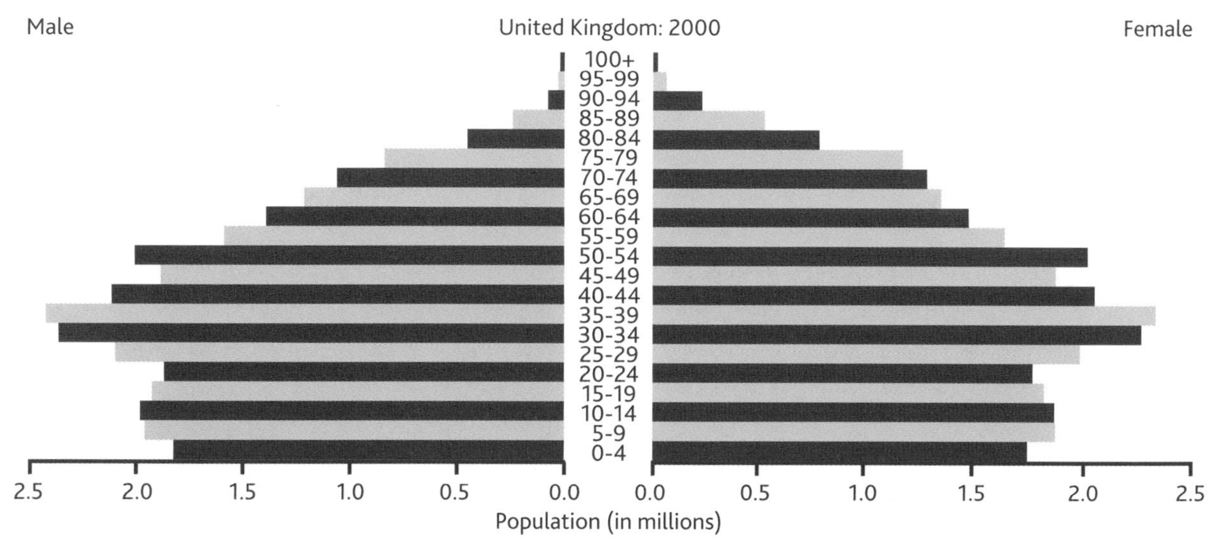

Source: U.S. Census Bureau, International Data Base

As you can see, there is a great deal of information on a population pyramid. For example, the general trend of females living longer than males can be easily seen. Similarly, it is possible to see that the age group with the most people is 35–39 years old. You can also see possible trends in the birth rate.

Exercise 4.4

1 The population pyramid shows the age distribution of the Isle of Man for 2009.

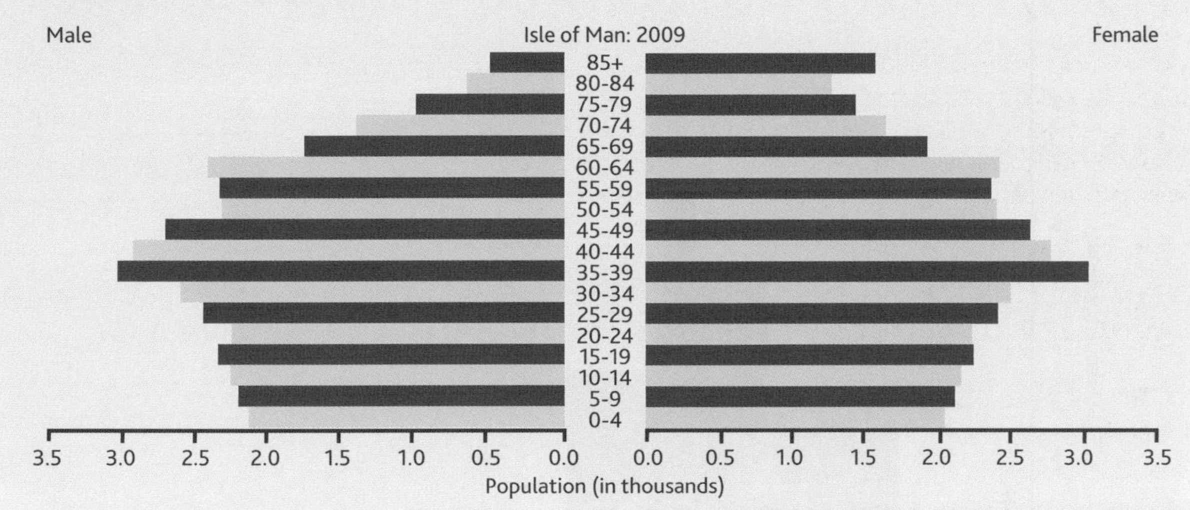

Source: U.S. Census Bureau, International Data Base

a Estimate the number of females under 5 on the Isle of Man.

b Estimate the total number of people aged 85 and over on the Isle of Man.

c Give a possible reason, in each case, why:

i there are fewer males aged 20–24 than males aged 15–19

ii there are more people aged 60–64 than 55–59.

2 The table to the right shows the age distribution of the workers in a factory.

Age	Frequency
$20 \leq x < 30$	18
$30 \leq x < 40$	38
$40 \leq x < 50$	12
$50 \leq x < 60$	8
$60 \leq x < 70$	4

a Explain why the middle of the class interval $20 \leq x < 30$ is 25.

b Draw a frequency polygon to show the data.

c Joe looks at the table and says that 25% of the workers are over 45 years old. Is Joe correct? You must explain your answer.

3 The length of time taken by 100 people to complete a questionnaire was monitored. The results are shown in the table to the right.

Length of time, l, minutes	Frequency
$2 \leq l < 5$	6
$5 \leq l < 7$	16
$7 \leq l < 8$	21
$8 \leq l < 10$	40
$10 \leq l < 20$	17

a Construct a histogram to show this information.

b Did the majority of respondents take between 8 and 10 minutes? You must explain your answer.

c The writers of the questionnaire wanted 75% of people to be able to complete it in between 6 and 12 minutes. Use the table or your histogram to estimate the percentage of people completing the questionnaire in between 6 and 12 minutes.

4 The histogram shows the weight of some suitcases weighed in at an airport check-in desk.

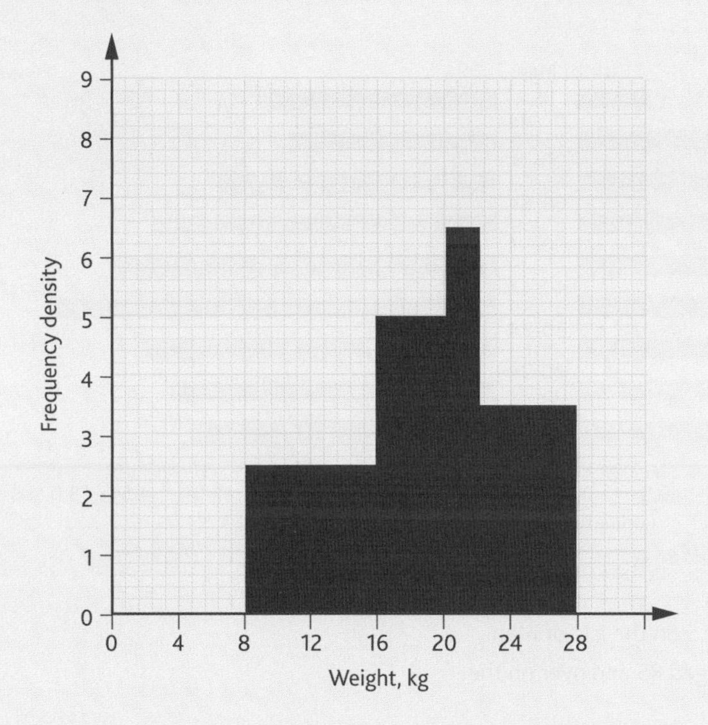

a Construct the frequency table from which the histogram was drawn.

b There is an excess charge for all suitcases above 25 kg.

Estimate the number of suitcases that had an excess charge to be paid.

5 The table shows the playing times of 100 'number one' hit records from the 1950s.

a Explain why the first class width is 20 seconds.

b Construct a fully labelled histogram for the data.

c Collect information about the playing times of the most recent 30 number one hit records. Group the data and draw a histogram. Compare this histogram with the one from the 1950s.

Playing times of hit record (seconds)	Frequency
90–109	6
110–119	7
120–129	22
130–149	47
150–199	16
200–299	2

Summary

You should:

- be able to use pictograms, bar charts or pie charts for qualitative data
- be able to use multiple or composite bar charts to compare two sets of qualitative data
- be able to use proportional pie charts to compare two sets of data using pie charts
- know that the area of the circle in a proportional pie chart is proportional to the total frequency it shows
- be able to use dot plots or vertical line diagrams for quantitative discrete data
- be able to use stem-and-leaf diagrams for discrete or rounded continuous data
- be able to use frequency polygons or histograms for grouped continuous data with equal-sized classes
- be able to use histograms for grouped continuous data with unequal-sized classes
- know that population pyramids are a form of histogram showing age distributions for males and females.

Summary of the diagrams to be used for different types of data:
the following table may help you to remember which type of diagram best suits which type of data:

Type of data	Most suitable diagrams	Other possible diagrams
Qualitative data	Pictograms Pie charts Proportional pie charts	Bar charts Dot plots
Quantitative discrete data	Bar charts Multiple bar charts Composite bar charts Dot plots Vertical line diagrams Stem-and-leaf diagrams	
Quantitative continuous data	Frequency polygons Histograms Population pyramids	Stem-and-leaf diagrams

AQA Examination-style questions

1. The percentage bar chart shows the proportion of votes cast in one constituency for each of three political parties, A, B and C, at the last general election.

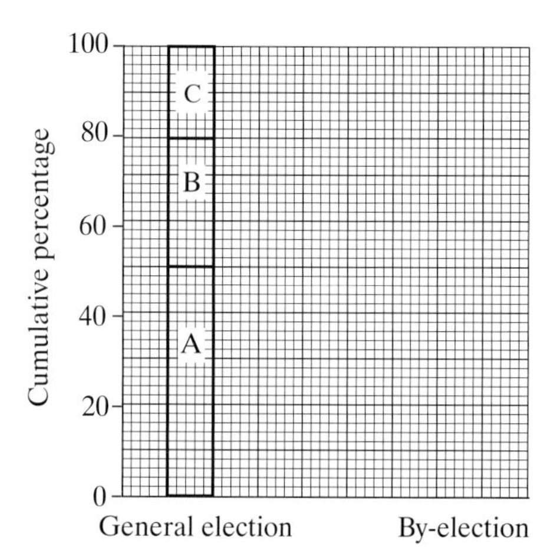

(a) What percentage of the votes were cast for party C at the election? *(1 mark)*

Later that year, a by-election was held in the same constituency.
The numbers of votes cast, to the nearest thousand are shown in the table.

Party	By-election votes
A	17
B	14
C	19

(b) Draw on the graph a percentage bar chart showing the votes cast in the by-election. *(5 marks)*

(c) State *two* differences between the general election and the by-election results. *(2 marks)*

AQA, 2002

2. A company makes rowing boats. The pictogram below shows the number of boats made in 1990 and 1995.

Year	Number of boats
1990	
1995	
2000	

represents 10 boats

(a) How many boats were made in 1990? *(1 mark)*

(b) How many boats were made in 1995? *(1 mark)*

(c) In the year 2000 the company made 43 boats. Complete the pictogram for the year 2000. *(2 marks)*

AQA, 2002

3. The population pyramids show the distribution of the population of Angola by age and gender for 2000 and a prediction for 2050.

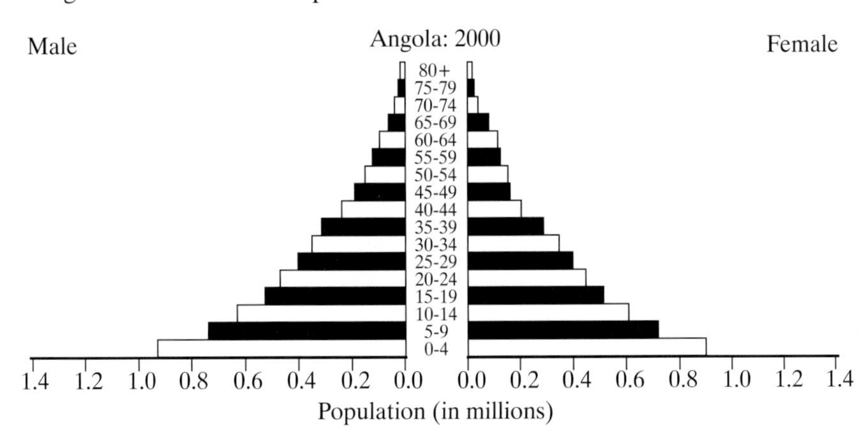

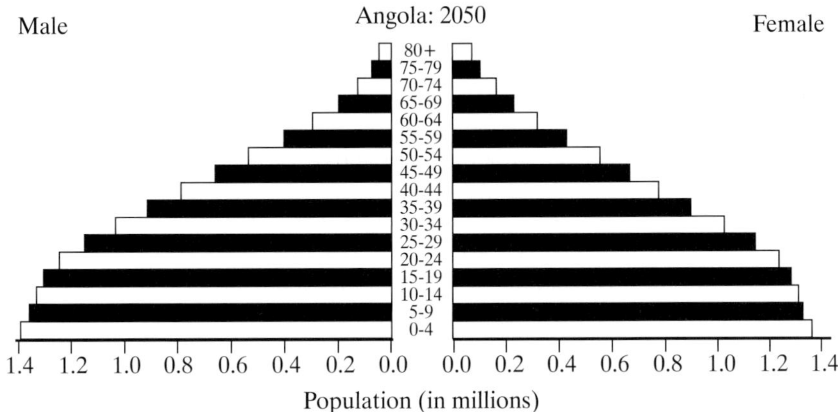

(a) Give one difference between the population for 2000 and the prediction for 2050. *(1 mark)*

(b) In the 2050 prediction which male age group will have approximately half of the number in the 30–34 male age group? *(1 mark)*

(c) Estimate the difference between the female population aged 10–14 for 2000 and the predicted value for 2050. *(2 marks)*

AQA, 2007

4. The number of matches in each of 20 boxes is given below.

42 42 43 42 41 41 42 41 43 41
42 42 41 43 41 41 42 41 43 41

(a) Complete the frequency chart.

Number of matches	Tally	Frequency
41		
42		
43		

(2 marks)

The data are to be displayed on a pie chart.

(b) Calculate the angle needed to represent 41 matches. *(2 marks)*

(c) Draw a fully labelled pie chart of the data. *(3 marks)*

(d) Represent the data on a fully labelled component bar graph. *(3 marks)*

(e) Give *one* advantage of displaying this information on a pie chart. *(1 mark)*

(f) Give *one* advantage of displaying this information on a component bar graph. *(1 mark)*

AQA, 2001

5. The following table shows the production in 1995 and 2001 at the Royal Wedgetown Pottery Company.

Production (millions of items)

Product	1995	2001
Cups	15.0	12.0
Mugs	18.5	20.0
Plates	24.0	26.5
Bowls	27.5	29.5

Draw a composite bar chart to show the data for 1995 and 2001. *(6 marks)*

AQA, 2003

6. The diagram shows the number of UK residents travelling abroad during 1991 and 2001.

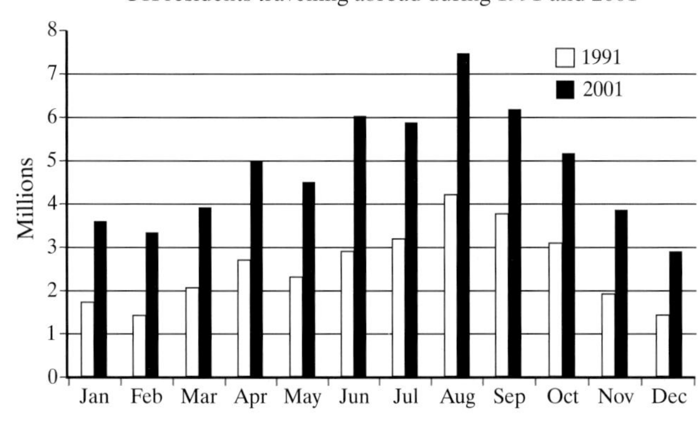

Source: Adapted from International Survey, Office for National Statistics, Social Trends

(a) Describe **two** features of the diagram. *(2 marks)*

(b) In May 1991, 2.2 million UK residents travelled abroad. Estimate from the diagram the number who travelled abroad in May 2001. *(1 mark)*

(c) In April 1991, 2.7 million UK residents travelled abroad. Find the percentage increase in the number of UK residents who travelled abroad in April 2001 compared to April 1991. *(4 marks)*

(d) A national newspaper reported that in the **first week** of June 2001 there were 600 000 UK residents travelling abroad. Explain why this report may be wrong. *(2 marks)*

AQA, 2005

7. A survey of 50 boys and 50 girls recorded the times they spent using the school computers last week. The time spent, t (hours), by each of the **boys** is given in the following table.

Time spent, t (hours)	**Frequency**
$1 \le t < 2$	1
$2 \le t < 3$	4
$3 \le t < 4$	5
$4 \le t < 5$	9
$5 \le t < 6$	12
$6 \le t < 7$	16
$7 \le t < 8$	3

The frequency polygon below shows the times for the 50 **girls**.

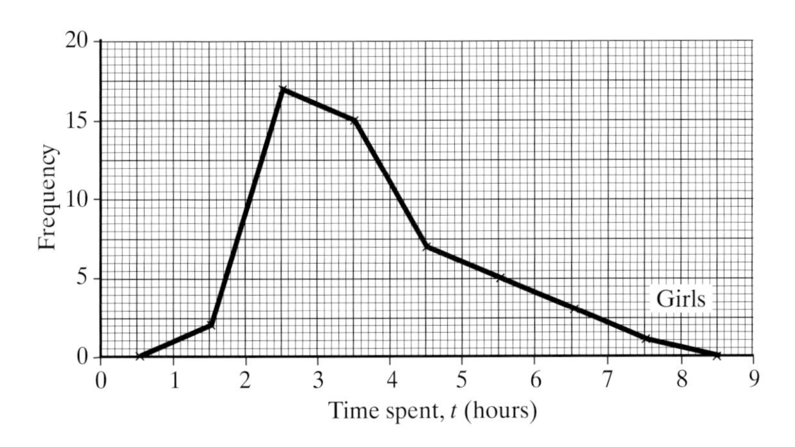

(a) Draw a frequency polygon for the **boys**' times on the same graph. *(3 marks)*

(b) What is the modal class for the **boys**' times? *(1 mark)*

(c) Write down:

(i) the number of **boys** who spent less than 4 hours last week using the school computers *(1 mark)*

(ii) the number of **girls** who spent less than 4 hours last week using the school computers. *(2 marks)*

AQA, 2004

8. The weights, w (grams), of 125 grapefruit are summarised in the table.

Weight, *w* (grams)	**Frequency**
$160 \le w < 180$	8
$180 \le w < 190$	15
$190 \le w < 200$	30
$200 \le w < 210$	35
$210 \le w < 220$	X
$220 \le w < 240$	Y
$240 \le w < 260$	Z

Part of a histogram is drawn to represent this data.

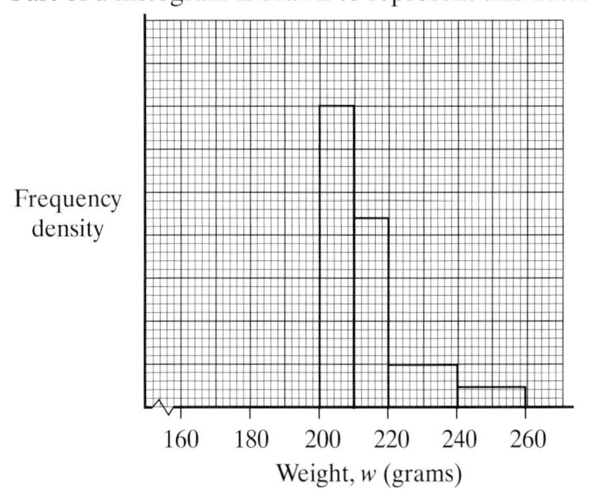

(a) Scale the frequency density axis. *(1 mark)*

(b) Calculate the value of:

(i) X *(2 marks)*

(ii) Y *(1 mark)*

(iii) Z. *(1 mark)*

(c) Complete the histogram. *(3 marks)*

(d) Calculate an estimate of the median weight. *(4 marks)*

AQA, 2005

9. The table shows some information about the number of schools by type in one local education authority (LEA) in the UK in 2006.

Type of School	Number	Angle on pie chart
Primary	63	
Comprehensive	45	120
Grammar		48
Others	35	
Total	135	360

(a) Copy and complete the table. *(4 marks)*

(b) A pie chart of radius 4.1 cm is to be drawn to illustrate these data. Copy and complete the pie chart shown. *(2 marks)*

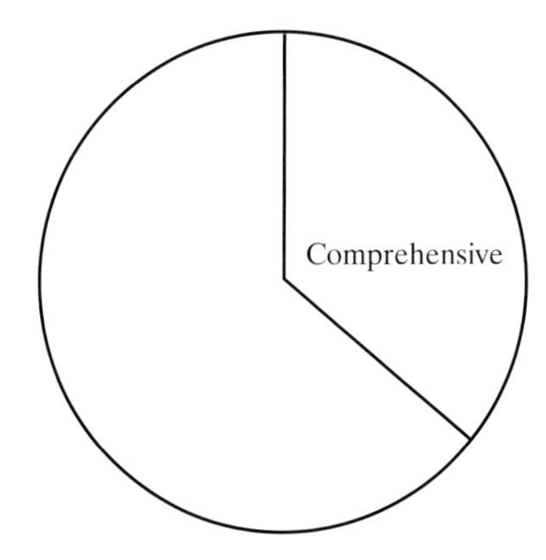

(c) A pie chart showing the number of schools by type, in another LEA is to be drawn for comparison. The radius for this pie chart will be 4.8 cm. Calculate the difference between the total number of schools in each of the two LEAs. (3 marks)

AQA, 2008

10. The journey times, in minutes, for a courier are given in the table.

Journey time (t minutes)	**Frequency**
$10 \le t < 20$	
$20 \le t < 25$	
$25 \le t < 30$	65
$30 \le t < 40$	58
$40 \le t < 60$	24

Some of the data is represented on the histogram drawn below.

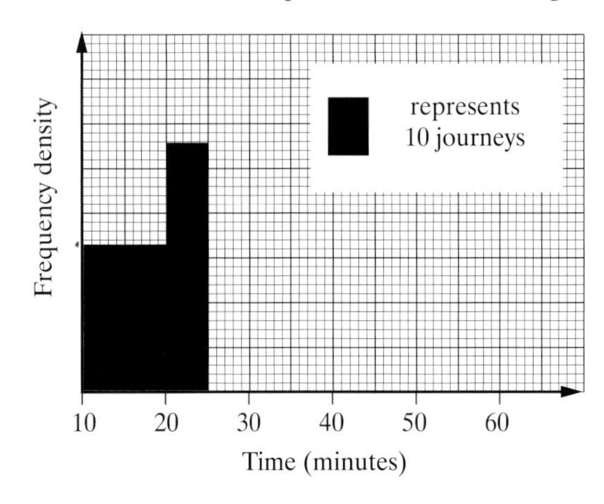

(a) Calculate the frequencies missing from the table for the intervals $10 \le t < 20$ and $20 \le t < 25$. *(3 marks)*

(b) Complete the histogram. *(5 marks)*

AQA, 2002

5 Methods for calculating average

In this chapter you will learn:

- how to calculate and interpret mean, mode and median for raw data
- what the advantages and disadvantages of the various measures of average are
- how to choose the best measure in a given situation
- how to use a change of origin or linear transformation to ease the calculation of mean and median
- how to calculate and interpret the measures of average for discrete frequency distributions
- how to find estimates of mean and median and find the modal group for grouped frequency distributions
- how to estimate the mode and median for grouped data using a histogram
- how to use Sigma notation for summary statistics
- how to calculate and interpret the geometric mean and know when it is used
- how to estimate the mean of a population from a sample
- about some of the issues that make an estimate more or less reliable
- how to estimate the proportion of a feature within a population.

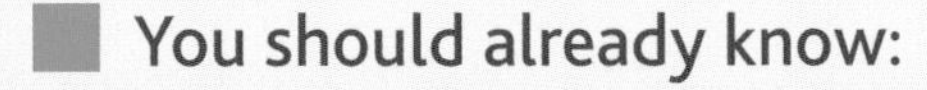

You should already know:

✔ how to find the lower and upper bounds for a class interval.

What's this chapter all about?

This chapter could be called 'everything you need to know about averages'! Once you have collected and displayed your data, the next natural thing to do with it is to work out some measure of average. This lets you compare your measure with previous averages, or with the average of another set of data. You probably already know the names mean, mode and median but there are things in here that you will not have learned about in your maths lessons.

They were the average family – 2.4 children

Focus on statistics

In this chapter you might be surprised to hear that there is a fair amount that you will not have met already in maths.

- In 5.1 you are asked to look in detail at why each measure of average may be better than another, you may have briefly met this in maths.
- Also in 5.1 is the change of origin, or linear transformation of data to make calculations easier. This is not done in maths, although you probably know that if you add the same number to each of the values, the mean or median goes up by the same.
- The geometric mean in 5.1 is not in maths.
- In 5.2, when dealing with grouped data in statistics, you are expected to be able to calculate an estimate of the median, whereas in maths you only have to be able to find the group in which the median lies.
- Section 5.3 is not explicitly tested in maths at all.

5.1 What are the mean, the mode and the median?

When looking at data it is useful to have some idea of a typical value for the data. This is usually called an **average**. An average is a summary statistic used to measure the central tendency of the data. Knowing the average for a set of data gives you some idea of what value or values you might expect to occur if further values from the same data source are to be obtained. If the data is from a sample, the average of the sample may well be used to estimate the average for the whole population. Using data from more than one sample, calculating averages enables simple comparisons to be made between the samples.

In this chapter you will meet three types of average (four for the higher tier). You have probably met them before but here is a reminder and a comparison of the different features of each one.

Objectives

In this topic you will learn:

- how to calculate and interpret mean, mode and median for raw data
- what the advantages and disadvantages of the various measures of average are
- how to choose the best measure in a given situation
- how to use a change of origin or linear transformation to ease the calculation of mean and median.

Key terms

Average: the general name given to the measures of central location – an idea of a typical value from the data.

Mean: the total of all the data values divided by the number of data values.

Arithmetic mean: the correct name for the mean, as it is usually worked out.

Median: the middle number in a list of ordered data.

Ordered: arranged from lowest to highest, or the other way around.

Mode: the number that occurs in a list most often.

Modal value: another name for the mode.

Bimodal: a set of data that has two modes is said to be bimodal.

The mean

When people talk about averages they are usually talking about the **mean** without actually calling it that. The full name of the mean is the **arithmetic mean**. The arithmetic mean is obtained by finding the total of all the data values and then dividing by the number of data values. This can be shown in a formula:

$$\text{mean} = \frac{\text{total of all the data values}}{\text{number of data values}}$$

At higher tier you are expected to know that the mean is often given the symbol $\bar{x}$ (you say 'x bar').

Example 1

The number of pupils absent in Year 10 for each day one week was 6, 9, 3, 9 and 7. Find the mean number of pupils absent for one day that week.

Solution

1 Add up all the values: $6 + 9 + 3 + 9 + 7 = 34$

2 Count the number of data values present $= 5$

3 Divide these answers $34 \div 5 = 6.8$

The mean number of pupils absent was 6.8.

The median

The **median** is the middle number of an **ordered** list of values.

If an ordered list has n values, the median is the $\left(\frac{n+1}{2}\right)$th value along.

Example 2

Find the median number of pupils absent in Example 1 above.

Solution

1 Write the list of data in order starting with the smallest: 3, 6, 7, 9, 9.

2 There are five data values so the middle number is the $\left(\frac{5+1}{2}\right)$th along = third along.

Alternatively, you can cross out one number from each end of the ordered list, until you reach the middle number. This gives the median number of pupils absent as 7.

If there is an even number of data values in the list, there is not actually one middle number – there are two. In this case, the median is taken to be halfway between those middle two numbers.

Example 3

Find the median of 5, 2, 1 and 8.

Solution

First put the data in order: 1, 2, 5, 8.

There are four values so the median is the $\left(\frac{4+1}{2}\right)$th along – the 2.5th along. This is halfway between the second and third values.

Halfway between 2 (the second value) and 5 (the third value) is 3.5 (or you can do $\left(\frac{2+5}{2}\right)$). Therefore the median is 3.5.

The mode (or modal value)

The **mode** or **modal value** is the value in the data that occurs most often.

For example, in the list of absent pupils the number 9 occurred twice, whereas all the other values only occurred once. So 9 is the modal number of pupils absent.

It is possible to have more than one mode if two items occur an equal number of times. Data like this is called bimodal. However, it is customary to stop at **bimodal**. If there are three or more numbers with the same frequency, it is usual to say that there is not a mode.

If all the data values are different to each other, there is no mode.

Example 4

Twelve teams entered a quiz to raise money for a hospital. The number of people in each of the 12 teams was 4 3 4 5 6 4 3 5 2 8 5 2. Find the modal number of people in a team.

Solution

Three teams had four people in; three teams had five people in. So 4 and 5 are both modes for this data. (This data is bimodal.)

It is also possible to have a mode for qualitative data. For example, the weather on the 10 days when Ellie was on holiday was: sunny, sunny, rain, cloudy, sunny, rain, sunny, sunny, sunny, cloudy. The most common weather was sunny, so sunny was the modal weather.

Spot the mistake

Toby counts the number of people in each car that passes his house during one minute. Here are his results.

3 1 1 3 2 4 2
1 1 1 4

He calculates the three measures of average for this data to give these results:

Mean 2.09

Median 1

Mode 1

1 Which **one** of his answers is wrong?

2 What is wrong with each of these statements made by Toby?

'The mean is 2.09 so I should expect 2.09 people in the next car.'

'The mode is 1 so most of the cars had 1 person in them.'

AQA Examiner's tip

The mean, median and mode are very commonly used and tested in examinations, but there are several very common errors students make when answering questions on them.

- Students mix them up! Remember MeDian = MiDdle, Mode = Most.
- Students frequently forget to put data in order before they try to find the median.
- It is common for students to miss out a number – take care when dealing with lists of data.

Choosing the best average

It is important to know which measure of average might serve you best in a given situation. After all, you want the average you work out to give you the idea of a typical value from the data and to be representative of the data. This table shows some of the issues connected with choosing the best measure of average.

Measure of average	Positive aspects	Negative aspects
Mean	uses all the data in obtaining the answer can be easily calculated using software packages and calculators	mean is distorted by particularly large or small values in the data answer may not be one of the data values so may not represent data very well
Median	easy to work out for small data sets or ordered data not affected by extreme data values	can be time consuming for large data sets, if appropriate software is not available answer may not be one of the data values so may not represent data very well does not use all the data values
Mode	usually easy to find will be a value of the data set the only measure of average that can be used for qualitative data	many sets of data do not have a mode at all there may be more than one answer

At foundation tier you need to know the advantages of each of these measures over the others.

At higher tier you can be asked to choose which average should be used.

AQA Examiner's tip

In exam questions, when talking about the advantages of a particular measure, make sure you discuss your answer in the context of the question.

Example 5

Alice notes the number of miles she travels in her car for 10 days.

The results are 12 8 12 34 23 12 12 5 12 135

a Calculate the mean, median and mode for the data.

b Which of these measures best represents this data?

c Give a reason for your choice in part (b).

Solution

a Mean $= (12 + 8 + 12 + 34 + \ldots + 135) \div 10 = 265 \div 10 = 26.5$

The *mean* is 26.5 miles.

To find the *median*, first order the data:

5 8 12 12 12 12 12 23 34 135

The median is halfway between the fifth and sixth values, which are both 12, so the median is 12 miles.

Mode. The most common value is 12, so the mode is 12 miles.

b The mode and the median both give 12, which better represents the typical value of the data compared to the mean.

Choose mode.

c The answer 'median' would have been fine but the mode probably occurs from Alice travelling to where she works. So in this context the mode gives a good average to use.

Be a statistician 

Zoe works in a clothes shop. She is paid £6.85 per hour. She finds out this information about workers in similar shops in the town where she works.

Mean pay per hour	£7.40
Median pay per hour	£7.10
Modal pay per hour	£6.95

Each year Zoe is given a pay rise. She asks for a 55p per hour rise to be brought in line with the 'average' as she sees it. The manager offers her a 10p per hour rise to be brought in line with the 'average' as she sees it.

1 Give reasons to support Zoe's claim.
2 Give reasons why the manager's offer might be fair.
3 What other factors are likely to be important, apart from the average pay elsewhere?

You could follow this up with some research of your own. Do males or females earn more from part-time jobs? Write a suitable hypothesis. Collect appropriate data. Show your findings on suitable diagrams. Calculate suitable measures for comparison. What are your conclusions? Prepare a short presentation to give to your class.

Change of origin and linear transformation of data

If a set of data is made up of particularly large or difficult values, it is possible to make calculations for the mean and median more easily by adjusting the values before finding the measure. This should lead to fewer errors in your calculations.

For example, finding the mean of 35 002, 35 005, 35 010, 35 007 and 35 006 can be made much easier by subtracting 35 000 from each value, finding the mean of the new data and then adding 35 000 to the mean, to give you the mean of the original data.

$$(2 + 5 + 10 + 7 + 6) \div 5 = 6$$

$$35\,000 + 6 = 35\,006$$

This is called **scaling** or **changing the origin** of the data. It is possible to go further and complete a **linear transformation** of data to make calculations more manageable. A linear transformation would mean that a value is added to or subtracted from the original data and the new data is then multiplied or divided to make it even easier to work with. These operations may also be carried out the other way around with multiplying or dividing first.

Linear transformations of data are useful in everyday life. For example, if everyone in a company receives a 10% plus £25 pay rise per week and the mean pay before the rise was £230 per week, it is not necessary to recalculate the mean from scratch. The new mean will be £230 with 10% added + £25 = £253 + £25 = £278

Key terms

Scaling: adding or subtracting the same value to or from each number in a set of data, to make it easier to calculate measures.

Changing the origin: another name for scaling.

Linear transformation: where data is divided or multiplied by a number, as well as possibly being scaled, in order to make it easier to calculate measures.

Example 6

Use a suitable linear transformation to find the mean of: 0.000078, 0.000082, 0.000073, 0.000084, 0.000077, 0.00008.

Solution

Working with very small values like this is difficult so firstly multiply each value by 1 000 000 to make each value into a whole number.

$\times$ 1 000 000 $\Rightarrow$ 78, 82, 73, 84, 77, 80

Now you can subtract 70 from each value to give numbers that you can add up in your head.

$-$ 70 $\Rightarrow$ 8, 12, 3, 14, 7, 10

Now find the mean of the new values.

$8 + 12 + 3 + 14 + 7 + 10 = 54$

$54 \div 6 = 9$

Now you must reverse the process.

9 add back on to 70 = 79

$79 \div 1\,000\,000 = 0.000079$

So the mean of the original data is 0.000079.

Geometric mean

The **geometric mean** is used for situations where values would naturally be multiplied together rather than added when combining them.

The geometric mean of n values is the nth root of all the values multiplied together. This might sound a bit complicated but look at the situation below.

By far the most common application of the geometric mean is in areas that use interest rates. So suppose that over three consecutive years you were lucky enough to have some money invested that gained 30% interest in year 1, 40% interest in year 2, and 20% interest in year 3.

The question, 'What is the average annual interest for these 3 years?' is not answered using the arithmetic mean and the answer is not 30% (if you don't believe this start with £100 and work through the percentage calculations). In fact the average you need to use in this question is the geometric mean, and the correct answer is $\sqrt[3]{1.3 \times 1.4 \times 1.2} = 1.2974$ (to four decimal places) giving an average percentage increase of 29.74% (to two decimal places).

Key terms

Geometric mean: an average where the n values in a data set are multiplied and the nth root is taken.

Real world

Try to find information about the monthly change in house prices in your area over one year. Use some diagrams to illustrate your data and use the geometric mean to find the average monthly increase (or decrease) in house prices for your area over the year.

Exercise 5.1

1 For each set of data find:

a the mean

b the median

c the mode or modes if one or more exist.

i 5 6 7 8 9

ii 12 11 8 15 11 16 14 9

iii 31 28 31 30 31 30 31 31 30 31 30 31

iv −6 −4 −7 −3 −4 −6 −5 −6 −8 −4

2 The mean of Monty's four rounds of golf is 71. What is his total score over the four rounds?

3 The median of Laura's four rounds of golf is 69. Her scores were (in order) 66 67 ** and 74. Find the missing score.

4 The amount of rain falling on London Heathrow Airport is recorded. The values for 10 days of June 2008 were (in mm): 0 3 17 0 0 2 28 0 1 6.

a Find the values of the three measures of average.

b Which value is the most useful in this case?

5 Look at each of these situations. Write down which of the measures of average is being used in each case. Explain your answer.

a The average number of children in a family is 2.2.

b The average number of arms a person has is 2.

c The average score Jean was awarded for her dive was 9.4.

d John looked on the shoe shop shelves and quickly decided that the average male shoe size must be 9.

6

a Write down four numbers with a mean of 5. They must not all be the same number.

b Write down five numbers with a median of 6. They must not all be the same number.

c Write down six numbers with a median of 7. They must not all be the same number.

d Write down four numbers with a median of 8 where none of the numbers are the number 8, and none of the numbers are odd.

7 A bag contains numbered discs. The numbers on the discs have a mean of 5, a median of 6 and a mode of 7. An extra disc with the number 5 on it is put in the bag. Which of the three measures of average:

a will not have changed

b might have changed

c will definitely have changed?

Explain your answers.

8 A bag contains numbered discs.

The numbers on the discs have a mean of 4, a median of 3 and a mode of 2.

An extra disc with the number 2 on it is put in the bag.

Which of the three measures of average:

a will not have changed

b might have changed

c will definitely have changed?

Explain your answers.

9 Find the mean and the median of

1008 1007 1006 1008 1003 1003 1010
1009 1011 1004

using a method of change of origin.

10 Use suitable linear transformations to find the mean of

a 23 045, 23 081, 23 054, 23 072, 23 063

b 0.00134 0.00128 0.0013 0.00129
0.00136 0.0014 0.00135 0.00137

11 The value of a house goes up by 8% one year, and 10% the next year.

a What is the average percentage increase in the house over the two years?

b Compare this value with the mean.

12 The geometric mean of three numbers is 6. Two of the numbers are 3 and 8, what is the third?

5.2 Measures of average for frequency distributions

Discrete data

If **discrete** data is presented in the form of a frequency table then it is possible to calculate the exact mean and median and simply read off the mode.

Mode: in a frequency table, the frequency column will tell you how many of each number has occurred. So the number corresponding to the highest frequency will be the mode.

Median: the median is the middle number when the data is in order. As the data is in a frequency table it is already effectively ordered. You need to add up the frequencies to find the total number of values. Use the fact that the median is the $\left(\frac{n+1}{2}\right)$th value along the data. Find the row of the table in which this halfway point occurs and the value of the data in that row will be the median.

Example 1

Find the mode and the median of the discrete frequency table showing the number of appliances left on stand-by in Mr Lazy's house per day during one month.

Number of appliances left on stand-by	Frequency
1	8
2	9
3	12
4	2

Solution

The mode is 3. This is because the value 3 has the highest frequency of 12. (Be careful, the mode is not 12.) For the median you have $8 + 9 + 12 + 2 = 31$ items. So the median is $\frac{31+1}{2}$ th $=$ 16th along. Looking at a running total of the values. The first eight values are 1s; the next nine values are 2s. This includes the 16th value, so the median is 2.

Mean: look at the example of the appliances on stand-by. What steps need to be taken to find the mean of that table? Firstly it is helpful to add an extra column to the frequency table for the total of each row.

Number of appliances left on stand-by	Frequency	Row total
1	8	$1 \times 8 = 8$
2	9	$2 \times 9 = 18$
3	12	$3 \times 12 = 36$
4	2	$4 \times 2 = 8$
	Total = 8 + 18 + 36 + 8 = 70	

Objectives

In this topic you will learn:

- how to calculate and interpret the measures of average for discrete frequency distributions
- how to find estimates of mean and median and find the modal group for grouped frequency distributions
- how to estimate the mode and median for grouped data using a histogram
- how to use Sigma notation for summary statistics
- how to calculate and interpret the geometric mean and know when it is used.

Key terms

Discrete: discrete values are exact with no rounding necessary to record them.

links

It is explained in Chapter 7 how the median can be found using a diagram for this type of data.

There are eight occasions where one appliance is left on: 8×1 (or 1×8) = 8. There are nine occasions where two appliances are left on: 9×2 (or 2×9) = 18, and so on to find the total of all the values.

Now you know that the mean is the total of all the values divided by the number of values. You have already seen that there are 31 values in this table (found by adding up the frequency column values), so the mean = 70 ÷ 31 = 2.26 appliances per day (to 2 d.p.).

Continuous grouped data (or discrete data that has been grouped)

If data is grouped into classes, this means that the exact values of the original data have been lost. Therefore it is not possible to calculate exact values for any of the measures of average. Mode: a mode cannot be found but the **modal group** or **modal class** can, if the classes are of equal width. It is the class with the highest frequency.

Median: the median can be estimated. This doesn't mean that you simply guess what it is. It means that you calculate an estimate based on finding the class in which the median lies and then work out approximately what its value will be based on how far into its class it lies. Note that if there is quite a lot of data (so n is large) there is no need to use $\left(\frac{n+1}{2}\right)$ for the position of the median. In this case it is quite acceptable to use the simpler formula $\left(\frac{n}{2}\right)$ as you are only providing an estimate of the median, because you cannot calculate the exact answer when the data has been grouped.

Example 2

A group of 40 students tried to estimate when a time of 30 seconds had lapsed. The actual length of time each student thought was 30 seconds is given in the grouped frequency distribution.

Actual length of time, x seconds	Frequency
$10 \le x < 20$	7
$20 \le x < 30$	8
$30 \le x < 40$	16
$40 \le x < 50$	6
$50 \le x < 60$	3

a Find the modal class.

b Calculate an estimate of the median.

Solution

a The highest frequency is 16. This belongs to the $30 \le x < 40$ class. The modal class is therefore $30 \le x < 40$ seconds.

b There are 40 values. The median is approximately the 20th. Remember that, as the data is tabulated, it is in order.

There are 15 (7 + 8) values before the $30 \le x < 40$ class so, as there are 16 values in the third class alone, the 20th value will be the fifth of those 16 into this class, or $\frac{5}{16}$ of the way into the class. Using this information you can estimate the median.

Estimate of median = $(\frac{5}{16} \times 10) + 30 = 33.125$ seconds.

(10: class width 40 – 30; 30: lower class bound)

Spot the mistake

This is one of the most common types of error seen in examinations from students who do not understand what a frequency table means.

The table shows the number of glass bottles recycled by a group of students in one day.

Number of bottles	Frequency
1	12
2	14
3	6
4	7
5	1

Abby and Barry both try to find the mean number of bottles recycled. They both make mistakes.

Abby does

12 + 14 + 6 + 7 + 1 = 40

40 ÷ 5 = 8

Why cannot Abby's answer possibly be correct?

Barry does

12 + 14 + 6 + 7 + 1 = 40

1 + 2 + 3 + 4 + 5 = 15

40 ÷ 15 = 2.67

Why is Barry's answer clearly a little too big? What have neither Abby nor Barry done with the data? What is the correct answer?

Key terms

Modal group: the class interval with the most values within it.

Modal class: another name for the modal group.

Histograms can be used to estimate the mode or median.

links

In Chapter 4 you learned how to draw a histogram.

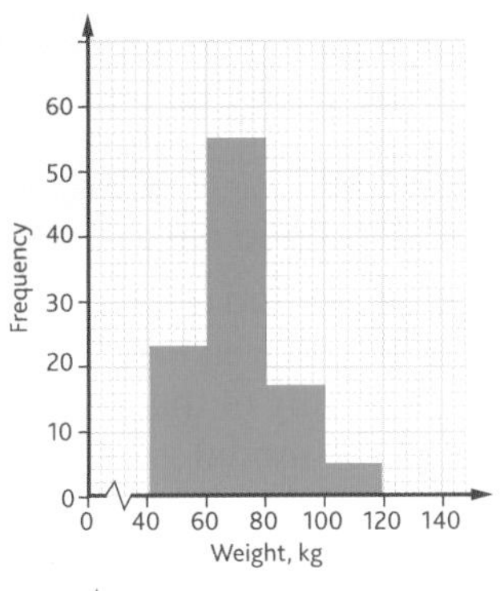

Example 3

Estimate:

a the mode and

b the median

from the histogram showing the weights of a group of people.

Solution

a Lines are drawn on the histogram to estimate the mode. The tallest bars inside top corners are joined diagonally to meet the point where neighbouring bars touch the tallest bar. Where these two lines cross, read from the horizontal axis to give an estimate of the mode.

Mode = 70

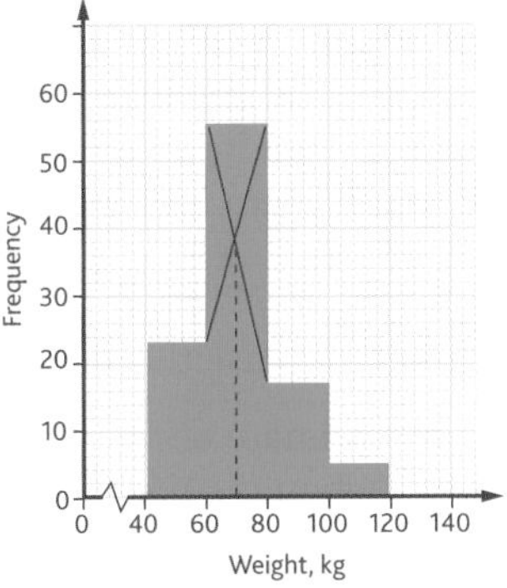

b The median will split the area of a histogram into two equal halves. The median can therefore be estimated by looking at the area of the histogram.

Area of histogram $= (20 \times 23) + (20 \times 55) + (20 \times 17) + (20 \times 5)$

$= 460 + 1100 + 340 + 100$

$= 2000$ sq units

The median will cut this area into '2 pieces' of area of 1000 sq units, so the median will fall somewhere in the second bar, $60 \leq x < 80$, because $460 + 1100 > 1000$

Median is then $60 + x$, where x is the distance it falls into the second bar.

For this second bar, we have $55x = 1{,}000 - 460$ where 55 is the height.

$$55x = 540$$

$$x = \frac{540}{55} = 9.818...$$

so median = 69.82 (2 d.p.)

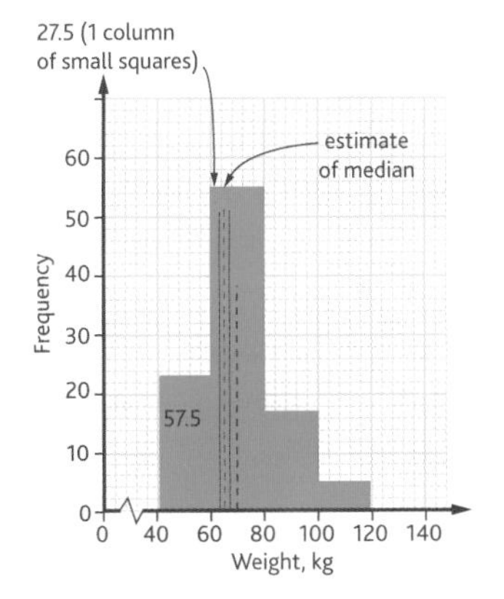

The mean can be estimated. Again this means that an estimate is calculated, not that there is any rounding or guessing taking place.

The method is similar to that for the discrete frequency distribution, except that you do not know the values in the class, so instead you use the midpoint to represent all the data in the class. This makes the assumption that the data is spread fairly evenly throughout the class, so the middle of the class is about the average (mean) of the values in it. So look again at Example 2. How would you estimate the mean for that data?

Add **two** columns to the original table, one for the midpoint values for each class and one for the total as before.

Actual length of time, x seconds	Frequency	Midpoint	Row total
$10 \le x < 20$	7	15	$7 \times 15 = 105$
$20 \le x < 30$	8	25	$8 \times 25 = 200$
$30 \le x < 40$	16	35	$16 \times 35 = 560$
$40 \le x < 50$	6	45	$6 \times 45 = 270$
$50 \le x < 60$	3	55	$3 \times 55 = 165$
			Overall total = 1300

Now the mean is the total of all the values (1300) divided by the total number of values (the total of the frequencies $7 + 8 + 16 + 6 + 3 = 40$).

Mean $= 1300 \div 40 = 32.5$ s

In a lengthy calculation like this, remember to check if your answer is sensible and around the expected value.

Finding midpoints for discrete data that has been grouped is slightly different. For example, data representing the number of people on a train could be grouped 10–19, 20–29, and so on. The midpoint for 10–19 would be 14.5, for 20–29 it would be 24.5, and so on. It is therefore important that you know whether your data is continuous, or whether it is discrete data that has been grouped.

On higher tier only these methods can be carried using a type of notation often called Sigma notation.

Sigma is the symbol Σ (the Greek letter 'S') meaning add up or find the sum of. (Sum begins with 's'.)

The mean can be denoted by the symbol $\bar{x}$ (you say 'x bar').

For a list of data $\bar{x} = \dfrac{\Sigma x}{n}$ where Σx is the sum of all the values and n is the number of values.

For a discrete frequency table $\bar{x} = \dfrac{\Sigma fx}{\Sigma f}$ where Σfx is the sum of all the frequency $\times$ data values.

Think back to the example of the appliances being left on (Example 1). The extended frequency table looked like this.

Number of appliances left on stand-by	Frequency	Row total
1	8	$1 \times 8 = 8$
2	9	$2 \times 9 = 18$
3	12	$3 \times 12 = 36$
4	2	$4 \times 2 = 8$
		Total = 8 + 18 + 36 + 8 = 70

Using this notation, column 1 is x, column 2 is f, and the row total column is fx giving $\sum fx = 70$, $\sum f = 31$ and $\bar{x} = \frac{70}{31} = 2.26$ as before.

For a grouped frequency table this formula is also used where the x values are the midpoints of the classes they represent.

AQA Examiner's tip

The focus on many questions in the examination will be on interpretation rather than calculation. Summary statistics such as $\sum fx$ could be given (or even the answers), with an interpretation then required. Look at Example 4.

Example 4

Two driving schools Passwell and Fastpass operate in a small town. Jodie is going to learn to drive and wants to know which school has a better record of passing drivers with fewer lessons. Passwell advertises that the mean number of lessons its drivers need before they pass is 22.

Fastpass do not give this information, so Jodie asks 10 people she knows, who passed their test with Fastpass, how many lessons, x, it took for them to pass. For this data $\sum x = 247$.

a Work out the mean for Fastpass and compare the two driving schools.

b Comment on the data Jodie collected and the reliability of the results.

Solution

a The mean $\bar{x} = \frac{\sum x}{n}$

$$\text{so } \bar{x} = \frac{247}{10} = 24.7$$

Therefore it seems to take a Fastpass driver longer, on average, than a Passwell driver.

b Sample size was small at 10. This was a convenience sample and may well have not been representative of all Fastpass drivers (for example, they might have only been female). Jodie also relied on people telling the truth. Altogether, there are doubts raised about the reliability of the results.

Exercise 5.2

1 The discrete frequency table shows the score made on each ball faced at cricket by Geoffrey.

Score	Frequency
0	42
1	20
2	8
3	2
4	3
5	6
6	1

a How many balls were bowled at Geoffrey in the whole innings?

b What was his total score in the innings?

c Use your answers to parts (a) and (b) to work out the mean number of runs scored per ball faced.

d Find the median number of runs per ball for Geoffrey.

e Find the modal number of runs per ball for Geoffrey.

f Ian says he averages 1.2 runs per ball in cricket when he is batting. Compare Ian and Geoffrey's figures.

2 Marlon compares the punctuality of two bus companies. He rounds times to the nearest 5 minutes. Company A has a mean number of minutes late of 8.2, with a median of 7.5. The data for company B is in the frequency table (again values are rounded to nearest 5 minutes).

Number of minutes late	Frequency
0	22
5	12
10	14
15	2

Compare the mean and median for the two companies.

3 The table shows the time in seconds taken for some children to solve a simple puzzle.

Time taken for some children to solve a simple puzzle	Frequency
$0 \leq t < 50$	15
$50 \leq t < 100$	28
$100 \leq t < 150$	20
$150 \leq t < 200$	27
$200 \leq t < 250$	10

a Calculate an estimate of the mean time taken to solve the puzzle.

b Calculate an estimate of the median time taken to solve the puzzle.

c Construct a histogram for the data.

d Use your histogram to estimate the modal time taken to solve the puzzle.

e Use your histogram to estimate the median time taken to solve the puzzle. Compare this answer to your answer in (b).

4 A company produces a bolt which is supposed to be 32 mm long. The machine producing them is old and now gives varied lengths. A sample of bolts is taken and measured, giving the results in the table below.

Length of bolt, l, millimetres	Frequency
$30 \leq l < 31$	15
$31 \leq l < 32$	28
$32 \leq l < 33$	20
$33 \leq l < 34$	27
$34 \leq l < 35$	10

a Bolts more than 1 mm away from the required length are scrapped. How many of the sample will need to be scrapped?

b If the mean of the sample is more than ½ mm away from the required length, the machine will also be scrapped. Use the data to find whether the machine will be scrapped.

5 Samples of consumers give two makes of jam marks out of 10 for taste. The summary statistics are given below.

Jam	$\sum fx$	$\sum f$
Toptaste	315	45
Proper preserve	724	91

a Calculate the mean score for each type of jam.

b Robert says, 'Toptaste has a higher mean score and so is definitely preferred to Proper preserve.' Comment on Robert's statement.

6 The mean of 47 values is 65. Find the value of $\sum fx$.

7 The weights of some plums are shown in the table.

Weight of plums, w, grams	Frequency
$8 \leq w < 10$	15
$10 \leq w < 12$	28
$12 \leq w < 14$	20
$14 \leq w < 16$	27
$16 \leq w < 20$	10

a Calculate an estimate of the mean weight of plums.

b Calculate an estimate of the median weight of the plums.

c Victoria plums have a mean weight of 13 g and a median of 12.5 g. State, giving two reasons, whether you think these plums are Victoria plums.

d Later, it was found that the scales used to weigh the plums were weighing items 2 g heavier than they actually were. Without further calculation, state how, if at all, your answers to parts (a), (b) and (c) change.

5.3 Estimation

In Chapter 2 you learned about useful and efficient sampling methods.

One of the key reasons for obtaining a sample from a population is so that you can make reliable and meaningful estimates of measures relating to the population such as the mean. The best way to estimate the mean of a population is to:

1 Find a representative sample.
2 Calculate the mean of this sample.
3 Use this value to estimate the measure for the population.

In statistics beyond GCSE, the sample mean is used to provide a range within which it is hoped the actual population mean lies.

The relevance of this here is that you need to be confident in your sampling technique if you are going to use a sample mean to give a specific single value estimate of the population mean. Also remember that different samples will, by their very nature, produce difference outcomes and therefore different sample means. So, if you take more than one sample you will be able to calculate more than one sample mean and therefore have more than one estimate of the population mean. Estimates that are otherwise comparable become more reliable as the sample size increases, so combining representative samples will often produce the best possible estimate of the population mean.

Objectives

In this topic you will learn:

- how to estimate the mean of a population from a sample
- about some of the issues that make an estimate more or less reliable
- how to estimate the proportion of a feature within a population.

links

See Chapter 2 to recap sampling methods.

Example 1

The weight of each the first 8 bags of grapes from a large crate is recorded in grams.

67.8 59.6 64.0 60.8 71.2 67.8 66.3 70.5

a Use the data given to obtain the best possible estimate for the mean of the population of bags of grapes of this type.

b Suggest two ways in which the estimate could be made more reliable.

Solution

a To estimate the population mean use the mean of the sample

$$\text{mean} = \frac{66.8 + 59.6 + \dots + 70.5}{8}$$

$$= \frac{528}{8}$$

$$= 66 \text{ g}$$

The estimate of the population mean of grapes is 66 g.

b Increase the sample size taken to produce the estimate. Ensure that the sampling method used is as reliable as possible in producing representative sampling units.

When a particularly small sample is collected, the effect of any outliers in that sample is greater in any calculation. Thus data sets with outliers, or more variable data, produce more variable or unreliable estimates.

There is a specific connection between the variability of a sample estimate and the sample size from which that estimate is taken. If you make a sample size n times bigger, the variability in the sample is divided by $\sqrt{n}$. This is perhaps easier to understand if stated as 'to halve the variability in a sample, you need to have a sample that is four times bigger'.

For example, if you have a sample of 20, you would have to collect an equivalent sample of 80 to halve the amount of variability in the results. Taking into consideration issues such as cost and time in collecting samples, it is therefore often not necessary to make slight increases in sample size as this has hardly any effect on the variability and therefore reliability of estimates. This is especially true where the data is unlikely to have outliers.

Estimating proportions

Suppose you wish to estimate the proportion of a particular feature within a population.

The best way to estimate this proportion is to:

1 Take a sample that is as representative as possible of the population.
2 Work out the proportion of the desired feature within the sample.
3 Use this value as your estimate of the population proportion of that feature.

This is often applied in opinion polls when seeking, for example, the proportion of the population who will vote for a particular political party.

Example 2

Sami is commissioned to find out the proportion of voters in Scotter who will vote for the Labour Party in an upcoming election. She goes to the centre of the village and asks the first 64 people she sees who they would vote for in the election. Of these people, eight say that they will vote for the Labour Party.

a Calculate the proportion of the sample who will vote Labour at the next election.
b Estimate the proportion of the village who will vote Labour at the next election.
c Give two reasons why the estimate in (b) may not be very reliable.

Solution

a $\frac{8}{64} = 0.125$

b 0.125

c She only sampled people from the village centre; they might not represent the population of the village very well. She is asking personal questions and some of the responses may not have been truthful.

AQA Examiner's tip

Sometimes you may be asked directly to estimate the proportion of a population, without first being asked to calculate the proportion of the sample. Remember that the way to answer this question is to work out the proportion of the sample.

Exercise 5.3

1 A strawberry producer is estimating the total yield of all his crops. He begins by estimating the mean number of strawberries produced by his plants. Over the summer he records the number of strawberries produced by the end plant on each of the first 9 rows of plants in his field. The results are given below.

9 12 8 13 10 17 12 15 16

a Calculate the mean number of strawberries for each of these plants.

b Estimate the mean number of strawberries per plant in the whole field.

c Give two reasons why this estimate may be quite unreliable.

d Suggest a better way to sample the crop yield.

2 A restaurant manager wishes to estimate the proportion of her customers who have a dessert at her restaurant. On one evening she asks her staff to count the number of customers sitting on the even numbered tables and the number of these who have a dessert. Of 85 customers, 51 have a dessert.

a Work out the proportion of customers who had dessert on even numbered tables that evening.

b Estimate the proportion of all customers at that restaurant who have dessert.

c Comment on the reliability of your estimate.

3 A researcher studies the number of different TV channels watched by families in the UK within one evening of viewing. For a random sample of 100 families, the table shows information about the number of different stations watched.

Number of stations watched	Frequency
1	38
2	43
3	13
4	4
5	2

Estimate the mean number of TV stations watched by families in the UK on one evening.

4 Sasha is investigating the pass rates on the driving test amongst 17-year-olds. Amongst her twenty 17-year-old friends, eight had passed their driving test.

a Estimate the proportion of all 17-year-olds who have passed their driving test.

b Explain why this answer is likely to be inaccurate.

c Suggest a better way of obtaining the sample data.

5 A component is produced to fit inside an engine. In order to fit, the component must be between 34 and 36 mm wide. The frequency table shows the widths of 80 components produced by one machine.

Component width, w (mm)	Frequency
$33 \le w < 34$	6
$34 \le w < 35$	31
$35 \le w < 36$	33
$36 \le w < 37$	5
$37 \le w < 38$	4
$38 \le w < 39$	1

a What proportion of components produced by this machine will fit inside an engine?

b Estimate the proportion of all components produced by this machine that will fit inside an engine.

c Estimate the mean width of a component produced by this machine.

6 The income of 45 working adults from a particular town is given in the frequency table.

Income (£)	Frequency
10 000–19 999	8
20 000–29 999	17
30 000–39 999	16
40 000–49 999	4

a Estimate the proportion of working adults in this town earning under £20 000.

b Estimate the mean income of working adults in this town.

c In order to halve the variability in the estimate obtained in part (b), what sample size should be taken?

d Discuss the appropriateness of using this data to estimate the mean income for the whole of the UK.

Summary

You should:

- know that the mean, mode and median are commonly used as measures of average
- know that sometimes, one or two of these measures may be unsuitable for use
- be able to use a change of origin or a linear transformation if desired to make numbers more manageable
- know that the geometric mean is used for averaging measures that change over time, such as interest rates
- be able to find row totals in discrete frequency tables in order to find the mean
- use midpoints in grouped frequency tables to represent the classes when finding an estimate of the mean
- know that the median is also only estimated for grouped data
- know that the modal group can be found for grouped data
- know that estimates of the mode and the median can be found using a histogram
- be able to use sigma notation where possible at higher tier
- know that the mean of a representative sample is a good estimate for the mean of the population from which the sample comes
- know that the proportion of a feature in a representative sample is a good estimate for the proportion of that feature in the population from which the sample comes.

AQA Examination-style questions

1. In May 2001 an estate agent sold nine three-bedroomed houses. The sale price in pounds were:

59 200 65 000 52 000
129 500 52 000 62 500
54 500 57 900 56 000

(a) Write down the mode of these prices. *(1 mark)*

(b) (i) Calculate the mean of these prices. *(2 marks)*

(ii) Give a *disadvantage* of using the mean to represent these prices. *(1 mark)*

AQA, 2003

2. Reuben read all 12 Sharren Day books. He gave each book an enjoyment score out of 10.
The scores were:

3 7 8 8 8 8 10 10 10 10 10 10

(a) For these scores work out:

(i) the mode *(1 mark)*

(ii) the median *(1 mark)*

(iii) the mean. *(2 marks)*

(b) Reuben says that the mean is the best average to use. Give a disadvantage of using the mean in this case. *(1 mark)*

AQA, 2006

3. A packet contains sweets. Some of the sweets are red.

A random sample of three sweets is chosen from the packet and the number of red sweets is recorded.

The sample is then replaced. Another sample is then taken.

The number of red sweets in each of 25 samples is given below.

1 2 2 1 2
1 2 2 2 2
2 0 2 0 1
1 0 0 1 1
1 1 1 1 3

(a) Complete the frequency table below.

Number of red sweets in each sample	Frequency
0	
1	
2	
3	

(2 marks)

(b) For this data:

(i) write down the mode *(1 mark)*

(ii) calculate the mean. *(3 marks)*

AQA, 2004

4. A class of pupils sat an examination and their marks were recorded. The mean, median, mode and range of the marks were calculated. Another pupil sat the examination at a later date. This pupil's mark was lower than any of the others. If this mark were to be included in the calculations, what effect would it have on:

(a) the mean *(1 mark)*

(b) the median *(1 mark)*

(c) the mode *(1 mark)*

(d) the range? *(1 mark)*

AQA, 2002

5. The frequency table shows the number of times members of a club attended monthly meetings in one year.

Number of meetings attended	Frequency
0	1
1	3
2	6
3	2
4	13
5	11
6	17
7	15
8	19
9	16
10	8
11	4
12	10

(a) What is the modal number of meetings attended by the members? *(1 mark)*

(b) How many members does the club have? *(2 marks)*

(c) The total attendance for the year was 875. Use this fact and your answer to part (b) to work out the mean number of meetings attended by each member. *(2 marks)*

(d) The data are put into a grouped frequency table.

Number of meetings attended	Frequency
0–3	12
4–6	
7–9	
10–12	

Complete the frequency column. *(3 marks)*

(e) Give one advantage of using:

(i) the original frequency table *(1 mark)*

(ii) the grouped frequency table. *(1 mark)*

AQA, 2008

6. The head teacher of a school recorded the number of pupils who were absent on the first twenty days of a school year. They were:

8	19	24	26	28
13	22	24	22	29
18	19	8	28	33
28	23	27	34	9

(a) Calculate the mean number of daily absences. *(3 marks)*

(b) Copy and complete the following grouped frequency table for these data:

Class	Tally	Frequency		
5–9				
10–14				
15–19				
20–24				
25–29				
30–34				

(3 marks)

(c) Using the grouped frequency table, calculate an estimate of the mean number of daily absences. *(4 marks)*

(d) Why are the means in parts (a) and (c) different? *(1 mark)*

AQA, 2002

7. **(a)** The table gives the price index for three years for the cost of a flat.

Price index for 2002 relative to 2001	105
Price index for 2003 relative to 2002	140
Price index for 2004 relative to 2003	130

Calculate the geometric mean of these three price indices. *(2 marks)*

(b) Write down the average annual rate of increase of the price of the flat over the three years. *(1 mark)*

AQA, 2004

8. The length of telephone calls is given in the table.

Duration (t minutes)	**Frequency**
$0 \leq t < 4$	8
$4 \leq t < 6$	14
$6 \leq t < 8$	9
$8 \leq t < 10$	7
$10 \leq t < 20$	10

(a) Draw a histogram to represent this data. *(5 marks)*

(b) Use your histogram to estimate the modal length of a telephone call in minutes. *(2 marks)*

(c) Calculate the mean of the duration of telephone calls. *(2 marks)*

(d) Explain why the mean may not be the best method of summarising the data. *(1 mark)*

AQA, 2001

6 Methods for calculating spread and skewness

In this chapter you will learn:

- to interpret and calculate the range
- to interpret and calculate quartiles and the interquartile range for discrete data
- to interpret and calculate deciles, percentiles, interdecile range and interpercentile range
- to calculate and interpret variance and standard deviation using, if appropriate, summary statistics
- to decide on the most appropriate measure of spread
- to recognise whether a distribution shape is symmetrical, or shows negative or positive skew
- to calculate measures of skew such as Pearson's
- about the Normal distribution and its basic properties
- how to compare distributions using measures of location, dispersion and skew
- how to compare distributions using standardised scores.

You should already know:

- ✓ how to calculate the mean of data in lists and frequency tables
- ✓ how to find the median for a discrete set of data
- ✓ how to round to decimal places and significant figures.

What's this chapter all about?

Knowing the average of a set of data is only a small part of the picture. There are many vastly different sets of data that have the same average values, so how do you find differences between them? This chapter looks at the way data can be spread using a number of different measures. Look out for skew and Normal distributions too – there's a lot more to the way data is distributed than you might think. Finally, does it mean you are better at a subject just because you get a higher score in its exam? Maybe it depends upon how hard the exam was and how everyone else does. This is what the chapter closes with – standardised scores, a genuine way of comparing distributions.

The Browns were a Normal type of family

Focus on statistics

Much of this chapter is not covered in maths. In 6.1 you will have met range and interquartile range before, but interdecile range and interpercentile range will be new to you. All of 6.2 and 6.3 will probably be new to you as these topics are not tested in maths.

6.1 Types of range

The range

The **range** is the most basic measure of **spread** or **dispersion**. It is simply the difference between the highest and lowest values in a set of data.

The interquartile range

The **interquartile range** (often written as IQR) measures the difference between the **upper quartile** (UQ) and the **lower quartile** (LQ).

As the upper quartile is three quarters along the ordered data, and the lower quartile is one quarter along the ordered data, the IQR measures the spread across the central 50% of the data, leaving out the top 25% and the bottom 25%.

Finding the upper and lower quartiles for discrete data is done much like the work for finding the median in Chapter 5.

If an ordered list has n values, the lower quartile is the $\left(\frac{n+1}{4}\right)$th value along. If an ordered list has n values, the upper quartile is the $3\left(\frac{n+1}{4}\right)$th value along.

Other similar methods are available.

Note that the lower quartile is sometimes denoted by Q_1, the median by Q_2, and the upper quartile by Q_3 as these three measures split the data into four equal groups.

Strawberry fields

Objectives

In this topic you will learn:

- to interpret and calculate the range
- to interpret and calculate quartiles and the interquartile range for discrete data
- to interpret and calculate deciles, percentiles, interdecile range and interpercentile range.

Key terms

Range: the highest value minus the lowest value.

Spread: a general term indicating the amount by which different values in a data set are close to each other or not.

Dispersion: another word for spread.

Interquartile range: the upper quartile minus the lower quartile.

Upper quartile: the upper quartile (Q_3) is the value ¾ along a set of ordered data.

Lower quartile: the lower quartile (Q_1) is the value ¼ along a set of data.

AQA Examiner's tip

Examiners often choose a value for *n* so that if you add one to the number of values you get a multiple of 4, therefore making the locating of the LQ and UQ relatively easy.

Example 1

The number of strawberries produced by 11 different plants is given below:

6 3 12 8 5 14 7 5 10 8 12

Find:

a the median
b the lower quartile
c the upper quartile
d the interquartile range.

Solution

All these parts require the data to be ordered, so you need to start by carefully putting the data in numerical order:

3 5 5 6 7 8 8 10 12 12 14

a There are 11 values. The median is the $\left(\frac{11+1}{2}\right)$th along, the sixth. The median is therefore 8.

b There are 11 values. The lower quartile is the $\left(\frac{11+1}{4}\right)$th along, the 3rd. The lower quartile is therefore 5.

c There are 11 values. The upper quartile is the $3\left(\frac{11+1}{4}\right)$th along, i.e. the 9th. The upper quartile is therefore 12.

d The interquartile range = upper quartile – lower quartile = 12 – 5 = 7

For grouped frequency distributions, estimates of the quartiles and therefore the IQR have to be made either by calculation or by graphical methods. The example from Chapter 5 can be used to show the calculation method.

links

See Chapter 7 for graphical methods of calculating or estimating the quartiles and IQR.

Example 2

A group of 40 students tried to estimate a time of 30 s. The actual length of time each student thought was 30 s is given in the grouped frequency distribution table.

Calculate an estimate of the interquartile range of the data.

Actual length of time, x seconds	Frequency
$10 \le x < 20$	7
$20 \le x < 30$	8
$30 \le x < 40$	16
$40 \le x < 50$	6
$50 \le x < 60$	3

Solution

There are 40 values. The lower quartile is approximately the tenth value. Remember that as the data is tabulated it is in order, and when n is large, $\frac{n}{4}$ is sufficient for the position of the lower quartile, as you are estimating the value anyway. (This method and the $\frac{(n+1)}{4}$ method are both acceptable for use in exams.)

There are seven values before the $20 \le x < 30$ class so, as there are eight values in that class alone. The tenth value will be the third of the eight in that class. Using this information you can estimate the actual lower quartile:

$$\text{estimate of lower quartile} = \left(\frac{3}{8} \times 10\right) + 20 = 23.75 \text{ s}$$

For the upper quartile, which is approximately the thirtieth value, there are 15 (7 + 8) values before the $30 \le x < 30$ class so, as there are 16 values in that class alone, the thirtieth value will be the fifteenth of the 16 in that class. Using this information you can estimate the actual upper quartile:

$$\text{estimate of upper quartile} = \left(\frac{15}{16} \times 10\right) + 30 = 39.38 \text{ (to 2 d.p.)}$$

So the IQR = 39.38 – 23.75 = 15.63 s

Interdecile and interpercentile range

The range can be criticised for using the extremes of the data and the interquartile range can be criticised for omitting 50 per cent of the data. The **interdecile range** and **interpercentile ranges** take into account more of the data than the interquartile range but do not use the extreme values as the range does.

Just as the quartiles split the data into four groups, **deciles** split the data into ten groups, and **percentiles** into 100 groups. Thus, for example, the third decile (or D_3) is $\frac{3}{10}$ or 30 per cent along the ordered data. The 83rd percentile (or P_{83}) is 83 per cent along the data.

The interpercentile range can be found for different percentiles, so no single formula exists.

For example, the 5 to 95 interpercentile range is $P_{95} - P_5$ giving the central 90 per cent of the data.

Key terms

Interdecile range: the ninth decile minus the first decile ($D_9 - D_1$). This gives the range of the central 80 per cent of the data.

Interpercentile range: the difference between two percentiles.

Deciles: deciles split ordered data into 10 equal groups.

Percentiles: percentiles split ordered data into 100 equal groups.

Spot the mistake

Mohammed said that $P_{50} = Q_2 = D_5$ and $P_{75} = Q_3 = D_7$

Where is the mistake he has made?

Exercise 6.1

1 Find:

a the range

b the interquartile range

for the following data sets:

i 5 3 7 8 3 2 8 4 5 1 6

ii 12 16 18 11 14

iii The number of letters in each of the words in the whole of this sentence.

2 The number of millimetres of snow falling during one week were as follows (rounded to the nearest mm):

17 0 5 54 1 22 13

a Find the range.

b Find the interquartile range.

c Which of these values is not affected by extreme values?

3 The interquartile range of a set of data is 12. The median is 36. The median is 4 below the upper quartile. Find the value of the lower quartile.

4 Use the data from the bolts question in Chapter 5 to calculate an estimate for the interquartile range.

Length of bolt, l, millimetres	Frequency
$30 \le l < 31$	15
$31 \le l < 32$	28
$32 \le l < 33$	20
$33 \le l < 34$	27
$34 \le l < 35$	10

5 Use the data in the table to answer the questions below about a set of discrete data.

Data Point	D1	D2	D3	D4	D5	D6	D7	D8	D9
Value	13	16	18	21	22	23	25	27	32

a Find the median.

b Find the Interdecile range.

c Find the 20–80 percentile range.

d Could the interquartile range be 9? Explain your answer.

6.2 Variance and standard deviation

Objectives

In this topic you will learn:

- to calculate and interpret variance and standard deviation using, if appropriate, summary statistics
- to decide on the most appropriate measure of spread.

Key terms

Variance: the average of the squared deviations from the mean.

Standard deviation: the square root of the variance.

The **variance** is a measure of spread that uses every value from a data set.

The **standard deviation** (s.d. or s) is the square root of the variance.

Consider the numbers 1, 2, 3, 4 and 5. You know that the mean of these values is 3.

The differences between the numbers and the mean are –2, –1, 0, 1 and 2; therefore the mean difference is 0, which is not much help. This will always be the case, can you explain why?

However, if you square these differences and then find the mean, you will have an idea of the spread of the data, and this in essence is the variance in its simplest form.

The differences squared make 4, 1, 0, 1 and 4; giving a total of 10 and a mean of 2. As you squared the data, you now find the square root of this answer, giving you 1.414 (to 3 d.p.), which is the standard deviation of 1, 2, 3, 4 and 5.

The higher the standard deviation, the higher the spread of the data. The lower the standard deviation, the more consistent the data is.

In practice for this course, you will nearly always use the standard deviation rather than the variance.

Discrete data

For a simple set of data use $s = \sqrt{\frac{\sum x^2}{n} - \bar{x}^2}$

where $\sum x^2$ means the sum of the squares of the data values, n is the number of data values and $\bar{x}$ is the mean of the data values. There is another equivalent formula but it using it can involve more work for a simple list of values.

Example 1

The number of strawberries produced by 11 different plants is given below.

6 3 12 8 5 14 7 5 10 8 12

Find the standard deviation of these values.

Solution

Find the mean of the values: 6 + 3 + 12 + ... + 12 = 90; 90 ÷ 11 = 8.18. . .

Find the sum of the squares of the values $6^2 + 3^2 + 12^2 + \ldots + 12^2 = 856$

Substitute these values and the value for n into the formula for standard deviation:

$s = \sqrt{\frac{856}{11} - (8.18\ldots)^2}$ (make sure you use the whole value for the mean – do not round it off)

$s = \sqrt{10.876033}$

$s = 3.2978831$

$s = 3.30$ (to 2 d.p.)

It is important to get a feel for the size of the answer as a means of checking that you have not made a silly mistake.

Frequency tables

For a discrete frequency table use $s = \sqrt{\frac{\sum fx^2}{\sum f} - \bar{x}^2}$

where $\sum fx^2$ is the sum of the values squared times their frequency, and $\sum f$ is the sum of the frequencies.

For a grouped frequency table, the standard deviation, much like the mean, has to be estimated, as the values in the original data are lost when grouping occurs. Again midpoints of the interval are used to represent the classes. It is perfectly possible to reuse the same formula as above for these calculations, but the alternative version of the formula tends to be easier to use, especially if the values are large or if the mean is an integer or a simple fraction.

The alternative formula is $s = \sqrt{\frac{\sum f(x - \bar{x})^2}{\sum f}}$

where $\sum f(x - \bar{x})^2$ is the sum of each value minus the mean, then squared, then multiplied by the frequency for that class. This is much like the method you saw when this section was introduced, using the values 1, 2, 3, 4 and 5. Obviously, this can be complicated for frequency tables and you are advised to use a table with additional rows added to it to obtain this information.

AQA *Examiner's tip*

All the formulae you will need to know for standard deviation are given at the beginning of the paper. You therefore don't need to learn the formulae but you do need to know how to use them.

Example 2

Find the standard deviation of the data in this grouped frequency table.

Data	Frequency (f)
$10 \le x < 20$	7
$20 \le x < 30$	8
$30 \le x < 40$	16
$40 \le x < 50$	6
$50 \le x < 60$	3

You need to add five extra columns for working (it is possible to use fewer but you risk making errors). Note that you need to use the totals from the $f \times x$ and f columns to find an estimate of the mean before you can find the values in the final three columns. (Mean = 1300 ÷ 40 = 32.5)

Data	Frequency (f)	Midpoint (x)	Row Total ($f \times x$)	$(x - \bar{x})$	$(x - \bar{x})^2$	$f(x - \bar{x})^2$
$10 \le x < 20$	7	15	105	–17.5	306.25	2143.75
$20 \le x < 30$	8	25	200	–7.5	56.25	450
$30 \le x < 40$	16	35	560	2.5	6.25	100
$40 \le x < 50$	6	45	270	12.5	156.25	937.5
$50 \le x < 60$	3	55	165	22.5	506.25	1518.75
Total of the column	40		1300			5150

Now substitute the relevant column totals into the formula

$$s = \sqrt{\frac{\sum f(x-\bar{x})^2}{\sum f}}$$

$$= \sqrt{\frac{5150}{40}}$$

$$= \sqrt{128.75}$$

$$= 11.346806$$

$$= 11.3 \text{ (3 significant figures)}$$

Although it is more difficult than for a list of data, you should try to check very roughly whether the size of the answer seems sensible for the data you have worked with.

Clearly this is quite a complex calculation and, although you will need to be able to do this, there are two issues to be aware of:

1 The emphasis in examination questions will be more on interpretation and comparison than in the past so it is more likely you will be given summary statistics such as the value of $\sum f(x - \bar{x})^2$ already calculated for you.

2 It is entirely appropriate to use the statistical functions on your calculator to do the working out for you. If you are required to do this calculation in an examination, you should be willing to do it twice as a check of accuracy. Calculators vary so much that it is not possible to give advice on how to use one here. However, always be aware, when using the statistical functions on your calculator, to make sure you have cleared all previous calculations from the memory and functions. Take time to become familiar with how to use your calculator long before the examination.

Which is the best spread?

Choosing the best measure of spread

It is important to be able to use the most appropriate measure of spread in different circumstances. The table shows the common measures you have just met and the positive and negative aspects of using each one.

Measure of Spread	Positive Aspects	Negative Aspects
range	easy to calculate	uses the two most extreme values in the data and so is badly affected if these are particularly large or small
interquartile range	excludes extreme data values	excludes 50% of the data
Interdecile range / interpercentile range	excludes extreme data values but can still use most of the data	can be difficult to calculate or estimate
standard deviation	uses all the data, useful for comparison and further statistical work	can be time consuming and is badly affected by extreme values

links

Extreme values are sometimes called outliers. More on how to find outliers in Chapter 7.

Be a statistician

The table contains real data from the US about fatalities due to being struck by lightning. The figures are broken down according to gender and age group.

You are a statistician who is writing a piece for a national magazine about the age and gender of people who are struck by lightning. Write a short piece that includes an estimate of the mean age and the standard deviation of the age of each gender making comparisons between the results. Make sure you explain your findings as if you were talking to a person who does not understand statistics.

www.weather.gov/os/hazstats.shtml – source of data from the NOAA website.

2006 Lightning fatalities by age and gender

Age	Female	Male	Unknown	Total	Percent
0 to 9	0	1	0	1	2
10 to 19	3	10	0	13	27
20 to 29	1	4	0	5	10
30 to 39	0	5	0	5	10
40 to 49	2	7	0	9	19
50 to 59	2	3	0	5	10
60 to 69	2	2	0	4	8
70 to 79	0	1	0	1	2
80 to 89	0	3	0	3	6
90 to –	0	0	0	0	0
Unknown	0	1	1	2	4
TOTAL	10	37	1	48	98*
PER CENT	21	77	2	100	

*Due to rounding total does not equal 100%

Exercise 6.2

1 Find:

a the standard deviation

b the variance for these sets of data.

i 5 4 8 3 10

ii 23 80 4 2040 33

iii

Score	Frequency
0	42
1	20
2	8
3	2
4	3
5	6
6	1

iv

Time taken for some children to solve a simple puzzle (seconds)	Frequency
$0 \le t < 50$	15
$50 \le t < 100$	28
$100 \le t < 150$	20
$150 \le t < 200$	27
$200 \le t < 250$	10

2 Which of the data sets in question 1 was the standard deviation least appropriate for? Explain your answer.

3 Given these summary statistics, find the mean and standard deviation for the data set.

$$\sum f = 50, \sum fx = 812 \text{ and } \sum fx^2 = 25\,080$$

4 Two employees, Alan and Richard, have applied for a job as the chief salesperson at a small firm. One of the criteria to be used in the process is to compare the size and consistency of the sales over the previous year. Alan has a monthly mean sales figure of £23 000 with a standard deviation of £8000. Richard's figures for the 12 months of sales are as follows (all in thousands of pounds):

28.3, 24.2, 19.2, 20.4, 24.5, 26.4, 18.2, 19.7, 20.0, 25.5, 26.3, 27.9.

By making suitable calculations, compare the performance of Richard and Alan over the year. Who should get the job based on these criteria?

6.3 Shapes of distributions and skew

Many naturally occurring sets of data come under distributions with one of three types of shape. **Symmetrical data** has a symmetrical shape of distribution and **skew data** has a skewed shape of distribution. The skew can be either positive or negative.

The shape of a distribution is easiest to see if a histogram is drawn with narrow groups. If the centre of the top of each bar is joined with a smooth curve you will get a distribution curve of one of these types:

1 Symmetrical distribution
2 Positively skewed distribution
3 Negatively skewed distribution.

1 Symmetrical distribution

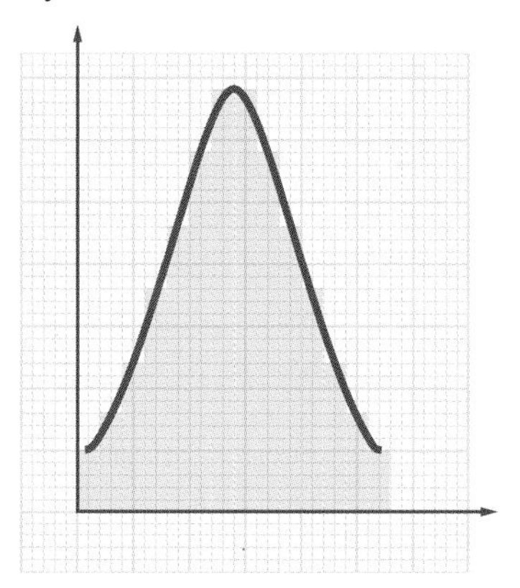

Mean = median = mode

2 Positively skewed distribution

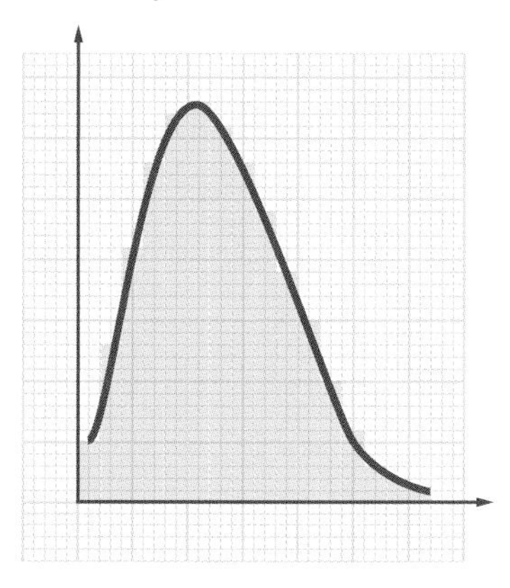

Mode < median < mean

3 Negatively skewed distribution

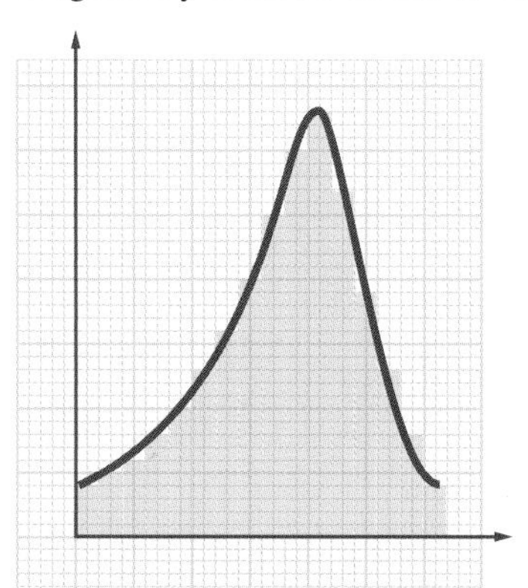

Mean < median < mode

Objectives

In this topic you will learn:

- to recognise whether a distribution shape is symmetric or shows negative or positive skew
- to calculate measures of skew such as Pearson's
- about the Normal distribution and its basic properties
- how to compare distributions using measures of location, dispersion and skew
- how to compare distributions using standardised scores.

Key terms

Symmetrical data: data that is equally spread either side of the centre.

Skew data: data that is unequally spread either side of the data.

Some examples of naturally occurring data that would take each of these shapes are shown in the table.

Distribution type	Symmetrical	Positive skew	Negative skew
Example 1	Heights of a random sample of men	Age at which a random sample of people first drove a car	The age at which a random sample of people retired
Example 2	Weight of all the apples from one tree	Temperature each day over one summer in England	Finishing times of people in a fun run

Calculating skew

There are several ways of calculating a measure of skew, which can then be used to compare distributions. The method you need to be aware of is called the **Pearson** measure of skew.

$$\frac{3(\text{mean} - \text{median})}{\text{standard deviation}}$$

Positive values of Pearson's skew indicate that the distribution is positively skewed. This is because, for positive skew, the mean is more than the median, giving a positive numerator for the formula.

Negative values of skew indicate that the distribution is negatively skewed. This is because for negative skew, the median is more than the mean, giving a negative numerator for the formula.

The size of the value for skew also shows the degree of skew present. For example, a value of 1.3 shows that a greater amount of positive skew is present than a value of 0.2, although both indicate that positive skew is present. Similarly, –1.5 indicates greater negative skew than –0.3, for example.

The Normal distribution

Continuous data that has a symmetrical distribution with the mean, mode and median all equal, follows a **Normal distribution**.

The Normal distribution is the most important distribution in statistics, as so much naturally occurring data follows a Normal distribution. It also has some unique properties.

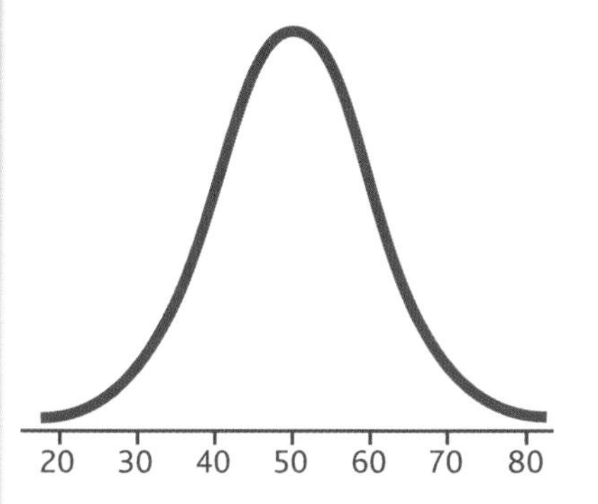

A Normal distribution

Real world

Can you explain why each of the items listed in the table has a symmetrical or skewed distribution as described?

Find two more examples of data that would fit into each category.

Key terms

Pearson's measure of skew: $\frac{3(\text{mean} - \text{median})}{\text{standard deviation}}$

Normal distribution: a symmetrical continuous set of data that has special features, often naturally occurring such as heights and weights.

links

There are other ways of seeing whether data is skewed – see Chapter 7 on box-and-whisker plots.

AQA Examiner's tip

You do not need to learn Pearson's formula; it will be given to you in the examination if needed. You need to know how to use it. Other measures of skew may be used but the required formula will always be given.

Properties of the Normal distribution

- It is symmetrical.
- The mean, mode and median are all equal.
- It has a value of skew = 0.
- About two-thirds of the distribution is within one standard deviation of the mean.
- Ninety-five per cent of the distribution is within two standard deviations of the mean.
- Virtually all the data from a Normal distribution is within three standard deviations of the mean.

Other features that you need to know relate to the drawing of Normal distribution curves. Normal curves are often drawn together on the same axes so you can compare distributions. The value of the mean determines the position of the peak of the curve on the x-axis. The value of the standard deviation determines the height and spread of the curve.

Example 1

On the same axes draw two Normal distribution curves.
Curve A should have a mean of 50 and a standard deviation of 10.
Curve B should have a mean of 60 and a standard deviation of 5.

Solution

Curve A (in brown) is flatter due to the larger standard deviation. Notice how the curve drops down to very close to the axis at about 20 and about 80 – these values represent + and – 3 standard deviations either side of the mean. Don't end your curves less than 3 standard deviations from the mean, or you will lose marks.

Curve B (in orange) is taller and positioned to the right of the brown curve due to the higher mean.

The area under each of the curves is equal.

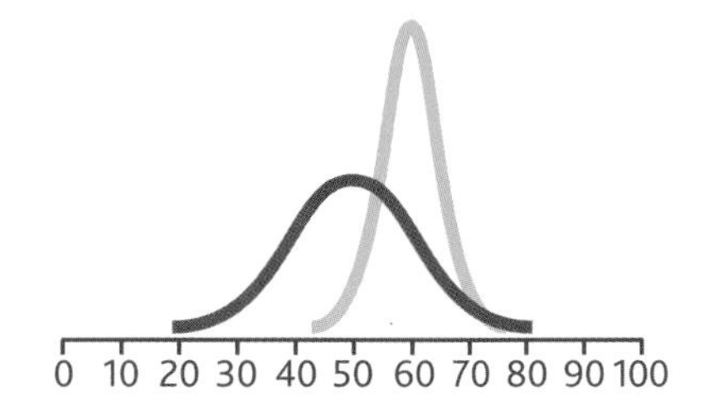

Comparing distributions using standardised scores

Using a method of **standardised scores** you can compare data from two or more different data sets. To do this you need to know the mean and standard deviation for each of the data sets, and use this formula:

$$\text{standardised score} = \frac{\text{actual score} - \text{mean}}{\text{standard deviation}}$$

Positive values of the standardised score indicate an actual score above the mean. Negative values of the standardised score indicate an actual score below the mean. The value of the standardised score is equal to the number of standard deviations between the value and the mean.

The above might seem obvious but the real value of finding standardised scores is when you have data from two different distributions with different means and different standard deviations and you are trying to compare these with each other. The classic case is when comparing scores in two different exams.

Key terms

Standardised score: $\frac{\text{actual score} - \text{mean}}{\text{standard deviation}}$

Example 2

The table shows Archie's marks in maths and statistics examinations, together with the mean and standard deviation for each examination. In which examination did Archie perform better?

	Archie's mark	Mean mark	Standard deviation
Maths	68	55	6.5
Statistics	74	59	10

Solution

Work out the standardised score for each subject. For maths, the standardised score $= \frac{68 - 55}{6.5} = +2$ (always put a + or – with a standardised score to emphasise its value).

For statistics, the standardised score $= \frac{74 - 59}{10} = +1.5$

Therefore, as Archie has a higher standardised score in maths, he has done better in maths relative to the group, even though at first sight he appears to have done better in statistics due to having a higher mark.

You now have many ways to compare different data sets using measures of average from Chapter 5 and measures of spread and standardised scores from this chapter. Examination questions tend to require use of one of each type of measures (a measure of average and a measure of spread), or will specifically ask you to compare using standardised scores.

Example 3

Here is an example of a question on this type of work.

The times taken by 300 Year 11 students to solve a simple puzzle are given in the table.

Time (seconds)	Frequency
$0 \le t < 40$	38
$40 \le t < 60$	36
$60 \le t < 80$	41
$80 \le t < 100$	58
$100 \le t < 120$	49
$120 \le t < 160$	48
$160 \le t < 200$	30

a Calculate an estimate of the mean and standard deviation.

b The mean and standard deviation of the times taken by some Year 13 students are 64 s and 35 s respectively. Comment on the differences in the times taken by Year 11 and Year 13 students.

Solution

a Use midpoints to represent each class interval. Midpoints are 20, 50, 70, 90, 110, 140 and 180. Then, using the statistical functions on the calculator, or by working out $\Sigma fx = 28\,160$, $\Sigma f = 300$ and $\Sigma fx^2 = 3\,281\,600$, you find that the estimated mean = 93.87 s and the estimated standard deviation = 46.13 s.

b When you are faced with a comment question like this, it is important you make one comment that compares the means in context and one comment that compares the standard deviations in context. The mean for Year 11 students is higher, therefore the Year 11 students tend to take longer to solve the puzzle. The standard deviation for Year 11 students is higher, therefore Year 11 students tend to have more varied times. (You could equally have commented, in relation to Year 13 students, that the mean for Year 13 students is lower so Year 13 students tend to solve the puzzle quicker. The standard deviation for Year 13 students is lower, therefor times for the Year 13 students tend to be more consistent or less widely spread.)

Exercise 6.3

1 Draw a sketch of each of the continuous distributions described below on separate sets of axes.

a A distribution with negative skew having a mode of 100.

b A Normal distribution having a mean of 5.

c A distribution with positive skew having a mode of 45.

2

a On the same axes, sketch these three Normal distributions, clearly labelling each one.

b Between what values does approximately 95% of each of these distributions lie?

Curve	Mean	Standard deviation
A	80	20
B	60	10
C	100	10

3 For each set of data in the table, state, with a reason, whether the data comes from a distribution that is Normal, positively skewed or negatively skewed.

Data set	Mean	Median	Mode
1	30	35	40
2	50	50	50
3	60	55	50

4 For each of the data sets in Question 3 you are now also given the standard deviation.

Calculate Pearson's measure of skew using the formula:

$$\frac{3(\text{mean} - \text{median})}{\text{standard deviation}}$$

Data set	Standard deviation
1	15
2	10
3	30

5 The table shows the mean and the standard deviation of the heights of a sample of adult males and a sample of boys aged nine.

	Adult males	Boys aged nine
Mean	180 cm	135 cm
Standard deviation	18 cm	10 cm

a David is a boy aged nine whose height is 120 cm. It is believed that the height of a nine-year-old boy gives a good indication of the adult height the boy will achieve.

i Calculate a standardised score for David.

ii Use the standardised score from part (i) to estimate David's likely adult height.

b Frank is an adult whose height is 2.025 m. How tall was Frank likely to have been at age nine?

6 When applying for a job Nasreen took a range of tests on memory, logic, numeracy and literacy. Use the information to the right to rank her performances in each of these tests.

Test	Nasreen's score	Test mean	Test standard deviation
Memory	35	40	5
Logic	48	46	8
Numeracy	50	60	5
Literacy	40	32	4

7 Anthony and Declan took the same test. Anthony scored 60 in the test, which gave him a standardised score of +1.5. Declan scored 50 in the test, which gave him a standardised score of –1. Work out the mean and standard deviation of the test scores.

Summary

You should:

- be able to use the range or interquartile range to find the spread of data
- be able to use the standard deviation, interdecile range or interpercentile range
- know that, if extreme values are present in the data, it is not wise to use the range or the standard deviation
- know that distributions can be symmetrical, or have positive or negative skew
- know that skew can be calculated, for example using the Pearson measure
- know that the Normal distribution is a commonly occurring symmetrical distribution
- know that 95% of a Normal distribution lies within two standard deviations of the mean and that virtually all the data within three standard deviations of the mean
- know that standardised scores can be used to compare distributions.

AQA Examination-style questions

1. Rod went fishing six times. The number of fish he caught was

3 5 6 8 11 x

The range of the number of fish caught is 10. Work out x, the largest number. *(2 marks)*

AQA, 2008

2. Two machines produce equal numbers of cartons of juice. Cartons are filled with apple juice by one machine and with blackcurrant juice by the other machine. The distributions of the volumes of juice are both Normal. The mean and standard deviation of each distribution are shown in the table.

	Mean (ml)	**Standard deviation (ml)**
Apple juice	100	10
Blackcurrent juice	110	2

(a) A carton is selected at random and contains at least 114 ml of juice. Is it more likely to contain apple juice or blackcurrant juice? You must support your answer with calculations. *(4 marks)*

(b) The two machines are equally likely to produce cartons that contain less than a certain volume of juice. What is this volume? *(4 marks)*

AQA, 2005

3. 100 cars of Type A and 100 cars of Type B were tested to see how far they could travel on one gallon of petrol. The distances travelled for each car are Normally distributed with means and standard deviations given below.

Car type	**Mean (miles)**	**Standard deviation (miles)**
A	45	2
B	43	4

On the grid below is a sketch of one of the two distributions.

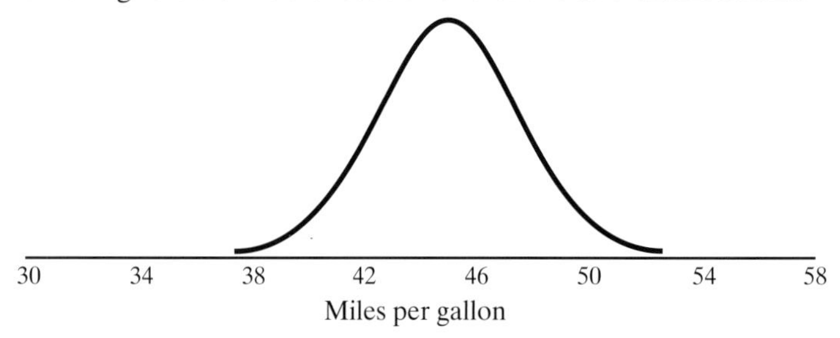

(a) (i) Which set of data is represented by the sketch? Give a reason for your answer. *(1 mark)*

(ii) Sketch the other distribution. *(3 marks)*

(b) What is the standardised score for a car of Type A that travels 50 miles on one gallon of petrol? *(2 marks)*

The standardised score for a car of Type B is –1.5.

(c) How far did this car travel on one gallon of petrol? *(3 marks)*

A man claims to travel 51 miles for each gallon of petrol used in his car.

(d) Which of Type A or Type B is most likely to be driving? Explain your answer. *(3 marks)*

AQA, 2001

4. Joan is a Road Safety Officer for a City council. Part of her work involves recording the number of vehicles exceeding the speed limit as they pass local schools. The following table gives the data recorded over a 120 day period. For example on 42 days two vehicles per day exceeded the speed limit.

Number of vehicles per day exceeding speed limit	2	5	6	10	14	15	20
Number of days	42	28	18	14	10	5	3

(a) Calculate the mean and standard deviation of the number of vehicles per day exceeding the speed limit. Give your answers to two decimal places. *(5 marks)*

(b) Due to a fault on the recording equipment, Joan's records did not show a further two vehicles each day which are exceeding the speed limit. What effect will this error have on the values for:

(i) the mean

(ii) the standard deviation? *(2 marks)*

(c) During the same period of time two other road safety officers recorded equal number of actual traffic speeds at different locations in the city. The speeds recorded by the first road safety officer are Normally distributed.

(i) On the grid below complete the diagram for this distribution. *(2 marks)*

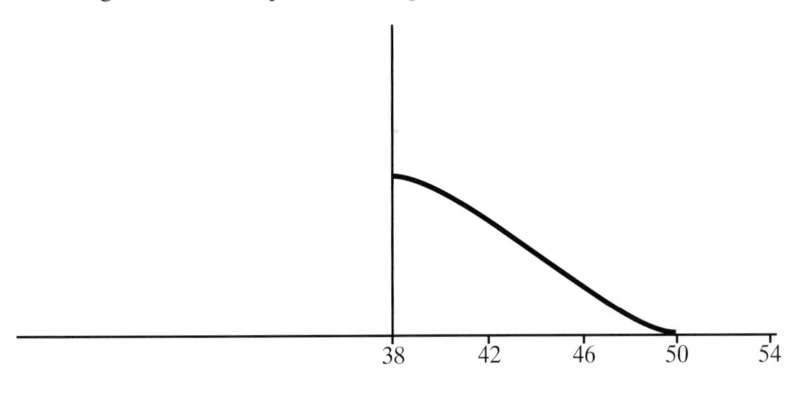

(ii) The speeds recorded by the second road safety officer were also Normally distributed. They had a mean of 44 mph and standard deviation of 2 mph. On the same grid draw a diagram to represent this distribution. *(3 marks)*

(iii) For the first road safety officer, what proportion of his records will show speeds of 38 mph or less? *(1 mark)*

(iv) For the second road safety officer what proportion of his records will show speeds above 52 mph? *(1 mark)*

AQA, 2006

5. A bank recorded the times taken to process an equal number of cheque errors at two of its branches, A and B.

The times taken in minutes at each of the two branches were Normally distributed with mean and standard deviation shown in the table.

	Mean (minutes)	**Standard deviation (minutes)**
Branch A	16.5	2.8
Branch B	14	4.5

To compare the performance of both branches it was agreed to standardise the times taken at each branch.

(i) What would be the standardised value for a cheque from Branch A taking 21 minutes to process? *(2 marks)*

(ii) A cheque processed at Branch B had a standardised time of 2.4. What was the actual processing time? *(3 marks)*

(iii) Between what limits would you expect approximately 99.9 per cent of the cheque processing time for Branch A to lie? *(3 marks)*

AQA, 2007

7 Cumulative frequency diagrams and box-and-whisker diagrams

In this chapter you will learn:

- what cumulative frequency is and how to calculate it
- how to display and interpret continuous data on a cumulative frequency diagram
- how to display discrete data on a cumulative frequency step polygon
- how to use cumulative frequency diagrams to estimate the median, quartiles and interquartile range
- how to use cumulative frequency diagrams to find or estimate deciles, percentiles and associated ranges
- how to draw and interpret box-and-whisker plots
- how to identify outliers and display them on a box-and-whisker plot.

You should already know:

- ✔ the difference between discrete and continuous data
- ✔ the meaning of median, quartiles, deciles and percentiles.

What's this chapter all about?

This chapter looks at the type of diagrams you can draw when you have a running total (cumulative frequency) of the data. These diagrams then give you an excellent way to find some of the measures of average and spread that you met in recent chapters. You might have been looking for this in Chapter 4, but the diagrams in that chapter were all for basic frequencies and we thought that the chapter was long enough already, so here it is now!

Focus on statistics

In this section there are some subtle but very important differences from what you might have met in maths. In 7.1 it is very important you know about cumulative frequency step polygons (at higher tier) and when to use them. The name sounds like the other types of diagram that you might know but they are something different and you will not have met them before in maths.

You will probably have heard of outliers before, but in 7.2 you need to know how to calculate whether a value is an outlier (at higher tier) and how to show them on a box-and-whisker plot. You need to know how a box-and-whisker plot can show skew. You won't have done that in maths.

7.1 Cumulative frequency diagrams

What is cumulative frequency?

When data is in a frequency table, the frequency column shows you how many times each data value or class has occurred.

Cumulative frequency gives you information about the total number of times data has occurred up to and including that value or class.

For example, this frequency table shows the weights of 100 people.

Weight, w kg	Frequency
$40 \le w < 60$	23
$60 \le w < 80$	55
$80 \le w < 100$	17
$100 \le w < 120$	5

If you wanted to calculate the cumulative frequencies, you would need to do two things:

1 Change the labelling of the data column to less than each value.

2 Add up a running total of the frequencies to get the cumulative frequencies.

The corresponding cumulative frequency table for the data on the weights of 100 people will then look like this:

Weight, w kg	Frequency	Cumulative Frequency
$w < 60$	23	23
$w < 80$	55	78 $(23 + 55)$
$w < 100$	17	95 $(23 + 55 + 17)$ or $(78 + 17)$
$w < 120$	5	100 $(23 + 55 + 17 + 5)$ or $(95 + 5)$

If the original data had been labelled using $40 < w \le 60$ or for a discrete distribution, the labelling of the data column would need to be $\le$ not $<$.

Consider the data obtained from rolling an ordinary die 50 times.

The frequency distribution on the left becomes the cumulative frequency distribution on the right.

Notice that there is no need to repeat the frequencies a second time.

Objectives

In this topic you will learn:

- what cumulative frequency is and how to calculate it
- how to display and interpret continuous data on a cumulative frequency diagram
- how to display discrete data on a cumulative frequency step polygon
- how to use cumulative frequency diagrams to estimate the median, quartiles and interquartile range
- how to use cumulative frequency diagrams to find or estimate deciles, percentiles and associated ranges.

Key terms

Cumulative frequency: the total frequency up to and including a particular value.

Score on a die	Frequency		Score on a die	Cumulative frequency
1	7		≤ 1	7
2	8		≤ 2	15
3	12	becomes →	≤ 3	27
4	5		≤ 4	32
5	9		≤ 5	41
6	9		≤ 6	50

AQA Examiner's tip

Many students make errors when adding frequencies to get cumulative frequencies. Always check that the final cumulative frequency gives the total you are expecting.

Cumulative frequency distribution tables can be used as a quick way of finding the median or quartiles.

Example 1

Use the data from the die cumulative frequency table to find the median, lower quartile, upper quartile and interquartile range.

Solution

The median is half way along the data; $n = 50$ so the median is the 25th value. The table shows that there are 27 values at 3 or less and 15 values at 2 or less, therefore the median is 3.

Lower quartile, $50 \div 4 = 12.5$th value.

The 12th and 13th values are 2, so the lower quartile is 2.

Upper quartile, $3 \times (50 \div 4) = 37.5$th value.

The 37th and 38th values are 5, so the upper quartile is 5.

Interquartile range = upper quartile – lower quartile

$= 5 - 2$

$= 3$

Values for these measures can also be obtained from the different types of cumulative frequency diagram you shall meet.

Cumulative frequency diagrams for grouped or continuous data

If you have the cumulative frequency table and you want to plot this information on a graph, then for grouped or continuous data, you need to produce a **cumulative frequency diagram**. The most important aspect of these diagrams is that you must plot the cumulative frequency values at the top end of the class interval that they represent.

The name 'cumulative frequency diagram' can refer to a **cumulative frequency curve**, where the points are joined by a curve (this can also be called an ogive), or a **cumulative frequency polygon**, where the points are joined by straight lines. Look out in exam questions to see whether they ask for a 'curve' or a 'polygon', or if they ask for a 'diagram', in which case either type can be drawn. You won't see the word 'ogive' used in an examination question.

Key terms

Cumulative frequency diagram: the name given to any diagram that shows the cumulative frequencies for a distribution.

Cumulative frequency curve: a cumulative frequency diagram for continuous data with the points joined by a continuous curve.

Cumulative frequency polygon: a cumulative frequency diagram for continuous data with the points joined by straight lines.

Example 2

The data shows the age distribution of workers in a factory.

a Complete a cumulative frequency table for these data.

b Draw a cumulative frequency polygon for the data.

c Use the cumulative frequency polygon to estimate:

- **i** the median age
- **ii** the lower quartile of the ages
- **iii** the upper quartile of the ages
- **iv** the range of the central 50% of the ages.

d Explain why your answers to part (c) are only estimates.

e Estimate the percentage of workers **over** 55 years old.

Solution

a The cumulative frequency for $30 \le x < 40$ is found by adding $18 + 38 = 56$, and so on to give the table on the right.

Age, x years	Cumulative frequency
$x < 30$	18
$x < 40$	56
$x < 50$	68
$x < 60$	76
$x < 70$	80

b The cumulative frequencies are now plotted on scaled axes at the top end of each class interval. For the interval $x < 30$, the upper bound is 30 as you are recorded as being 29 years old up to the moment you become 30. The remaining upper bounds are 40, 50, 60 and 70. The coordinates that need to be plotted are therefore (30, 18), (40, 56), (50, 68), (60, 76) and (70, 80). The point (20, 0) can also be plotted as the original table tells you that there was no one under 20 years old. As a cumulative frequency polygon has been asked for, these points are now joined with straight lines. (A cumulative frequency curve should have the points joined with a smooth curve, a cumulative frequency diagram or graph can be either a smooth curve or a set of straight lines.)

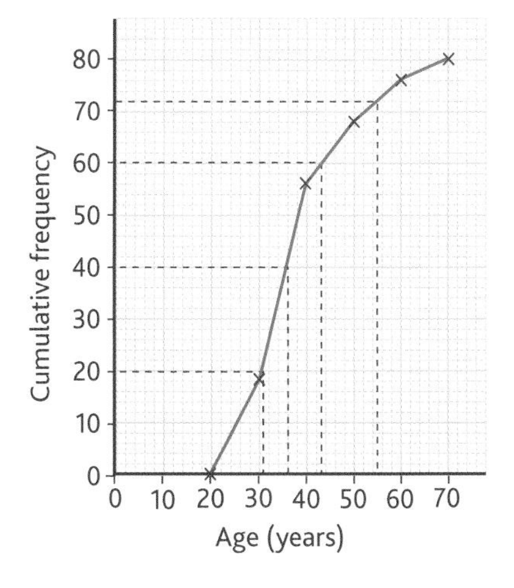

c **i** The median of 80 numbers is the 40th along. It represents half way up the polygon vertically. Draw a dotted line from 40 on the vertical axis across to meet the graph. Drop down to the axis and read off the median. Estimate of median = 36 years.
(If you use the median as $\frac{(n+1)}{2}$ along, which is 40½ up, this would be fine in the examination.)

ii For the lower quartile draw a line from the cumulative frequency of 20 ($80 \div 4$) across and read off from the x-axis. Estimate of lower quartile = 31 years.

iii For the upper quartile draw a line from the cumulative frequency of 60 ($3 \times (80 \div 4)$) across and read off from the x-axis. Estimate of upper quartile = 43 years.

iv The central 50 per cent of the data is covered by the interquartile range. Use upper quartile – lower quartile = 43 – 31 = 12 years.

d These are estimates as the data has been grouped so you do not know any actual values within the groups. The fact that, on this occasion, the cumulative frequencies were joined with straight lines on the graph assumes that the data is evenly spread within the class interval in which it lies.

e Draw a line up from 55 years to the graph. This meets at a cumulative frequency of 72. Therefore 80 – 72 = 8 workers are over 55 years old.
The percentage over 55 years old is $\frac{8}{80} \times 100 = 10\%$

Example 3

The length of 50 rumbles of thunder was timed. The table shows the frequency distribution for these times.

List the coordinates that you would need to plot in order to draw a cumulative frequency diagram for this data.

Length of thunder rumble (s)	Frequency
1–3	4
4–6	12
7–10	19
11–20	15

Solution

The cumulative frequencies are 4, 16 (4 + 12), 35 (4 + 12 + 19) and 50. The top end of the class intervals are 3.5, 6.5, 10.5 and 20.5. So the coordinates required are (3.5, 4), (6.5, 16), (10.5, 35) and (20.5, 50). Note that it is also possible to plot a point at (0.5, 0) as you know there were no rumbles of less than 0.5 s duration – the lower limit of the first class interval.

Cumulative frequency diagrams for discrete data (step polygons)

Discrete data can only take specific values so it would not be sensible to draw the type of cumulative frequency graph seen so far. Instead a **cumulative frequency step polygon** is required for this type of data. As the name suggests, the graph looks like a series of steps with the increases only occurring as each data value is reached, instead of the continuous increase seen before.

Key terms

Cumulative frequency step polygon: a cumulative frequency diagram for discrete data with the points joined in steps.

Example 4

The cumulative frequency table shows the die scores for 50 rolls. (This is the same data as on p.135.)

Score on a die	Cumulative frequency
≤ 1	7
≤ 2	15
≤ 3	27
≤ 4	32
≤ 5	41
≤ 6	50

a Draw a cumulative frequency step polygon of the data.

b Confirm that the graph gives the same values for the median, lower quartile, upper quartile and interquartile range as the calculation method did.

c Are these values estimates? Explain your answer.

d Work out the interdecile range.

Solution

a The cumulative frequencies are already calculated. The plotting points are the values 1, 2, 3, 4, 5, 6 as there are no upper bounds. There is no data before the score 1 so the point (0, 0) can also be plotted.

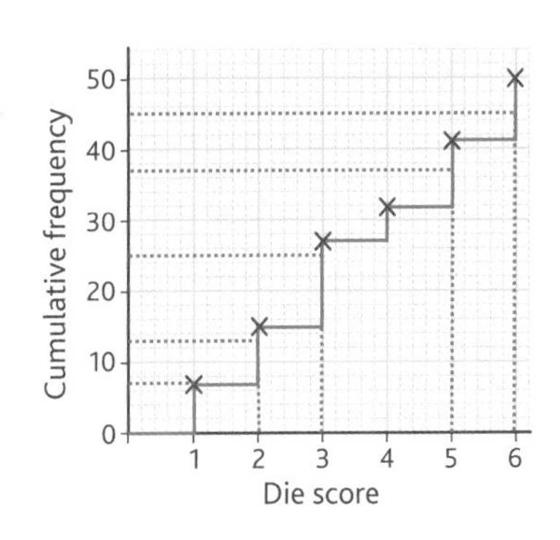

b Median line from cf of 25 to polygon and read down. Median = 3.

Lower quartile line from cf of 12.5 to polygon and read down. Lower quartile = 2.

Upper quartile line from cf of 37.5 to polygon and read down. Upper quartile = 5.

Interquartile range = upper quartile – lower quartile

= 5 – 2

= 3

These values are the same as were achieved by the calculation method.

Spot the mistake

Clive owns a corner shop. He is thinking of employing more staff, so counts the people in the queue every 5 minutes for 2 hours.

He draws this graph of the data.

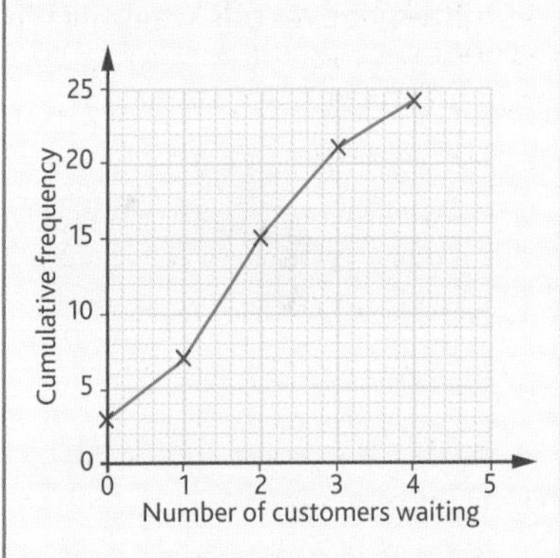

Without knowing the data values, spot the mistake he has made.

c These values are not estimates: they are exact because you know all the individual values (the data has not been grouped and is discrete).

d The interdecile range is the value of the ninth decile (D_9) – the value of the first decile (D_1).

D_9 = the value 90 per cent along the data. 90 per cent of 50 = 45 so D_9 = 45th value.

Draw a line from 45 on the cumulative frequency axis to the graph and read down to give $D_9 = 6$.

D_1 = the value 10 per cent along the data; 10 per cent of 50 = 5 so D_1 = fifth value.

Draw a line from 5 on the cumulative frequency axis to the graph and read down to give $D_1 = 1$.

Therefore interdecile range = 6 – 1

= 5

(Interpercentile ranges are found in a similar ways – finding the appropriate percentages of the cumulative frequencies and reading off the graph at these values before subtracting.)

Exercise 7.1

1 Construct an appropriate cumulative frequency table for each of these frequency tables. Remember to change the data labels.

a A survey of 50 adults recorded the time they spent on the internet over one weekend. The information is given in the table.

Time spent, t (hours)	Frequency
$1 \le t < 2$	1
$2 \le t < 3$	4
$3 \le t < 4$	5
$4 \le t < 5$	9
$5 \le t < 6$	12
$6 \le t < 7$	16
$7 \le t < 8$	3

b

Length of time, l, minutes	Frequency
$2 \le l < 5$	6
$5 \le l < 7$	16
$7 \le l < 8$	21
$8 \le l < 10$	40
$10 \le l < 20$	17

2 Construct a frequency table from this cumulative frequency table given that there were no values below 490 ml.

Remember to change the data labels.

Volume of liquid in bottle, ml	Cumulative frequency
$v < 500$	14
$v < 510$	29
$v < 520$	65
$v < 530$	87
$v < 540$	100

3

a Use the data from Question 1 (a) to draw a cumulative frequency polygon.

b Use the polygon to find an estimate of:

i the median

ii the lower quartile

iii the upper quartile

iv the interquartile range

v the percentage of the adults who spent longer than five and a half hours on the internet.

4

a Use the data from Question 2 to draw a cumulative frequency curve for the data.

b Use your graph to estimate:

i the interquartile range

ii the percentage of bottles with a volume between 515 and 525 millilitres.

c Explain why your answers in part (b) are estimates.

5 The table shows the length of 100 number one hit records from the 1950s.

Length of hit record (s)	Frequency
90–109	6
110–119	7
120–129	22
130–149	47
150–199	16
200–299	2

a Explain why the first coordinate that can be plotted in order to draw a cumulative frequency graph is (89.5, 0).

b Draw a cumulative frequency diagram for this data.

c Use the diagram to find

i the median

ii the interquartile range

iii the percentage of these records under 3 minutes.

6 The table shows information about the number of people (including the driver) in each of 200 cars sampled on the main road into a city during the rush hour.

Number of people	Frequency
1	121
2	48
3	13
4	16
5	2

a Draw a cumulative frequency step polygon for the data.

b Use the graph to find the median and interquartile range.

c Use the graph to find the interdecile range.

d Why are your answers to parts (b) and (c) so similar for these data?

7 The table shows the time taken for runners to finish a marathon.

Finishing time, t minutes	Frequency
$130 \le t < 140$	18
$140 \le t < 150$	66
$150 \le t < 160$	86
$160 \le t < 170$	58
$170 \le t < 190$	42
$190 \le t < 220$	16
$220 \le t < 260$	14

a Draw an appropriate cumulative frequency graph for this data.

b Use the graph to estimate:

i the median finishing time

ii the interquartile range of finishing times

iii the 35th percentile.

c Runners who finished in under three hours are automatically invited back to the next year's race. Estimate the number who were invited back to next year's race.

d Describe the skew of the distribution.

7.2 Box-and-whisker diagrams (box plots)

An effective way to show the median, quartiles and extreme values of a distribution is through a **box-and-whisker diagram** (often referred to just as a **box plot**).

The box plot also shows information about the possible skew of a set of data and its spread. It is particularly effective to use two or more box plots for comparing two or more sets of data.

Objectives

In this topic you will learn:

- how to draw and interpret box-and-whisker plots
- how to identify outliers and display them on a box-and-whisker plot.

Key terms

Box-and-whisker diagram: a diagram that shows the position of the minimum, the lower quartile, the median, the upper quartile and the maximum values from a set of data.

Box plot: the short name for a box-and-whisker diagram.

Structure of a box plot

There are five measures shown on a box plot:

1 the minimum
2 the lower quartile
3 the median
4 the upper quartile
5 the maximum.

The whiskers go from 1 to 2 and 4 to 5. The box goes from 2 to 4 with a vertical line at 3 as shown.

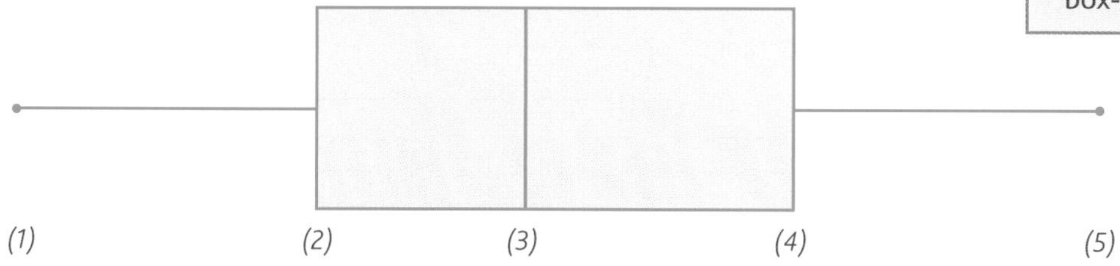

Later, for higher tier only, you will see how outliers can also be shown on a box plot.

Box plots also need a numbered and labelled scale added below (or occasionally above) the box plot, and they should always be drawn using a pencil and ruler.

Showing skew on a box plot

It is a relatively easy task to look at a box plot and determine whether a distribution is symmetrical or has any skew. If the median is equally spaced between the lower and upper quartile, the distribution is largely symmetrical.

If the median is nearer the lower quartile than it is to the upper quartile, the distribution is positively skewed.

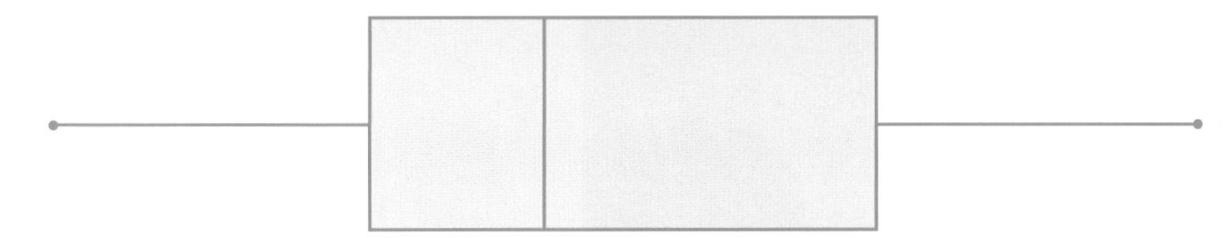

If the median is nearer the upper quartile than it is to the lower quartile, the distribution is negatively skewed.

Example 1

Over one month Seve and Monty played golf several times. They always record their scores for the whole round of golf.

Seve's scores are summarised in the table.

	Minimum	Lower Quartile	Median	Upper Quartile	Maximum
Seve's scores	63	69	72	74	79

Monty's scores over the month are:

71 72 68 73 70 67 72 77 75 70 69 68 73 73 73 70 67 71 74

a Work out the five measures needed for a box plot for Monty's data.

b On the same axes draw box plots for Seve's and Monty's data.

c Describe any skew present in the two distributions.

d Compare the performances of Seve and Monty.

Solution

a First the data needs to be ordered.

67 67 68 68 69 70 70 70 71 71 72 72 73 73 73 73 74 75 77

It is now easy to see that the minimum is 67 and the maximum is 77. There are 19 values, so the lower quartile is the fifth value, the median is the tenth value and the upper quartile is the 15th value, giving the lower quartile = 69, the median = 71 and the upper quartile = 73.

b You need to draw a scale running from slightly below 63 (the overall minimum) to slightly above 79 (the overall maximum). Then draw the box plots neatly one above the other, to make the later comparisons easier, clearly labelling which is which.

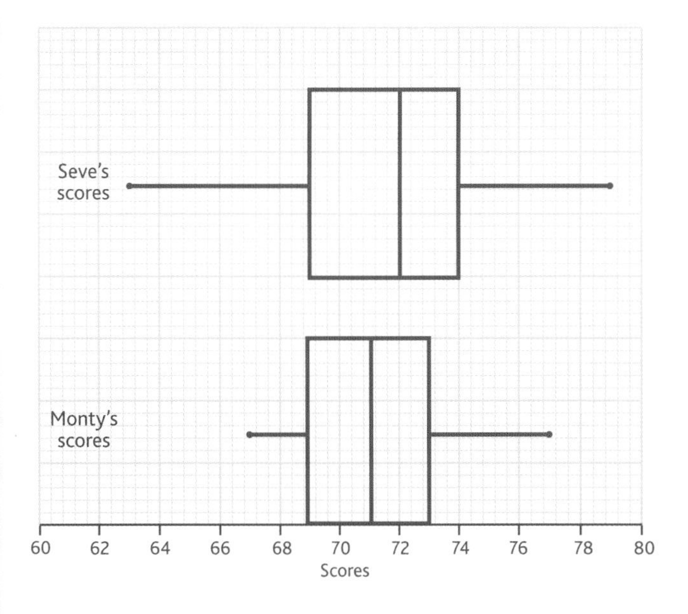

c Seve's scores are negatively skewed, as the median is closer to the upper quartile than the lower quartile. Monty's scores are symmetrical, as the median is equidistant between the two quartiles (but notice it is not symmetrical beyond the quartiles).

d Monty's scores are on average lower than Seve's (which is better in golf) as the median is lower than Seve's. Monty's scores are also more consistent than Seve's, as the box part of the box plot, which represents the interquartile range, is smaller for Monty than for Seve.

AQA Examiner's tip

Remember, when answering questions like part (d), to make a comment about a measure of average and a comment about a measure of spread.

Finding outliers

The general definition of an **outlier** is a data value that does not seem to fit the pattern or size of other data values from the same set. Sometimes other terms are used, including extreme, rogue or freak data, but the word 'outlier' also has a technical meaning in statistics. An outlier is any value that is either:

- less than the lower quartile − 1.5 × the interquartile range or
- more than the upper quartile + 1.5 × the interquartile range.

For example, look again at Seve's golf data. What range of values would be outliers?

1 Low outliers would be less than 69 – 1.5 × (74 – 69)
= 69 – 7.5
= 61.5

So all scores of 61 or below would be outliers.

2 High outliers would be more than 74 + 1.5 × (74 – 69)
= 74 + 7.5
81.5

So all scores of 82 or above would be outliers.

Key terms

Outlier: any value that does not fit in well with the rest of the data: the value is either less than the lower quartile – 1.5 × the interquartile range; or more than the upper quartile + 1.5 × the interquartile range.

Dealing with outliers

Finding outliers is not difficult but it can be very difficult to decide what to do with them. Outliers can greatly affect some of the measures you have met in this book so far, such as the mean and standard deviation – but is it a good idea to ignore them just to make answers to measures more predictable? There is no right or wrong answer to this and it will depend a great deal on the context of the problem being studied. Be aware that you need to be very careful before you discard data, unless you have serious doubts about the reliability of the particular value or reading in question.

For example, if one of Seve's scores had been recorded as 174, it is fairly obvious that this will be some kind of recording error and that value should be discarded. However, if one of his scores had been 84 (which is a statistical outlier) then there is a good reason to keep this value and use it in any comparison, for example with Monty's scores.

Displaying outliers on box plots

If outliers are present in data, the box plot drawn under normal circumstances for the data would have extremely long whiskers, which would dominate the diagram. It is usual to mark outliers with a cross (one for each outlier) at the correct point on the scale, and only join whiskers from the appropriate quartile to the lowest/highest value that is **not** an outlier. This then gives instant information about the presence of outliers allowing judgements to possibly be made about whether they could be ignored in interpreting the overall information.

Example 2

A laboratory measures the vitamin C content of different varieties of apples.

The number of milligrams of vitamin C per 100 grams of 15 varieties of apple rounded to the nearest integer is given below.

11 10 14 7 25 13 14 11

11 15 8 10 12 12 8

a Find any outliers present in the data.

b Display the data on a box-and-whisker plot.

c The mean of this data is to be found. Discuss whether any outliers found should be left out of the calculation.

Solution

a Firstly you need to find the values of the lower and upper quartiles. The lower quartile of 15 values is the fourth in the ordered data = 10. The upper quartile of 15 values is the 12th in the ordered data = 14. Thus low outliers are below $10 - (1.5 \times 4)$ = below 4. High outliers are above $14 + (1.5 \times 4)$ = above 20. Therefore we have one high outlier of 25.

b Minimum = 7, lower quartile = 10, median = 11, upper quartile = 14, highest non outlier = 15, one outlier at 25. This gives the box-and-whisker plot as drawn below.

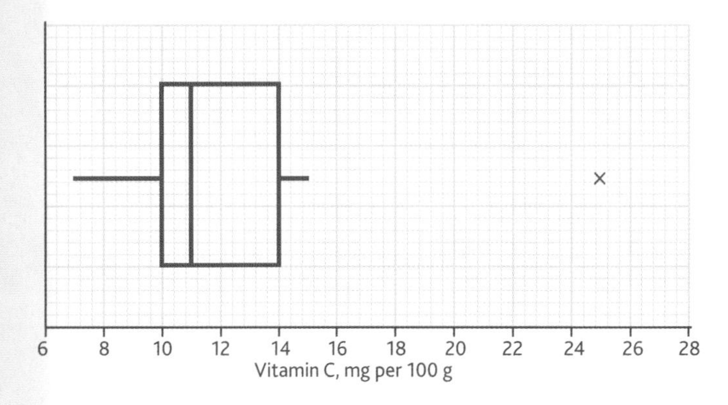

c The mean would be quite distorted by this outlier but there is no good reason to simply disregard it. It might be better to consider using a different measure of average such as the median, which is unaffected by the actual size of the largest value. (The value 25 is not a wrongly recorded value: it is the value for the variety of apple called 'Sturmer'.)

Real world

Vanessa decided to open her corner shop an hour earlier as a trial. During that hour she had 15 customers. The amount each one spent is given below:

£2.05 75p £3.98 £5.00 45p £19 £6 £1.78
£28.40 £1.85 £7.12 £3.61 £1.95 20p £15.07

a Identify any outliers in this data.

b Draw this information on a box-and-whisker plot.

LQ = £1.78 med = £3.61 UQ = £7.12 IQR = £5.34 lower bound <0
upper bound = 7.12 + 1.5 x 5.34 = 7.12 + 8.01 > 15.13

c Identify any skew in the data.

d Discuss whether Vanessa should disregard the outliers when deciding whether the early opening was worthwhile.

Be a statistician!

If you collect the value of the maximum temperature each day over a month in your nearest town or city, would you expect:

- symmetrical or skewed data
- any outliers?

Would this be any different if you considered minimum temperatures? Would the results to the two questions vary according to the time of year? Collect some relevant data to investigate some of these issues. Draw box-and-whisker plots as part of your findings. Can you extend this investigation to include other countries and parts of the world?

Exercise 7.2

1 A set of data has the following measures: minimum value = 35; lower quartile = 48; median = 66; upper quartile = 76; maximum = 89.

a Draw a box-and-whisker plot to illustrate this data set.

b Describe the skew present in the data. How did you come to this decision?

2 The two box-and-whisker plots below show the times taken to get home by 100 football supporters who used the underground and 100 football supporters who used the bus.

Compare these times, commenting on at least four differences between the two distributions.

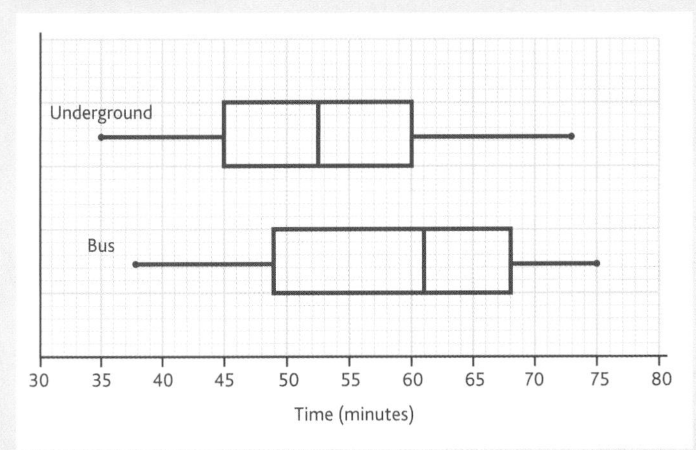

3 Here is a set of data about the time in seconds spent by 19 cars at a red traffic light:

30 22 45 67 12 35 60 55 16 44 48 37 33 28 65 49 61 30 44

a Construct a box-and-whisker diagram to illustrate the data.

b Comment on whether the distribution is symmetrical or has skew.

4 Look at these situations. Imagine you are presented with this data (you do not record the data yourself). In each case make a judgement about the outlier. Should it be kept as part of the data in analysis, or disregarded for the purposes of analysis? In which cases would further information be helpful in your decision making? What would your decision be? Give a full discussion of your decision.

a Daily number of visitors to a small museum. Lower quartile = 141, interquartile range = 40, outlier: on Wednesday there were 56 visitors.

b Journey time on a weekday morning from Scunthorpe to Manchester by car.
Upper quartile = 152 minutes, interquartile range = 18 minutes, outlier: on Monday the journey took 3½ hours.

c Weights of packets of dried fruit labelled as containing 80 g.
Upper quartile = 81.4 g, interquartile range = 1.17 g, outlier: one bag recorded as 104.3 g.

d Recorded monthly rainfall. Lower quartile = 11.3 mm, interquartile range = 6.8 mm, outlier: last month no rainfall at all.

5 Two taxi firms are competing for a contract to take employees from a company to the airport for business flights. For one week Topcab provide the taxis. They make 11 journeys to the airport. The data shows how late each taxi was in minutes. A value of –5 means the taxi was 5 minutes early.

–2 0 24 –1 –3 –1 0 1 –2 11 3

For the second week Fastcar provide the taxis. They make 15 journeys to the airport. The data shows how late each taxi was in minutes.

5 –2 –1 0 3 2 4 –2 –4 3 6 7 3 0 10

a Determine whether there are any outliers in either set of data.

b On the same axes, draw box-and-whisker plots to illustrate the data.

c Compare the two taxi companies. Use your comparisons to state which company you would use.

Summary

You should:
know that cumulative frequency is a running total of frequency
know that grouped/continuous data is shown on a cumulative frequency curve or polygon
know that discrete data is shown on a cumulative frequency step polygon
know that all these diagrams can be used to find or estimate quartiles, medians, deciles or percentiles
know that box-and-whisker plots are useful to show spread, skew and location and are particularly good for comparisons
know that outliers are more than 1½ times the interquartile range below the lower quartile or above the upper quartile.

AQA Examination-style questions

1. Stephanie works at a local garage. The table shows the cumulative frequencies for the number of hours of overtime worked by Stephanie each week over a period of 60 weeks.

Number of hours, *t*	Cumulative frequency
less than 1	2
less than 2	7
less than 3	10
less than 4	22
less than 5	41
less than 6	58
less than 7	60

(a) Draw a cumulative frequency polygon for these data. *(3 marks)*

(b) Estimate on how many weeks Stephanie worked between 4½ and 5½ hours overtime. *(3 marks)*

2. The cumulative frequency polygon shows the distribution of weekly earnings of a sample of 120 male manual workers in the ceramics industry.

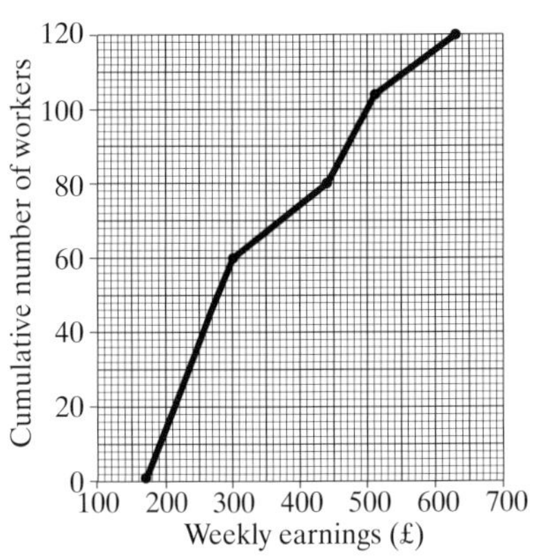

(a) Use the graph to estimate:

(i) the median *(1 mark)*

(ii) the interquartile range *(2 marks)*

(iii) the percentage of workers earning under £320 per week. *(3 marks)*

(b) The following information was found from a sample of 120 female manual workers in the ceramics industry. The median of the weekly earnings was £230. Twenty-five percent of the sample had weekly earnings more than £280. The interquartile range was £100. No one earned less than £120 per week or more than £420 per week. Six workers earned more than £390 per week.

(i) Use your answers to part (a) and the information on female earnings to make two statements that support the following hypothesis:

Female workers in the ceramics industry have lower and less variable weekly earnings than male workers in the ceramics industry. *(2 marks)*

(ii) Describe another source of data that could be used to expore this hypothesis. *(1 mark)*

AQA, 2006

3. As part of a school project, Paul carried out two surveys on the ages of passengers using his local train service. The surveys were undertaken at 10 am and 5 pm on a Tuesday. There were 100 passengers in each survey. The results for the 10 am survey were as follows:

Age, x (years)	Frequency	Cumulative Frequency
$0 \le x < 10$	14	14
$10 \le x < 20$	41	
$20 \le x < 30$	13	
$30 \le x < 40$	19	
$40 \le x < 50$	9	
$50 \le x < 60$	4	

(a) Complete the cumulative frequency column *(2 marks)*

(b) Draw a cumulative frequency polygon.

(c) Use your diagram to estimate:

(i) the median *(1 mark)*

(ii) the lower quartile *(1 mark)*

(iii) the upper quartile. *(1 mark)*

(d) The youngest passenger was 3 years old. The oldest passenger was 57 years old. Find the range. *(1 mark)*

(e) Paul drew a box-and-whisker plot for the 5 pm survey.

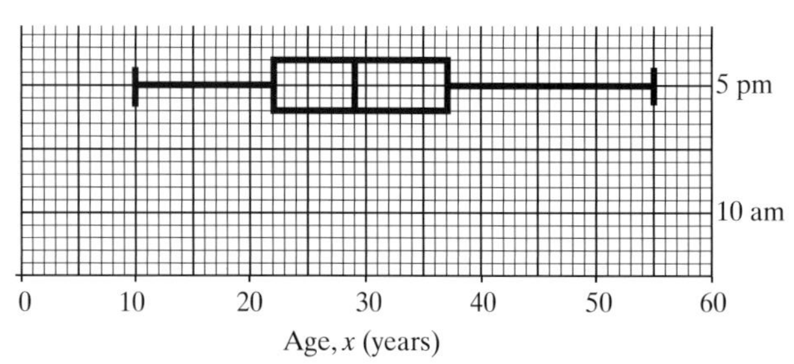

(i) Draw a box-and-whisker plot for the 10 am survey. *(4 marks)*

(ii) Write down two differences between the ages of passengers in the two surveys. *(2 marks)*

AQA, 2005

4. The times taken for all 5500 competitors to complete a cross-country race are summarised in the table.

(a) Draw a cumulative frequency graph to illustrate this data. *(3 marks)*

Time (minutes)	Position
83.2	1st
84.4	1000th
84.8	2000th
85.3	3000th
86.4	4000th
88.0	5000th
89.1	5500th

(b) From your graph estimate:
(i) the median time *(1 mark)*
(ii) the upper and lower quartiles. *(2 marks)*
(c) Draw a box-and-whisker diagram to illustrate this data. *(3 marks)*
(d) The box-and-whisker diagram for the race the previous year is shown below.

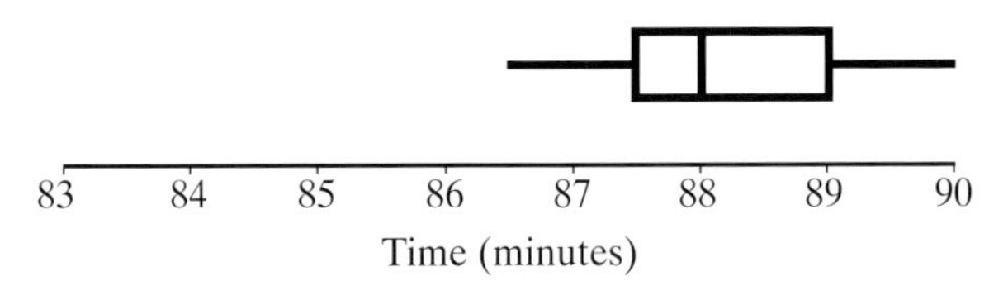

Give two comparisons between this year's and last year's race times. *(2 marks)*

AQA, 2001

5. The table shows the annual wage of women working for a company.
(a) Draw a cumulative frequency graph for these data. *(4 marks)*

Wages (£ per annum)	Frequency
wage 8,000	0
8000 ≤ wage < 10 000	14
10 000 ≤ wage < 12 000	33
12 000 ≤ wage < 15 000	38
15 000 ≤ wage < 20 000	30
20 000 ≤ wage < 25 000	17
25 000 ≤ wage < 35 000	11
35 000 ≤ wage < 60 000	7

(b) Estimate the proportion of women with an annual wage of more than £31 000. *(2 marks)*
(c) Use your graph to estimate the median annual wage. *(1 mark)*
(d) (i) Use your graph to estimate the range between the first and ninth decile. *(3 marks)*
The range between the first and ninth decile for the male workers at this company was £24 000.
(ii) What does this tell you about the wages of men and women working for this company? *(1 mark)*
(iii) Give one advantage of using an interdecile range. *(1 mark)*

At Christmas all the women receive the same bonus of £300.

(e) What effect will this bonus have on:

(i) the median *(1 mark)*

(ii) the range between the first and ninth decile? *(1 mark)*

AQA, 1999

6. **(a)** The length of reign of each of the last 19 monarchs is given in the table.

George VI	16 years	George IV	10 years	James II	3 years
Edward VIII	0 years	George III	60 years	Charles II	25 years
George V	26 years	George II	33 years	Charles I	24 years
Edward VII	9 years	George I	13 years	James I	22 years
Victoria	64 years	Anne	12 years	Elizabeth I	45 years
William IV	7 years	William III	14 years	Mary	5 years
				Edward VI	6 years

(b) Find the median and quartiles of the length of reign of these 19 monarchs. *(3 marks)*

(c) Write down the name of any monarch whose length of reign is an outlier.
You must show calculations to support your answer. *(3 marks)*

(d) The box-and-whisker plot shows the length of reign of the last 19 popes.

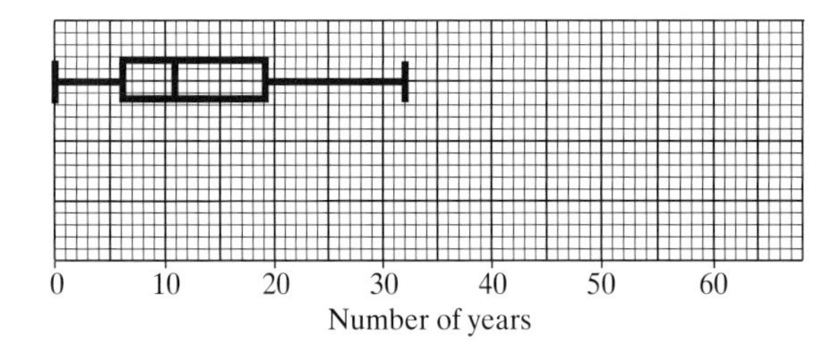

Draw a box-and-whisker plot for the length of the last 19 monarchs. *(4 marks)*

(e) Compare the length of reign of monarchs and popes. *(2 marks)*

AQA, 2004

7. Fifteen teams took part in a quiz. Their scores are as follows:

81	64	75	70	68
78	74	69	76	72
62	82	53	75	69

(a) Find the median and quartiles of the scores. *(3 marks)*

(b) (i) Find the interquartile range. *(1 mark)*

(ii) Explain why 53 is an outlier. *(2 marks)*

(c) Draw a box-and-whisker plot to illustrate these data. *(4 marks)*

AQA, 2003

8 Scatter diagrams, correlation and regression

In this chapter you will learn:

- how to draw a scatter diagram
- the different types of correlation and how to recognise them
- that just because two variables have a correlation this does not necessarily mean that one causes changes in the other
- when and how to draw line of best fit using a double mean point
- how to calculate the equation of a scatter diagram and draw the regression line for it
- that correlation may be nonlinear
- how to use lines for interpolation and extrapolation and the dangers of the latter
- how to calculate and interpret values of Spearman's rank correlation coefficient including tied ranks
- to interpret the product moment correlation coefficient values and compare it with Spearman's correlation.

You should already know:

- ✔ how to plot coordinates
- ✔ how to square numbers
- ✔ how to find the equation of a line of best fit.

What's this chapter all about?

Lots of data comes in pairs: your height and weight, your maths grades and the amount of revision you did, the time you spent getting ready for school and the time it takes you to get to school. Some of these things may be connected and some may not. In this chapter, you will see that there are graphs and calculations available to help you decide if two variables are connected and you should learn that, even if two things are connected, it doesn't mean one going up or down caused the other to go up or down!

Focus on statistics

There are some important differences to how things are done in statistics and in maths and there are some new topics here as well.

In statistics, when you draw a line of best fit you must use a double mean point (it is more accurate). You are shown how to do this in 8.1. If you have done it before in maths you might have just tended to guess where the line should go.

None of 8.2, which is all about correlation, is covered in maths – so we like asking about it in statistics!

8.1 Scatter diagrams and correlation

A **scatter diagram** is a good method for displaying two sets of data at the same time (**bivariate data**). It consists of a series of plotted points and can reveal relationships between variables. Points are best plotted with a cross, for reasons to be discussed later.

Any connection between variables is called **correlation**, and will be shown by patterns emerging in the plotted points. There are three situations that can occur (with two different types of correlation).

Positive correlation

In this situation, as the values of one of the variables increase, the values of the other variable also increase.

The scatter graph will look like one of the three following graphs, depending upon whether the **positive correlation** is weak, moderate or strong. Perfect positive correlation would see a perfect straight line of crosses from the bottom left of the graph to the top right of the graph. An example of positive correlation might be the relationship between the length of a person's foot and the same person's shoe size.

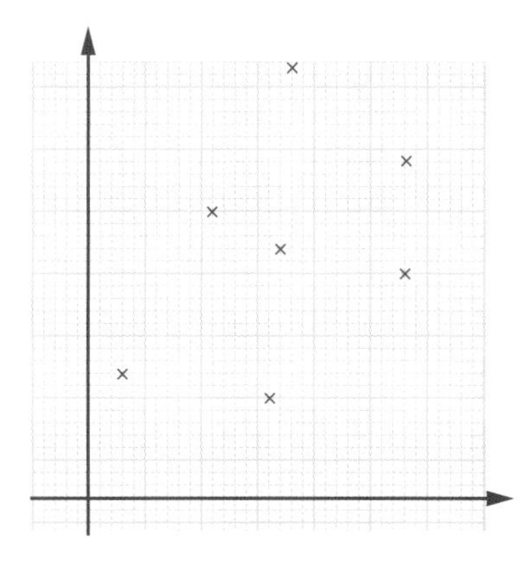

Weak positive correlation

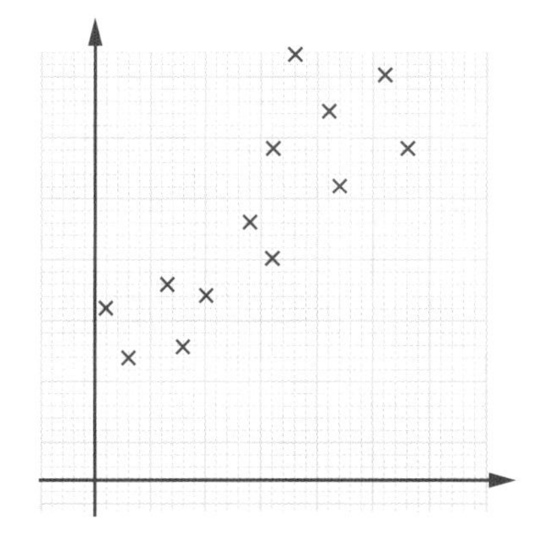

Moderate positive correlation

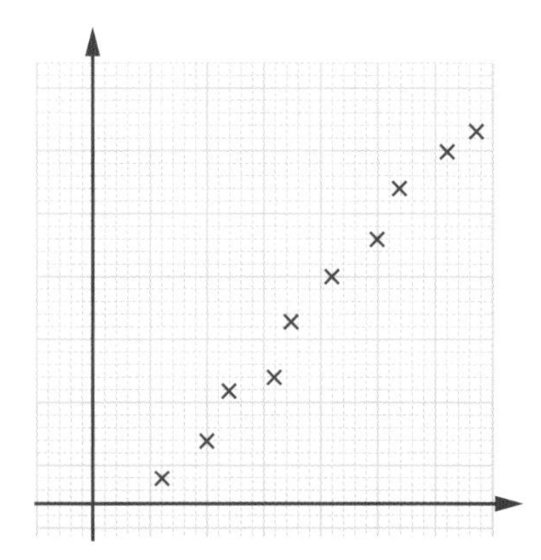

Strong positive correlation

Objectives

In this topic you will learn:

- how to draw a scatter diagram
- the different types of correlation and how to recognise them
- that just because two variables have a correlation this does not necessarily mean that one causes changes in the other
- when and how to draw line of best fit using a double mean point
- how to calculate the equation of a scatter diagram and draw the regression line for it
- that correlation may be nonlinear
- how to use lines for interpolation and extrapolation and the dangers of the latter.

Key terms

Scatter diagram: a visual way of showing bivariate data.

Bivariate data: data that has two variables.

Correlation: the statistical word for a connection between two variables.

Positive correlation: positive correlation exists when as one variable increases the other variable also increases.

Negative correlation

In this situation, as the values of one of the variables increase, the values of the other variable decrease.

The scatter graph will look like one of the following three graphs, depending upon whether the **negative correlation** is weak, moderate or strong. Perfect negative correlation would see a perfect straight line of crosses, from top left of the graph to the bottom right of the graph.

An example of negative correlation might be the year group of a student and the money raised for the school in a sponsored event (the higher year groups tend to raise less money).

Key terms

Negative correlation: negative correlation exists when as one variable increases the other variable decreases.

No correlation: there is no correlation between two variables when the value of one seems to have no connection to the value of the other.

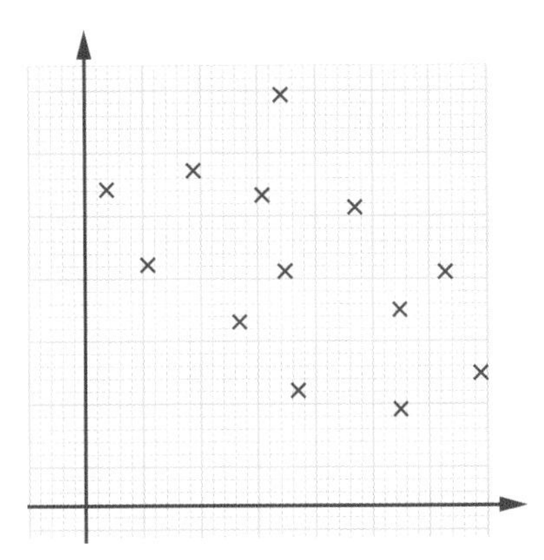

Weak negative correlation

Moderate negative correlation

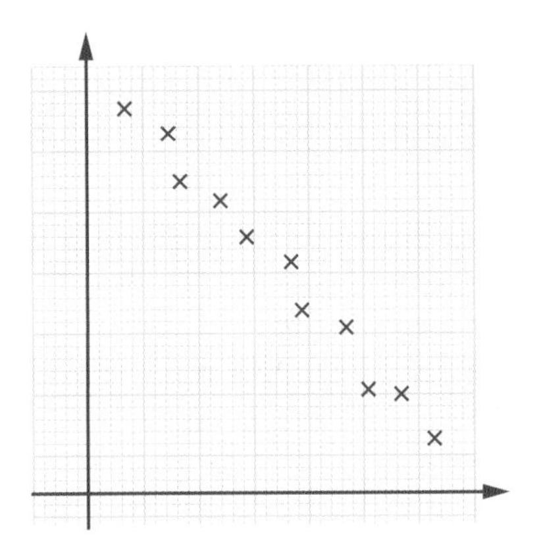

Strong negative correlation

No correlation

Sometimes there is **no correlation** between two variables. In this situation there is no apparent pattern in the positions of the plotted points, and no apparent connection between the variables.

An example of no correlation might be the number of bedrooms in a house and the last digit in the telephone number for the house.

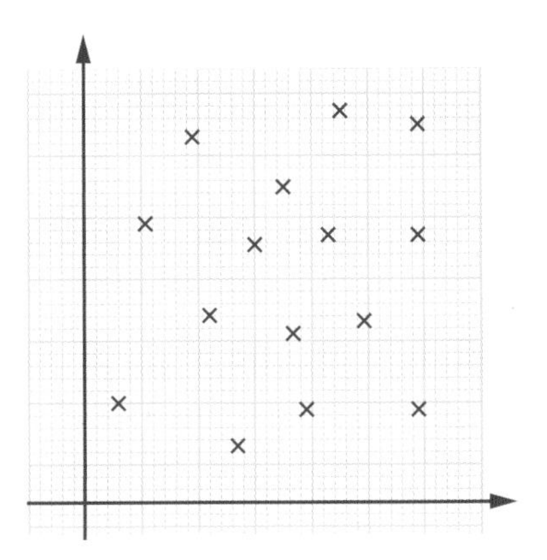

No correlation

Nonlinear correlation

There is the possibility, however, that although there is a connection between the data, it is not a linear (straight line) connection. The points in the scatter diagram below have a connection, but they certainly do not have simple positive or negative correlation.

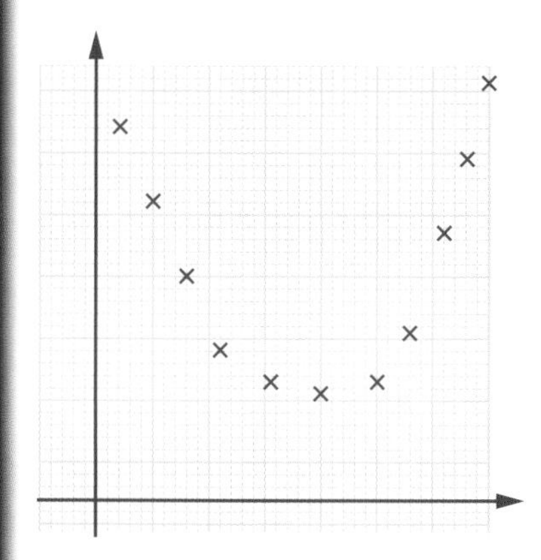

Nonlinear correlation

Lines of best fit

Where there is a clear correlation between the bivariate data shown on a scatter diagram, you can draw a **line of best fit** through the data to show the pattern that exists more clearly and, if required, to make predictions.

There are some clear rules that you need to follow when deciding whether to draw a line of best fit and when completing the line:

- only draw a line of best fit for moderate and strong correlation
- the line must be ruled straight
- the line must pass through a plotted double mean point
- the line should be long enough to reach the outer values of the plotted data
- the line should not normally go beyond the plotted points.

A **double mean point** is a plotted point using a dot in a circle (to avoid confusion with ordinary plots which are crosses) with coordinates (mean of first x variable, mean of second y variable). Using a double mean point enables you to position a line of best fit more accurately than just guessing where its position should be, as a correct line of best fit would be expected to pass through the double mean point for the data sets.

Key terms

Line of best fit: a straight line drawn through a double mean point to represent the apparent connection between two variables.

Double mean point: a point plotted at the values of the means of the two variables – the point through which the line of best fit must pass.

AQA Examiner's tip

Judgements on whether correlation is weak, moderate or strong, are sometimes difficult to call and, unless it is absolutely obvious how strong the correlation is, examiners will often accept two possible answers. For example, the answers 'moderate' or 'strong' would each get a mark in the graph shown in Example 1 – but the correlation is clearly not weak.

Example 1

The table shows the number of rooms and the number of paying guests in 10 hotels one Friday night in Torquay.

Number of rooms	32	15	108	45	55	20	65	28	72	85
Number of paying guests	45	22	121	40	76	24	96	31	99	124

a Draw a scatter diagram to illustrate the data.

b Describe the strength and type of correlation present in your scatter diagram.

c Is it appropriate to draw a line of best fit? Explain your answer.

d If appropriate, draw a line of best fit showing any calculations made.

No correlation between hotel brochure and customer satisfaction

Solution

a Plot each point using an x. The data from the top row of a data table always goes on the x-axis.

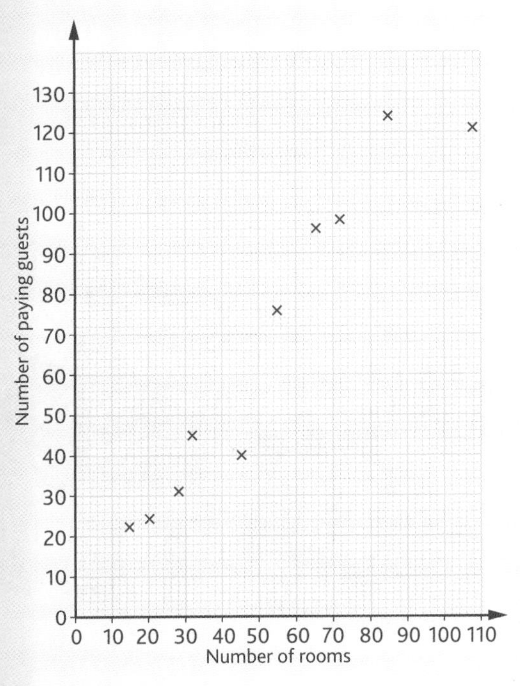

Part (a)

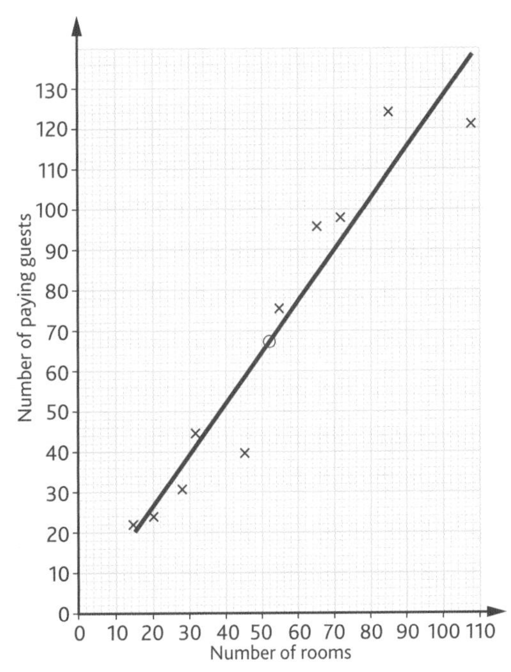

B *Part (d)*

b The correlation is positive and moderate.

c The correlation is moderate and possibly fairly strong so there is justification for drawing a line of best fit as there appears to be a fairly close relationship between the two variables.

d The double mean point needs to be calculated.

For the rooms data: $(32 + 15 + 108 + \ldots + 85) \div 10 = 525 \div 10 = 52.5$

For the guests data: $(45 + 22 + 121 + \ldots + 124) \div 10 = 678 \div 10 = 67.8$

So plot (52.5, 67.8) clearly with a dot in a circle and make sure that the line of best fit goes through it. You also need to draw the line of best fit so that it has roughly the same number of crosses on either side of it and is long enough to go as far as the highest and lowest data but does not unnecessarily go longer than that requirement.

Once a line of best fit has been drawn it can be used to make estimates of values of one variable, given the value of the other variable. These estimates can vary in their likely accuracy according to these rules:

- Estimates from lines of best fit where the correlation is strong are more likely to be reliable than those made from lines of best fit where the correlation is less strong.
- Estimates made from within the range of the data (**interpolation**) are more likely to be reliable than those made from outside the range of the data (**extrapolation**) as there is no evidence that any relationship between the data continues beyond the range of the plotted data.
- Consider whether there a genuine connection between the variables, or whether it is simply a coincidence?

To illustrate the third point further, look at the feature 'Spot the mistake'.

Even when a genuine connection exists, you have to very careful when attaching **causality** to the situation. Causality means that the increase (or otherwise) in one variable is caused by the increase (or otherwise) in the other variable.

For example, it is probably likely that an increase in exercise would cause a reduction in weight, especially in a person who was somewhat overweight in the first place, so there might be a causal relationship present here (although other factors could be present too).

However, although there would be some connection between the number of donkey rides taken on the beach one day and the number of cans of cold drink sold at the shop on the edge of the beach, one will not be the direct cause of the other. The reality is that a third variable (the number of people at or around the beach) is the important factor here.

AQA Examiner's tip

One of the most common sources of lost marks on the whole examination paper is candidates not *plotting* the double mean point and not making sure the line passes through it.

Key terms

Interpolation: an estimate often made using a line of best fit from within known data values.

Extrapolation: an estimate made often using a line of best fit from outside the data values known.

Causality: causality means that the changes in one variable are as a direct result of changes in the other variable and not some other factor.

Spot the mistake

The scatter graph shows information about the number of letters in 11 of the months of the year and the average daily maximum temperature for that month.

Isobel says: 'There is a moderate correlation present, so I can use it to estimate the average daily maximum temperature for March, the month with data missing.'

What two mistakes has Isobel made?

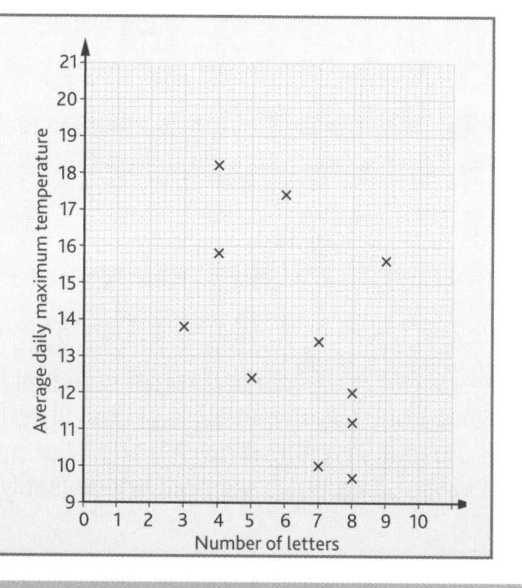

Example 2

For Example 1:

a Use your line of best fit to estimate the number of paying guests in Torquay that night for a hotel with 60 rooms.

b Estimate the number of paying guests in Torquay that night for a hotel with 10 rooms.

c State, with a reason, which of your estimates above will be the more reliable.

d Is any relationship here likely to be a causal relationship?

Solution

a Draw a line from the x-axis up to the line of best fit. Read off from the y-axis. Estimate = 78 paying guests.

b Extend the line of best fit backwards a little. Now draw a line from the x-axis up to the line of best fit and again read off on the y-axis. Estimate = 14 paying guests.

c The first will be the more reliable estimate, as it is interpolation, an estimate from within the data range, whereas the second was extrapolation, from outside the original data range. This means that in part (b) you needed to extend the line of best fit to produce the estimate.

d The number of paying guests is limited by the number and nature of the rooms available at a given hotel (the fourth hotel in the original table of data might only have single rooms). However, there are too many other variables at work here, and if it was the middle of winter, no matter how many rooms there were in the hotel, it is likely that a third variable (the weather/season/location) could be the controlling factor here.

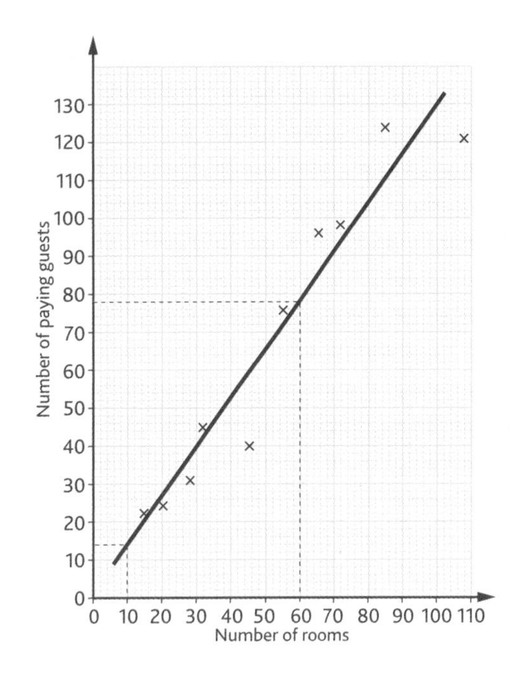

Equation of the line of best fit

A major area in statistics called regression analysis is devoted to the use of lines of best fit, which are properly known as least squares regression lines. A line of best fit is an accurately drawn line that keeps the vertical distances from the points to the line at a minimum. This goes beyond what you will need to know for GCSE but you can be expected to calculate the equation of a regression line, or line of best fit using the techniques below.

You might know from GCSE mathematics that the equation of a straight line has the general form:

$$y = mx + c$$

where m is the **gradient** of the line, and c is the **intercept**, the point where the line cuts, or would cut, the y-axis.

Key terms

Gradient: the steepness of a line.

Intercept: the value at which a line passes through the y-axis.

m can be calculated by choosing two points on the line (with 'nice' coordinates if possible). Then,

$$m = \frac{\text{change in the } y\text{-coordinates}}{\text{change in the } x\text{-coordinates}}$$

Note that you should **not** use coordinates from any tables of values, as the line of best fit is highly unlikely to pass exactly through two such points.

In exam questions, you might be asked to state what the values of m and c mean in the context of the question.

Example 3

The scatter diagram shows scores out of 100 awarded to 14 competitors in the 'best kept village' competition from two different judges.

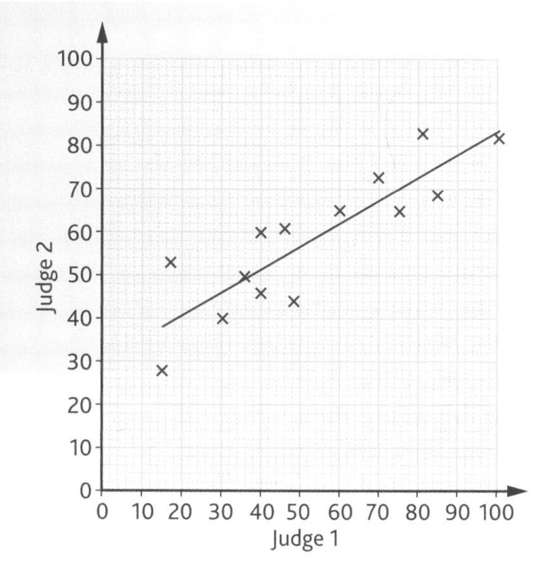

a What is missing from this scatter diagram?

b Find the equation of the line of best fit on the scatter diagram.

c What is the interpretation of the values of m and c in this case?

d Use the equation to estimate the score given by Judge 2 when Judge 1 awards 55.

Solution

a The label for the vertical axis is missing, and a plotted double mean point for the line of best fit (though the line is accurate). The drawn line also extends beyond the data.

b Find the equation in the form $y = mx + c$.

c the intercept, which is 30.

For m choose two coordinates on the line; (0, 30) and (90, 78) are chosen as they are whole numbers and easier to work with.

$$m = \frac{78 - 30}{90 - 0} = \frac{48}{90}$$

$$= 0.53 \text{ (2 d.p.)}$$

So the equation of the line of best fit is $y = 0.53x + 30$

c The value of c actually gives the mark Judge 2 would give if Judge 1 gave a score of zero. The value of m is the number of extra marks Judge 2 would give for each 1 extra mark Judge 1 gave.

d Now you have the equation of the line you can use it to say that

when $x = 55$, $y = 0.53 \times 55 + 30 = 59.15$

So, Judge 2 would award around 59 marks.

In some instances it might not be possible to extend the line of best fit to read off the intercept. If this is the case, you need to find the gradient first as above.

You would then have $y = 0.53x + c$

Next use the coordinates of a point on the line to substitute in a y and an x value.

Suppose you used (90, 78) you would get $78 = 0.5333333 \times 90 + c$

(Notice that it is better to use the full decimal for any previously rounded value, and also that you can use the double mean point here if you wish.)

Rearranging this gives $c = 78 - 0.533333 \times 90 = 30$ as before.

It is then wise to use a second point to check that the values obtained work. The double mean point can be used for this, though the values are not always that friendly.

Be a statistician 

Road safety campaigners were concerned that the older the car the longer it would take to stop safely at 30 miles per hour.

1 If their concerns were true what type of correlation would exist?

The campaigners commissioned a garage to test 12 cars of various age to see how many metres it was before they stopped when travelling at 30 miles per hour. The results are in the table.

Age of car (months)	6	12	18	24	30	42	54	70	78	88	96	102
Stopping distance (metres)	21.4	22.7	23.6	23.4	24.3	24.6	25.8	27.3	27.1	29.5	31.2	33.0

2 Draw a scatter diagram of these results (make the graph large enough to answer Question 4 (b)).

3 Draw a line of best fit on your graph.

4 Use your line of best fit to estimate:

a the stopping distance in metres of a 5-year-old car

b the stopping distance in metres of a 10-year-old car.

5 Which of your answers in Question 4 is more reliable? Give a reason.

6 Calculate the equation of the line of best fit from question 3.

7 Use your answer to question 6 to verify your estimate made in Question 4 (a).

8 Write a short report for the road safety campaigners from the viewpoint of being the garage commissioned to do the tests clearly stating your findings.

Exercise 8.1

1 State the strength and type of correlation in each of these diagrams.

a

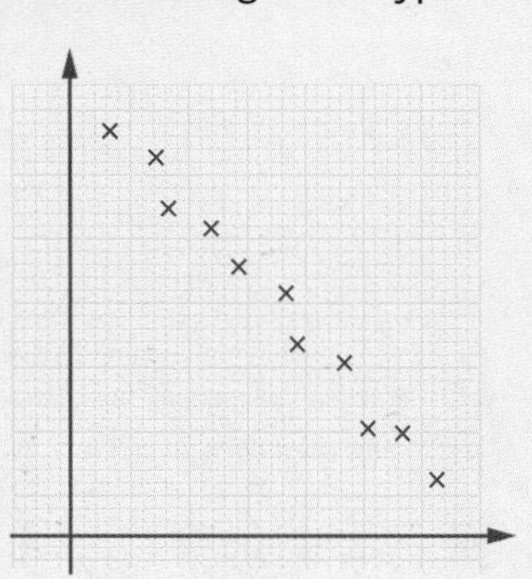

b

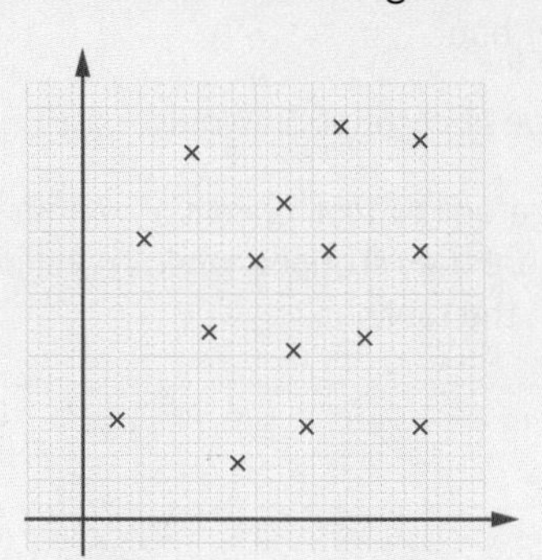

c

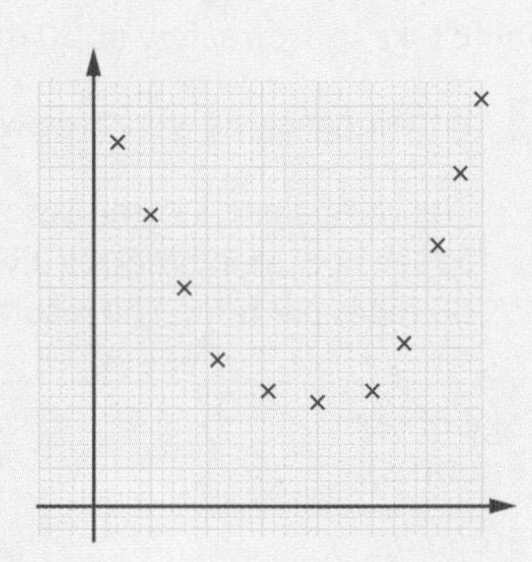

2 Gabriel was asked to estimate the size of eight angles. The table shows the results.

Actual angle	66	93	115	165	207	240	289	323
Gabriel's estimate	60	100	125	160	195	240	300	335

a Plot the data as a scatter diagram.

b Describe the strength and type of correlation present in your diagram.

c Draw a line of best fit.

d Use your line of best fit to estimate the size of Gabriel's angle estimate for an angle of:

i 140 degrees

ii 30 degrees.

e Which of your estimates in part (d) are likely to be more reliable? Explain your answer.

3 Give a reason why an estimate found using extrapolation is likely to be unreliable.

4 The table shows the height of 10 athletes from different sports and the time taken to run 100 metres.

Height (cm)	193	195	188	186	175	180	172	183	168	168
Time (seconds)	56	63	65	66	69	71	74	76	77	81

a Plot these data as a scatter diagram.

b Describe the strength and correlation present in your diagram.

c Draw a line of best fit.

d Use your line of best fit to estimate the time taken to run 100 m for a person of height:

i 177 cm

ii 165 cm.

e Which of your estimates in part (d) is the more reliable? Explain your answer.

f Why are both your estimates in part (d) likely to be fairly unreliable?

g Calculate the equation of the line of best fit.

h Give an interpretation of the values of m and c from your equation of the line of best fit.

5

a Calculate the equation of the line of best fit for your line in Question 2.

b Interpret the value of m for this line of best fit in the context of the question.

8.2 Calculating and interpreting values of correlation

Interpreting Spearman's rank correlation coefficient

It is possible to calculate measures of correlation that indicate whether positive or negative correlation is present in bivariate data and the strength of such correlation. At foundation tier you need to know how to interpret values of **Spearman's rank correlation coefficient (SRCC)**. You do not need to know how to calculate these measures.

The SRCC is a measure often denoted by the symbol r_s.

–1 represents perfect negative correlation.

+1 represents perfect positive correlation.

The labelled scale below shows the interpretations that you can make for values between these extremes.

-1		0		+1
strong moderate weak	←	no correlation	→	weak moderate strong
negative correlation				**positive correlation**

The use of 'strong', 'moderate' and 'weak' also depends on other factors such as the number of pairs of data being considered. For example, a typical moderate correlation of 0.75 would actually be a very strong correlation if there were 100 pairs of data on which that value had been calculated. You should also be able to estimate values of the SRCC by looking at scatter diagrams.

Spearman's rank coefficient of correlation uses the **rank** order of the values for the x and y coordinates and so any graph be it a curve or straight line, where each point is above and to the right of the previous point would have a SRCC value of 1. Likewise, values of –1 will be from continuously decreasing data.

Objectives

In this topic you will learn:

- how to calculate and interpret values of Spearman's rank correlation coefficient including tied ranks
- to interpret the product moment correlation coefficient values and compare it with Spearman's correlation.

Key terms

Spearman's rank correlation coefficient (SRCC): the SRCC is a way of calculating a value for correlation that uses the ranks of the two variables in a formula.

Rank: the rank of a piece of data is its position in the ordered list of that data.

Example 1

Estimate the value of Spearman's rank correlation coefficient for each of these scatter diagrams.

a

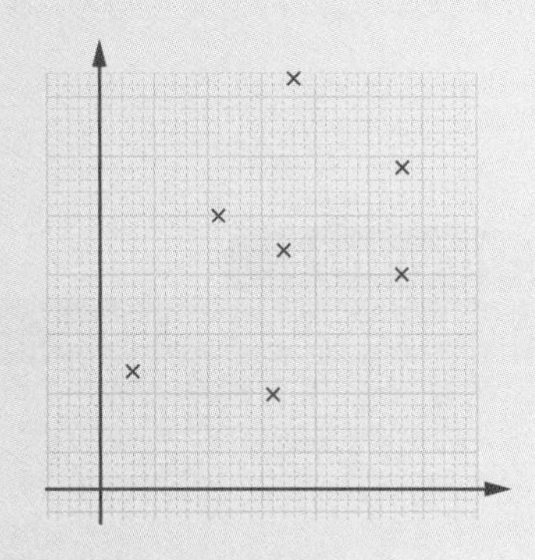

b

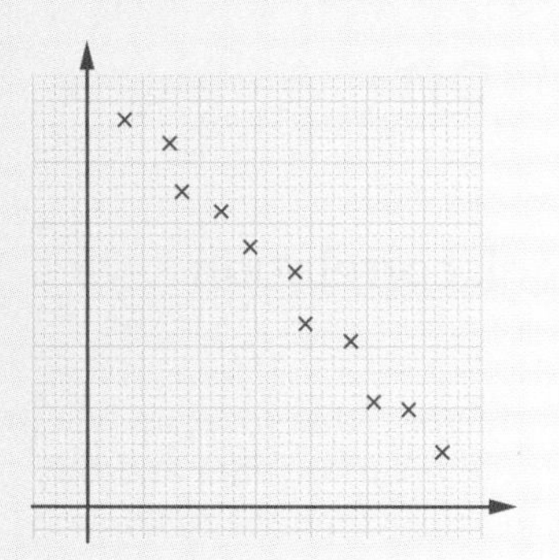

c

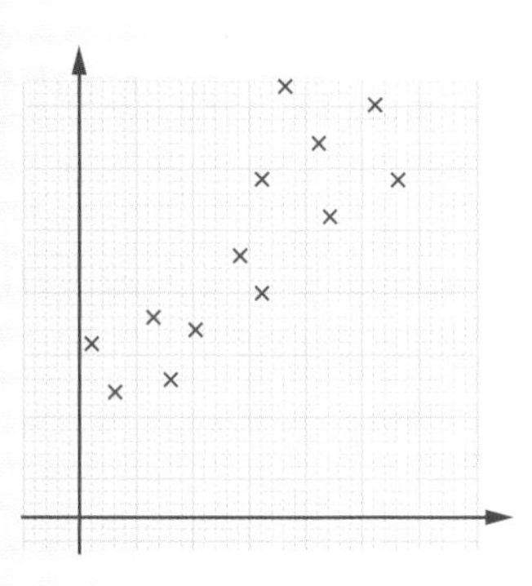

Solution

a This looks like very weak positive correlation, possibly about 0.5.

b This looks like quite strong negative correlation, possibly about –0.9.

c This looks like moderate positive correlation, possibly about 0.7.

Calculating Spearman's rank correlation coefficient

At higher level you need to be able to calculate values of SRCC. The example shows how this can be done. You need to pay particular attention to:

- the absolute need to rank the data before anything else is done
- be consistent when ranking – work from lowest to highest on both data sets
- how to deal with tied ranks when 2 (or more) items of data have the same value
- the structure of the formula for SRCC
- the fact that the formula is given so you do not need to learn it off by heart, just know how to use it.

The formula for Spearman's rank correlation coefficient is $1 - \frac{6\Sigma d^2}{n(n^2-1)}$

d is the difference between the ranking of the data values

n is the number of pairs of data.

Example 2

For nine months Carlo kept a record of the number of accidents on the roads in his village and whether it was raining at any point each day. The results are shown in the table.

Number of accidents	3	2	5	1	7	4	0	5	6
Number of dry days in month	21	24	18	26	16	20	26	21	19

a Draw a scatter diagram to illustrate the data.

b State the type and strength of the correlation.

c To what extent do you feel there is a causal relationship present? Explain your answer.

d Calculate the value of SRCC.

e Interpret the value obtained in part (d) in the context of the question.

Solution

a

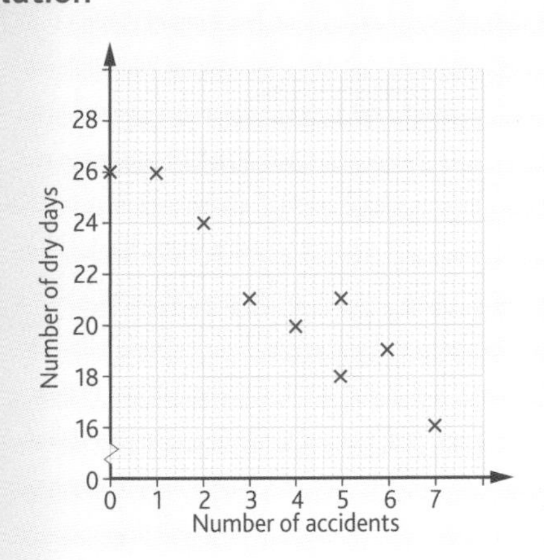

b Correlation is negative and appears to be quite strong.

c There would seem to be a fair chance that wetter weather would lead to a greater risk of accidents, hence leading to a negative correlation between the number of dry days and the number of accidents. However, a linked third variable is also likely, such as in wet weather more traffic is on the road, especially around the time of the start and end of school.

d The formula for SRCC is given at the front of the examination paper and is $1 - \frac{6\Sigma d^2}{n(n^2 - 1)}$

When calculating SRCC, it is recommended that four extra rows are added to the table as shown. One for the ranks of each of the data sets, one for d, the difference in the ranks and one for d^2. This final row is then totalled to give $\sum d^2$ in the formula.

Number of accidents	3	2	5	1	7	4	0	5	6
Number of dry days in month	21	24	18	26	16	20	26	21	19
$r_{accidents}$	4	3	6.5*	2	9	5	1	6.5*	8
r_{days}	5.5	7	2	8.5	1	4	8.5	5.5	3
d	−1.5	−4	4.5	−6.5	8	1	−7.5	1	5
d^2	2.25	16	20.25	42.25	64	1	56.25	1	25

* The rank of 6.5 on the accidents row comes from the fact that the value 5 occurs twice. These would be the sixth and seventh values so they are averaged to give each one a rank of 6.5 A similar occurrence causes ranks of 8.5 on the dry days' data.

The total for the bottom row gives $\sum d^2 = 228$.

You can now use the formula, substituting $\sum d^2 = 228$ and $n = 9$.

$$\text{Spearman's rank correlation coefficient} = 1 - \frac{6 \times 228}{9(9^2 - 1)}$$

$$= 1 - \frac{1368}{720}$$

$$= 1 - 1.9$$

$$= -0.9$$

AQA Examiner's tip

Each year thousands of students do not rank their values when attempting to calculate Spearman's and immediately score zero out of 5 or 6 marks as a result. Therefore you *must* remember to rank your data first.

e The strong negative correlation seen between number of accidents and number of dry days would indicate that the more dry days there are the fewer accidents there tend to be.

Interpreting product moment correlation coefficients

Another measure of correlation is called the **product moment correlation coefficient**, or PMCC for short.

The main difference between PMCC and SRCC is that the PMCC uses the data values themselves to calculate the value of correlation. This means it is more sensitive a measure than the SRCC, which relies solely on the order that each set of data falls in. It is not necessary to be able to calculate the PMCC in GCSE statistics.

The requirements are that you understand the outcomes that you know the difference between the PMCC and the SRCC. This key difference is the use of ordinal (ranked) data for the SRCC but numerical or interval data for the PMCC.

The PMCC produces a result between –1 and +1 just as the SRCC does, with the same interpretation of the values along the scale.

Key terms

Product moment correlation coefficient (PMCC): the PMCC is a way of calculating a value for correlation that uses the actual size of the data for the variables.

Example 3

A teacher sets and marks a piece of homework from a class. He also asks the pupils to state how long the work took them, and how many televisions they have in their house.

The teacher calculates the value of the product moment correlation coefficient between homework mark and time taken and finds the value is 0.77.

The teacher also calculates the value of the product moment correlation coefficient between homework mark and the number of televisions and finds the value is –0.16.

Interpret, in context, these two values of the PMCC.

Solution

The value of 0.77 between mark and time taken indicates a moderate to quite strong correlation between the two variables. This shows that the longer a pupil spends on the work, the higher the mark tends to be.

The value of –0.16 between mark and televisions indicates that there is no correlation and therefore no connection between those two variables.

Real world

Search the internet for examples of values of correlation coefficients between variables. Choose your five favourite examples, displaying the results on a poster with suitable artwork relevant to the contexts. Use ICT if possible.

Exercise 8.2

1 The table shows values of Spearman's rank correlation coefficient for some sets of bivariate data.

Part	First variable	Second variable	Value of SRCC
a	Age of child	Weekly pocket money received	0.58
b	Score in examination	Shoe size	0.09
c	Time taken to complete a run	Length of run	1
d	Age of car	Price of car	-0.74

Comment on the values of Spearman's correlation coefficient in each case in the context given.

2 Estimate the value of Spearman's rank correlation coefficient for each of these diagrams.

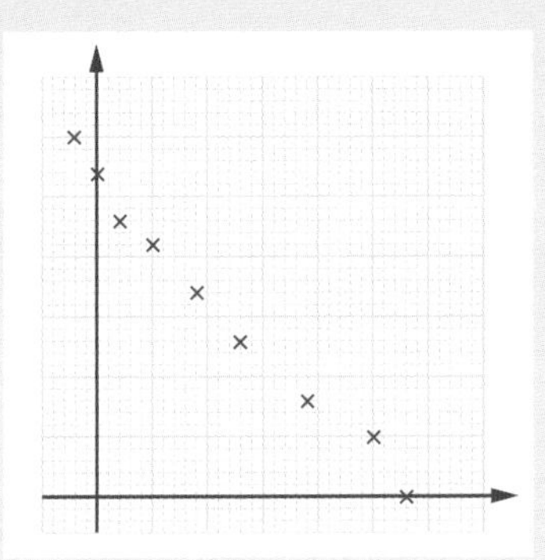

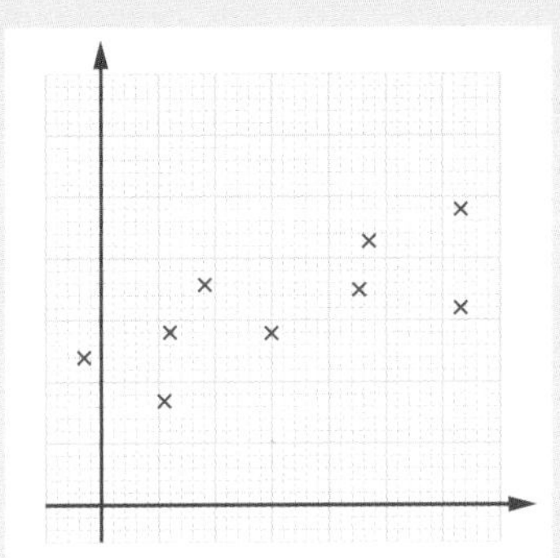

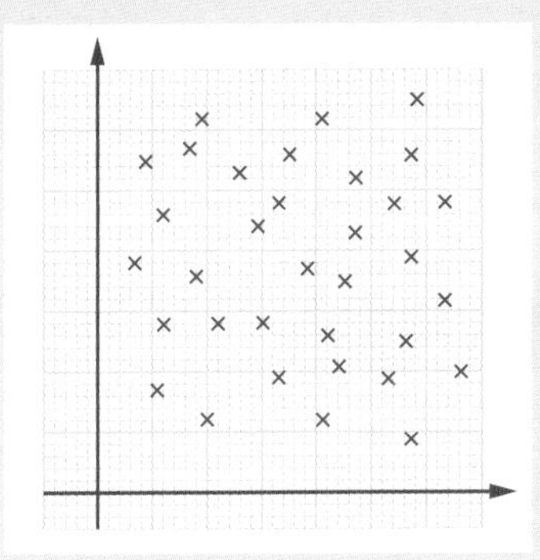

3 Draw a scatter diagram with 10 plotted points, so that the value of Spearman's rank correlation coefficient is –1.

4 For each of these sets of bivariate data, calculate the value of Spearman's rank correlation coefficient.

a

Height (m)	1.54	1.46	1.60	1.87	1.49	1.79	1.65
Weight (kg)	78	64	61	89	66	84	60

b

Number of pets	4	0	1	2	9	3
Number of cars	2	2	1	1	0	3

c

Age	18	21	22	25	25	34	41	55	63
100 m time (seconds)	11.3	10.8	10.6	11.0	11.5	11.1	11.8	12.6	14.2

d

Engine size	1.6	1.6	1.6	1.4	1.1	2.1	1.8	2.8
Miles per gallon	46	59	44	54	59	31	35	28

5 Ten friends who graduate from the same course at university compare their final examination percentage with their salary one year after leaving university. The results are shown in the table.

Person	A	B	C	D	E	F	G	H	I	J
Final %	78	63	68	96	63	45	80	73	61	92
Salary £	32 000	26 000	34 000	35 000	33 000	27 500	38 000	45 000	30 000	39 000

a Calculate the value of Spearman's rank correlation coefficient.

b Interpret the value obtained in part a in the context of the question.

6 Which of the measures product moment correlation coefficient and Spearman's rank correlation coefficient is affected by outliers in the data? Explain your answer.

7 Put these values of product moment correlation coefficient in order, starting with the value that shows the highest correlation: 0.35 –1 –0.67 0.94 0.06 –0.44

8 A researcher was looking at the number of pedestrians and the number of shops in fixed areas of a town centre. Once the data was collected, the value of the product moment correlation coefficient was calculated to be 0.76 between the variables. Interpret this result in the context of the problem.

Summary

You should:

- know that scatter diagrams are used to shows patterns in bivariate data
- know that the diagrams will show either positive, negative or no correlation and that correlation can vary in strength
- know that, for moderate or strong correlation, a line of best fit can be drawn through a double mean point
- know that these lines can be used for estimation of values of one variable given the value of the other
- know that estimates made within the data (interpolation) are more reliable than estimates made outside the data range (extrapolation)
- know that the equation of the line of best fit can be calculated
- know that Spearman's rank correlation coefficient can be calculated and interpreted to give a measure of correlation
- know that product moment correlation coefficients use the actual values of the data and can be interpreted to give a measure of correlation.

AQA Examination-style questions

1. A student was shown the line AB and asked to adjust the length of the line CD so that the length of CD appeared to be the same as the length of AB. When the line AB was 8 cm long the student made the length of CD equal to 10.3 cm.

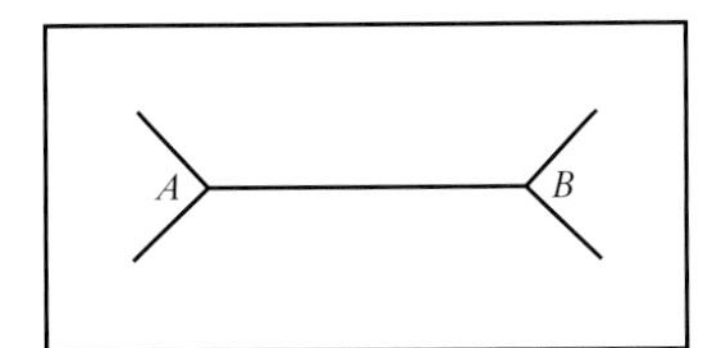

Not to scale

C
D
Move left or right to change the length of CD

The experiment was repeated using different lengths for AB and the results are given the in the table.

AB cm	5	6	7	8	9	10	11	12
CD cm	6.5	7.8	8.6	10.3	11.4	12.7	14.2	15.1

(a) Plot a scatter graph to illustrate these results. *(2 marks)*

The mean length of CD is 10.8 cm.

(b) Calculate the mean length of AB. *(1 mark)*

(c) Draw a line of best fit on the scatter graph. *(2 marks)*

(d) Use your line to estimate the length of CD when:

(i) the length of AB is 6.5 cm *(1 mark)*

(ii) the length of AB is 14 cm. *(1 mark)*

(e) Which of these two answers is the most reliable? Give a reason for your answer. *(1 mark)*

AQA, 2000

2. **(a)** The table shows the number of people in each of 12 households and number of items of post they received one Monday.

Number in household	1	2	2	3	3	3	4	5	5	5	7	8
Number of items of post	8	3	9	7	6	10	3	9	8	11	4	7

(i) Complete the scatter diagram. The information for the smallest six households has been plotted. *(2 marks)*

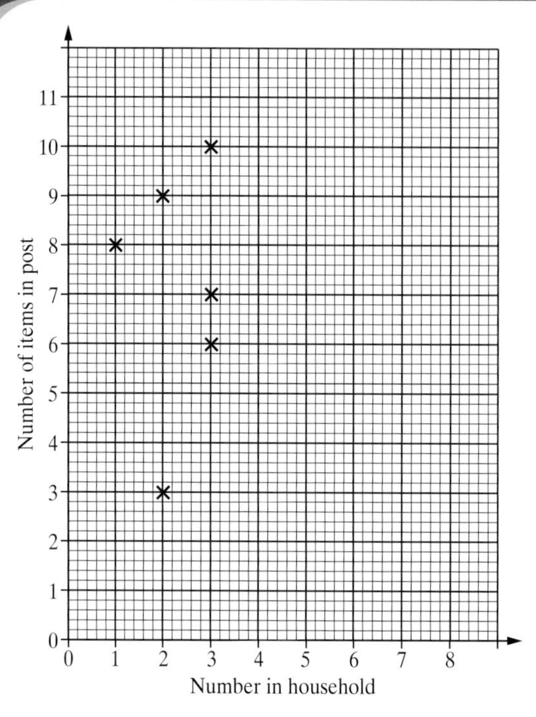

(ii) Explain why it is not appropriate to draw a line of best fit for these data. *(1 mark)*

(b) The number of phone calls received by the same households that Monday is shown in this scatter diagram.

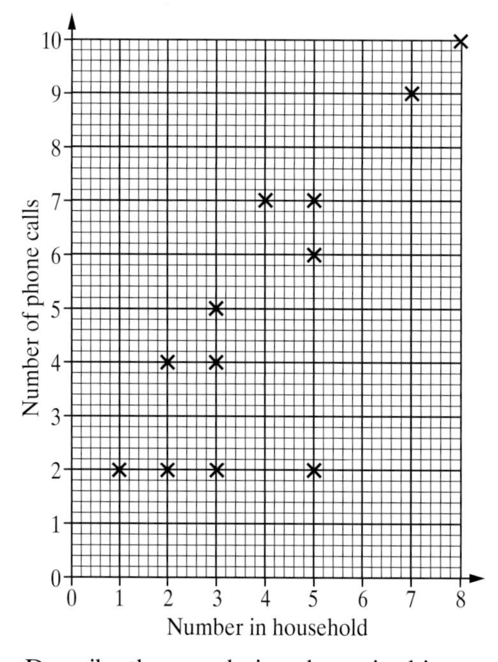

(i) Describe the correlation shown in this scatter diagram. *(1 mark)*

(ii) The mean number in a household is 4.

The mean number of calls received is 5.

Use this information to draw a line of best fit on the graph. *(2 marks)*

(iii) Use your line to estimate the number of phone calls received by a household of six. *(1 mark)*

AQA, 2008

3. The floor areas of some houses and the cost of each house are given.

(a) Plot a scatter graph of these data.

Floor area (m^2)	156	183	180	185	190	200	210
Cost (£000s)	48	54	50	52	57	59	58

(2 marks)

The mean cost of the houses is £54 000.

(b) Calculate the mean floor area. *(2 marks)*

(c) Draw a line of best fit on your scatter graph. *(2 marks)*

(d) A family wants to buy a house with a floor area of 170 m^2.
How much might it cost? *(1 mark)*

(e) Explain why the house with floor area 180 m^2 seems to be good value for money. *(1 mark)*

AQA, 2001

4. A survey was carried out at eight retail stores on one morning. Information was recorded about the number of employees and the percentage of employees absent. The information is shown in the table.

Store	A	B	C	D	E	F	G	H
Number of employees	60	75	100	50	100	130	200	80
Percentage of employees absent	5	8	4	2	12	10	8.5	2.5

(a) Calculate the value of Spearman's rank correlation coefficient for the two sets of data. *(6 marks)*

(b) Use your answer to part (a) to comment on the stateent, 'The absentee rate increases with the size of the workforce'. *(1 mark)*

(c) Write down the most likely value of Spearman's rank correlation coefficient for each of the following graphs.

Select from the following list:

-0.45 0.75 -1.2 -0.98 0.96 *(2 marks)*

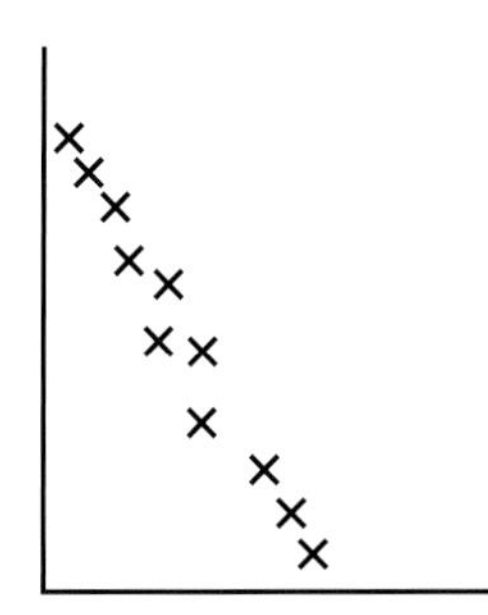

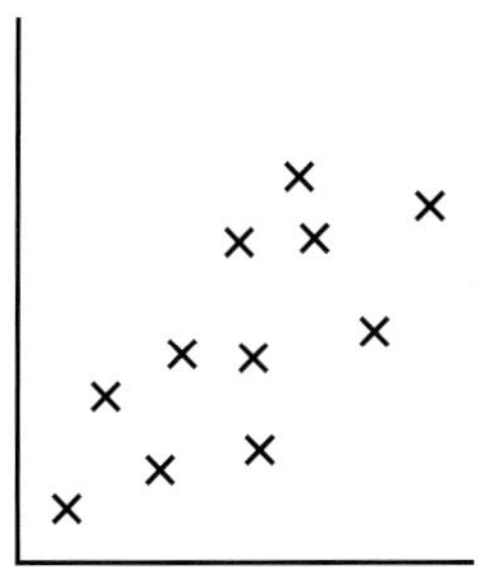

AQA, 2007

5. The scatter diagrams shows the floor area (100m²) and the daily takings (£1000) for a chain of supermarkets.

(a) The mean floor area is 1700 m².

The mean daily takings are £184 000.

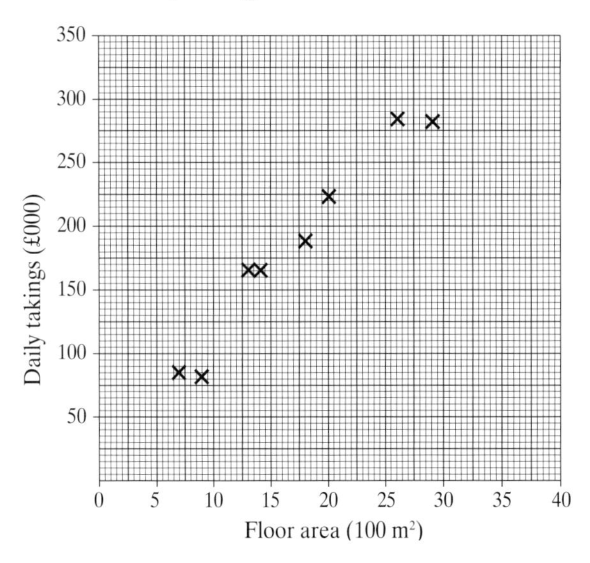

Draw a line of best fit on the scatter diagram. *(2 marks)*

(b) It is proposed to build two new supermarkets.

Use your line of bet fit to estimate the daily takings for a supermarket with floor area:

(i) 2200 m²

(ii) 3500 m²

(c) Which of these estimates is more reliable?

Give a reason for your answer.

(d) What is the expected increase in daily takings for each additional 100 m² of floor area?

(e) The equation of the line of best fit for a chain of DIY stores passes through the points (6, 25) and (20, 250).

Draw this line on the scatter diagram. *(1 mark)*

(f) Compare the daily takings of the chain of supermarkets and the chain of DIY stores. *(2 marks)*

Supermarket	A	B	C	D	E	F	G	H
Floor area (100 m²)	7	6	13	14	18	20	26	29
Daily takings (£1000)	85	82	165	165	188	223	283	281

(g) The floor areas and the daily takings of the supermarkets are given in the table.

(i) Calculate the value of Spearman's rank correlation coefficient. *(6 marks)*

(ii) Interpret in context the value of Spearman's rank correlation coefficient. *(1 marks)*

(h) Write down the least and greatest daily takings of supermarket F that will not change the value of Spearman's rank correlation coefficient. *(2 marks)*

AQA, 2005

9 Time series, moving averages and seasonal effects

In this chapter you will learn:

- how to draw and interpret time series graphs
- how to choose and calculate appropriate moving averages
- how to draw and interpret trend lines
- to identify seasonal effects
- how to calculate average seasonal effects and use them for prediction of future values
- how to draw and interpret Z-charts.

You should already know:

✓ how to plot coordinates

✓ how to evaluate and plot cumulative frequencies.

What's this chapter all about?

There is a vast amount of data in the real world that is collected over periods of time. If you ran a business such as a shop, you would keep a careful record of takings each day, and possibly each hour, to inform your opening times; if you were on a diet, you would weigh yourself every week or so to keep track of changes. Government data is also produced monthly and annually. To be able to deal with this wealth of time related data, you need special methods. This is what this chapter is all about – graphs and measures looking for patterns in the data and searching for periods when things don't seem to go as expected!

Focus on statistics

Most of this chapter is covered in maths but some of it is done differently in statistics and some of it is not in maths at all! In 9.2 the work on seasonal effects is only in statistics and this leads to a better method of predicting future values than the one used in maths. In statistics it is important to use the average seasonal effects when predicting – do not use the method you would use in maths, which is to base your prediction on the next moving average value.

In 9.2 Z-charts are found only in statistics.

9.1 Time series graphs

A **time series** is simply the name given to a graph or set of data where the data is measured over a period of time. The time periods can be, for example, every minute, day, month or year. Quarterly data is popular in finance and retail – as the name suggests, it is collected four times per year, each quarter being three months' data.

The raw data is plotted at the appropriate time values and then the plots are joined with straight lines. Time series are used widely in real life, as much real data such as prices, temperatures, earnings, profits and so on, change with time.

Objectives

In this topic you will learn:

- how to draw and interpret time series graphs
- how to choose and calculate appropriate moving averages
- how to draw and interpret trend lines.

Example 1

The table shows the number of UK females (in thousands) not in full time education aged 16 or 17, for each quarter over three years.

a Draw a time series graph to show the data.

b Make two comments about the graph of the data.

2004 Q1	133
2004 Q2	122
2004 Q3	135
2004 Q4	126
2005 Q1	128
2005 Q2	122
2005 Q3	113
2005 Q4	106
2006 Q1	102
2006 Q2	101
2006 Q3	107
2006 Q4	115

Source: www.statistics.gov.uk

Solution

a The time scale is always placed horizontally.

The vertical scale is for the data. Note that the lowest value on the scale is 100(000), and the highest 135(000). If you were to begin the scale at 0, a large part of the graph would be blank. Therefore, you should use a break mark in the graph to begin the scale close to the lowest value. However, you must be aware that this will exaggerate the changes in the data.

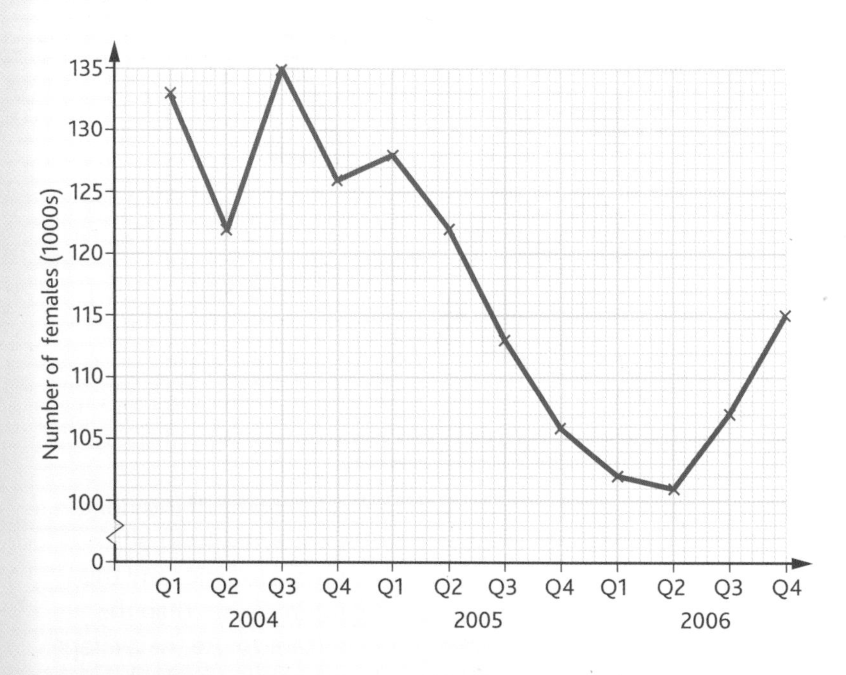

b In 2004 the data fluctuates before falling quite significantly during 2005. There was some evidence of an increase in the data towards the end of 2006.

Key terms

Time series: any set of data that has values recorded at different times.

Many time series graphs show fluctuations due to changes in data values that occur in some kind of cycle, usually throughout a week or a year. These are called **seasonal effects**.

Examples would be the size of gas bills, where in winter they are always higher than in summer; items like sales at a fish-and-chip shop, where traditionally more people used to eat fish on a Friday than any other day; or stocks of Wellington boots, which are probably bought more often in autumn, ready for winter, than at other times. Perhaps you could research whether these assumptions are actually true.

It is perfectly possible to look for trends using the entire data set by drawing a line of best fit for the times series data. However, there is a strong chance that these seasonal effects mask the genuine trend of the data.

Moving averages

A good way of taking into account seasonal effects or patterns in time series data when trying to find trends is to find **moving averages**.

Moving averages take the mean of the data over one whole period, such as a week if the data is daily, or a year if the data is quarterly. Changes in the values of the moving average then represent overall trend changes taking into account seasonal fluctuations.

For example, for the data 45, 53, 61, 60, 47, 57, 66, 62, suppose three-point moving averages were required.

The first three-point moving average would be $(45 + 53 + 61) \div 3 = 53$

The second three-point moving average would be $(53 + 61 + 60) \div 3 = 58$, and so on.

Notice how, each time, one extra data point is used in the calculation and one is removed from the calculation. To avoid errors it is important to be systematic in your approach.

Moving averages can be plotted on to time-series graphs to give a clear idea of overall trends within what is often quite widely fluctuating data. The method for producing such plots has two key features:

- The moving average is plotted at the centre point of the time interval it represents.
- The **trend line** through the moving average plots should be a line of best fit, drawn by eye and not by plotting the double mean points.

Key terms

Seasonal effects: the fluctuations in data that occur in any kind of regular time pattern, such as particular times of a week or year.

Moving average: the mean of one whole cycle of data in a time series that moves through the data one value at a time.

Trend line: a line of best fit drawn by eye through the plotted moving average points.

Spot the mistake

Syd and Hattie run a cafe. It opens on Thursdays, Fridays, Saturdays and Sundays. They collect data on takings each day for six weeks, and want to work out moving averages to look at trends in their takings. They work out seven-point moving averages as they know the data is collected on a week-by-week basis.
Spot their mistake.

Example 2

The data shows daily sales at a small sandwich shop over three weeks. The shop opens four weekdays only.

Week	1				2				3			
Day	Mon	Tues	Weds	Thurs	Mon	Tues	Weds	Thurs	Mon	Tues	Weds	Thurs
Sales (£)	140	128	82	154	134	126	76	152	122	120	74	146

a Plot the data as a time series graph.

b In order to determine trends in the data, four-point moving averages are calculated. Explain why four-point moving averages are used.

c Calculate the set of four-point moving averages.

d Plot your four-point moving averages on the time series graph.

e Draw a suitable trend line on the graph.

f Describe two patterns in the data.

Solution

a Use a cross to represent each data point and join up with straight solid lines.

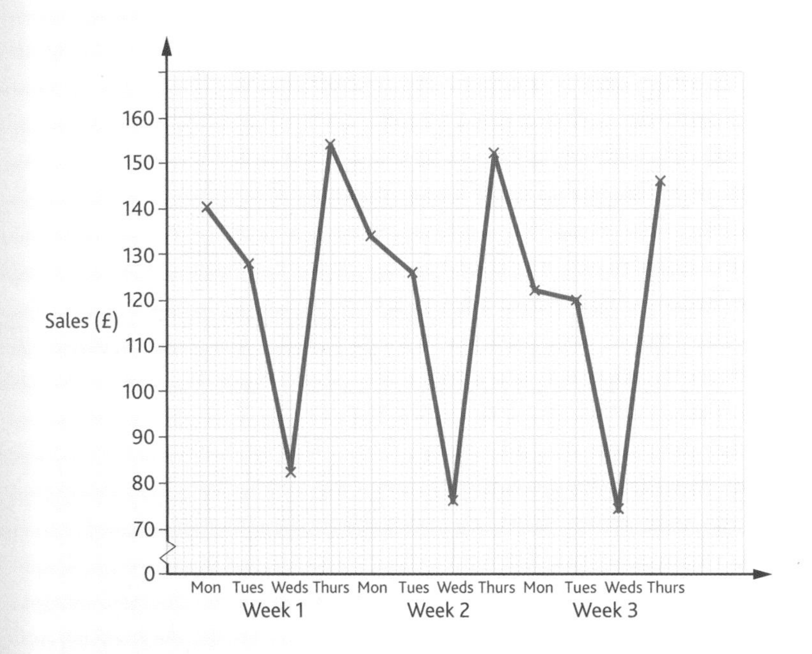

b Each cycle has four data values as the shop is open for four days each week.

c The first moving average is $(140 + 128 + 82 + 154) \div 4 = 126$

The second moving average is $(128 + 82 + 154 + 134) \div 4 = 124.5$

The third moving average is $(82 + 154 + 134 + 126) \div 4 = 124$

Continuing these calculations gives the remaining four-point moving averages as: 122.5, 122, 119, 117.5, 117 and 115.5. Once there are fewer than four data values remaining, no further four-point moving averages can be calculated.

d The values are now plotted at the centre of the period they each represent.

So, the first four-point moving average is plotted half way between Tuesday and Wednesday for Week 1, the second is plotted halfway between Wednesday and Thursday for week 1, and so on. Use a dot with a circle around it to make the plots look different from the actual data plots.

The graph will now look like this:

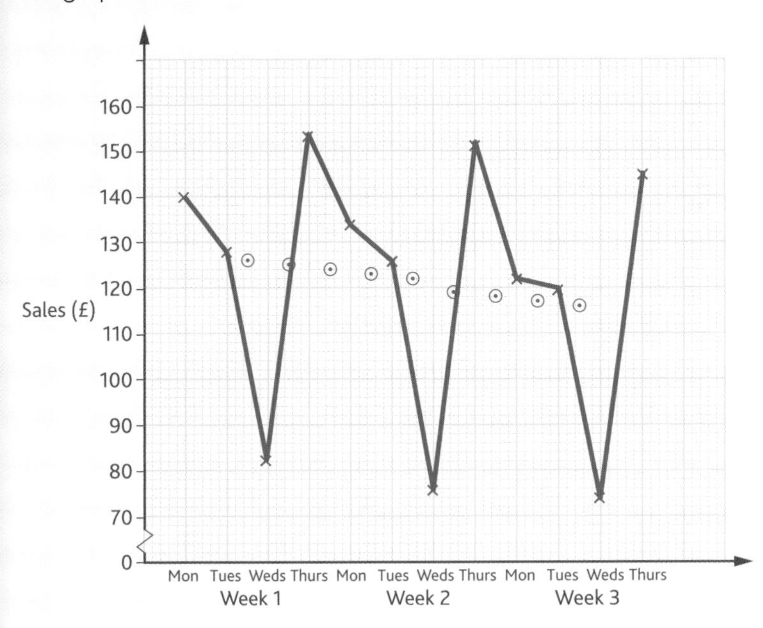

e The trend line is now a line of best fit drawn using judgement (by eye).

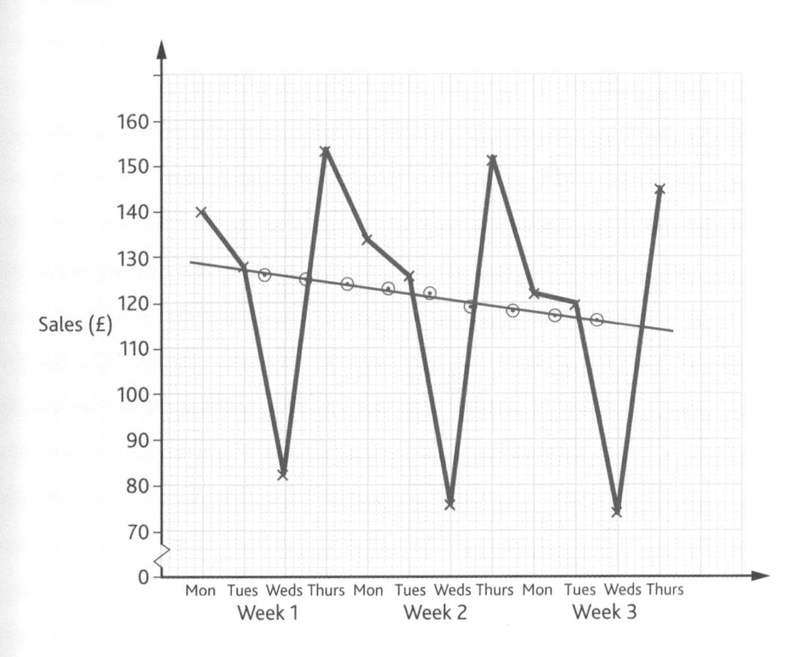

f The weekly pattern sees the highest sales on a Thursday and the lowest sales by far on a Wednesday. The general trend is for a slight decrease in sales over the period as shown by the trend line.

AQA Examiner's tip

Questions on time series often ask for two patterns in the data. One mark is for a comment about the data *within* one cycle (here that is the comment about the highest or lowest day for sales) and one mark is for a comment about the *trend* over the whole period. You cannot score full marks by making two similar comments about one aspect of the data.

Be a statistician

Population changes within a country can be shown in time series. Research the total UK population over the last 10 or 20 years. Use time series to display the data and moving averages to show any trends. Make some basic predictions about the future based on your findings. Present your work using ICT if possible.

Exercise 9.1

1 The time series graph shows stocks of three types of fish in the North Sea.

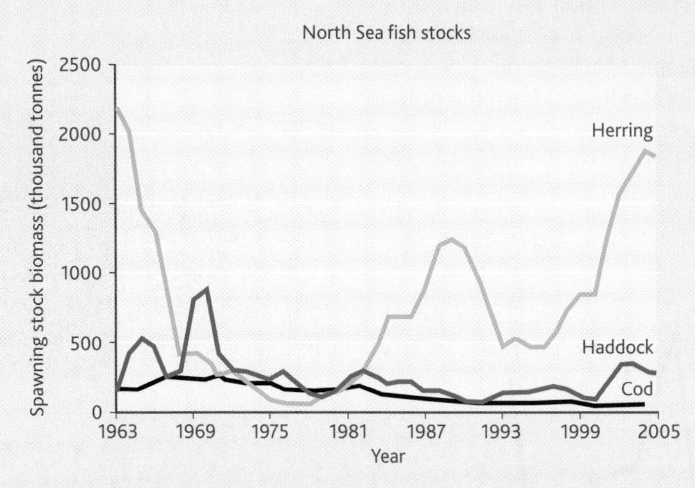

Source: The Centre for Environment, Fisheries and Aquaculture Science (CEFAS), International Council for the Exploration of the Sea (ICES), Department for Environment, Food and Rural Affairs (Defra)

Comment on the stocks of each of the three types of fish shown in the time series graph.

2 For each of these sets of data, calculate the full set of moving averages as indicated.

a 56, 58, 62, 64, 60, 66, 72, 74, 74, 76, 80 (four-point moving averages)

b 93, 87, 90, 81, 78, 75, 78, 72, 66, 69, 63 (three-point moving averages)

c 143, 165, 173, 153, 148, 170, 153, 168, 154, 177, 154, 168, 173 (five-point moving averages)

3 The table shows the number of books borrowed from the town library on each of the three days it is open over four weeks.

Week	1			2			3			4		
Day	Mon	Weds	Fri	Mon	Weds	Fri	Mon	Weds	Fri	Mon	Weds	Fri
Number borrowed	36	21	54	33	24	57	36	30	66	45	36	72

a Draw a time series graph to show the data.

b Explain why three-point moving averages would be the useful in analysing the data.

c Calculate the set of three-point moving averages.

d Plot the three-point moving averages on the time series graph.

e Draw a trend line.

f Describe two patterns in the data.

4 The data to the right is about pupil numbers post 16 in Devon.

a Draw a time series graph of this data.

b Calculate and plot four-point moving averages for the data.

c Draw in a trend line.

d Describe some features of the data as shown by the graph and the moving averages.

Year	Number of pupils
1999	3455
2000	3860
2001	3806
2002	3874
2003	4130
2004	4392
2005	4470
2006	4810
2007	4826
2008	4666

9.2 Calculating seasonal effects

Seasonal effects are changes in data that occur in a regular pattern or cycle, often in a fairly predictable manner.

At foundation tier you are expected to be able to identify seasonal effects when they occur. In the example that follows this means that only parts (a) to (d) could be asked at foundation tier.

At higher tier, average seasonal effects should be calculated in order to make predictions about the future as accurate as possible.

The **size of a seasonal effect** is the difference between the observed value and the prediction of that value from the trend line – seasonal effect equals the actual value less the value on the drawn trend line. The drawn trend line shows what the value would have been if there were no seasonal effect.

In most cases there will be more than one reading from each part of the 'cycle', giving several values for the seasonal effect at each point. The mean of these values gives the **average seasonal effect**.

Objectives

In this topic you will learn:

- to identify seasonal effects
- how to calculate average seasonal effects and use them for prediction of future values
- how to draw and interpret Z-charts.

Key terms

Size of seasonal effect: actual value less the value on the drawn trend line.

Average seasonal effect: the mean of the calculated seasonal effects from the graph.

Example 1

The data shows the size of Javed's gas bill each quarter over the three years 2006–2008.

Year	2006				2007				2008			
Quarter	1	2	3	4	1	2	3	4	1	2	3	4
Gas Bill (£)	184	142	88	126	240	166	94	148	306	212	124	178

a Plot the data as a time series.

b Find and plot appropriate moving averages.

c Draw a trend line.

d Describe the trend and any apparent seasonal effects.

e Calculate the average seasonal effect for Quarter 1.

f Use your answers for parts (c) and (e) to estimate Javed's gas bill for Quarter 1 of 2009. (This is called a seasonally adjusted forecast.)

Solution

a See the graph to the right.

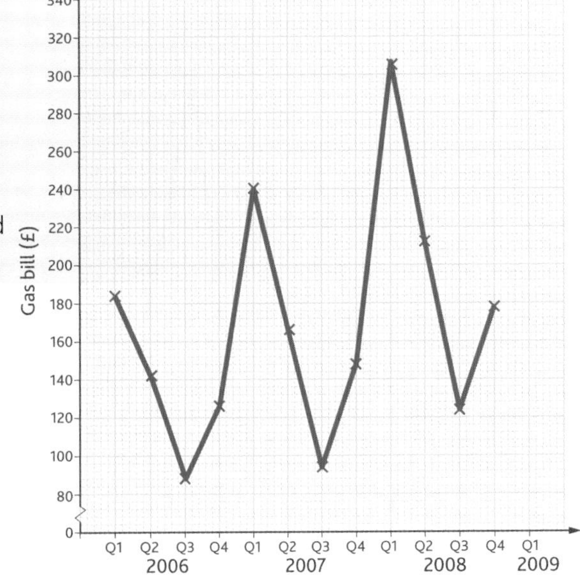

A *Part (a)*

b The only appropriate moving average here are four-point moving averages, as there are clearly four parts to each cycle (one year), as can be seen from looking at the shape of the graph. The first four point moving average is $(184 + 142 + 88 + 126) \div 4 = 135$. The remaining four-point moving averages are: 149, 155, 156.5, 162, 178.5, 190, 197.5 and 205. These are now plotted in the middle of the period each covers – the first one halfway between Q2 and Q3 for 2006, and so on.

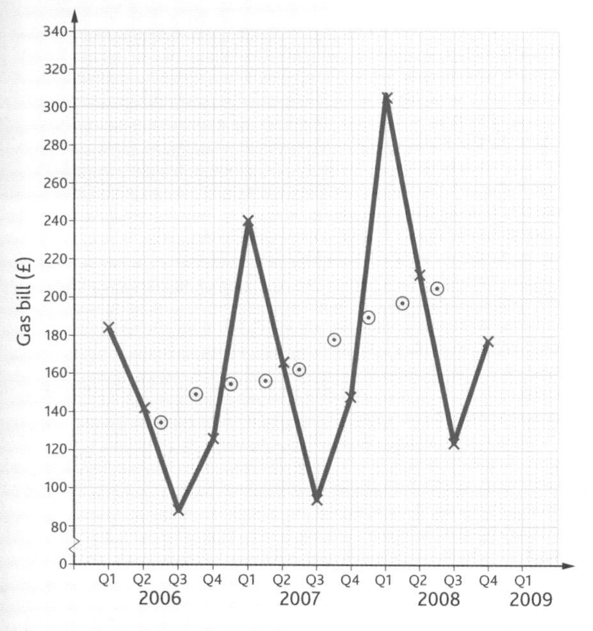

B Part (b)

c The trend line is drawn below. Make it long enough to be able to be used for later parts of the question.

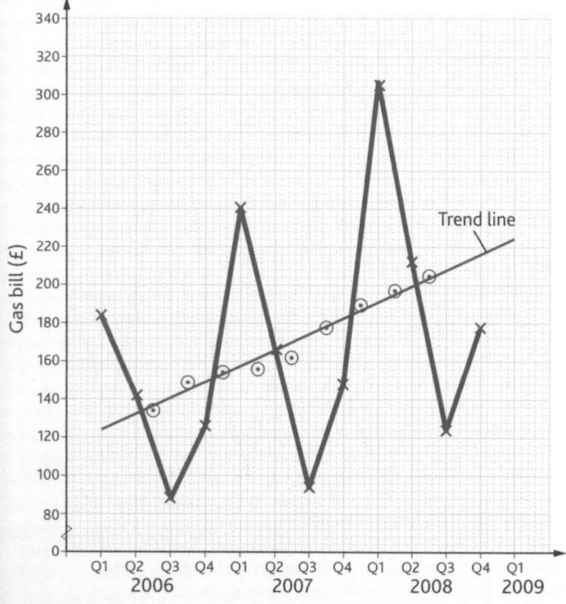

C Part (c)

d The trend is definitely for quite a large rise in gas bills. Seasonal effects are a peak in Quarter 1 and a trough in Quarter 3 and to a lesser extent Quarter 4.

e For Quarter 1 there are three readings available. Each will have a seasonal effect equal to the gap between the actual reading and the position on the trend line. As the data is above the trend line for Quarter 1, the seasonal effect is positive. The three seasonal effects are shown on this copy of the time series with double-headed arrows.

The three seasonal effect values are

$184 - 124 = 60$

$240 - 158 = 82$

and $306 - 192 = 114$

The average seasonal effect for Q1 based on the data available is $(60 + 82 + 114) \div 3 = 85.333...$

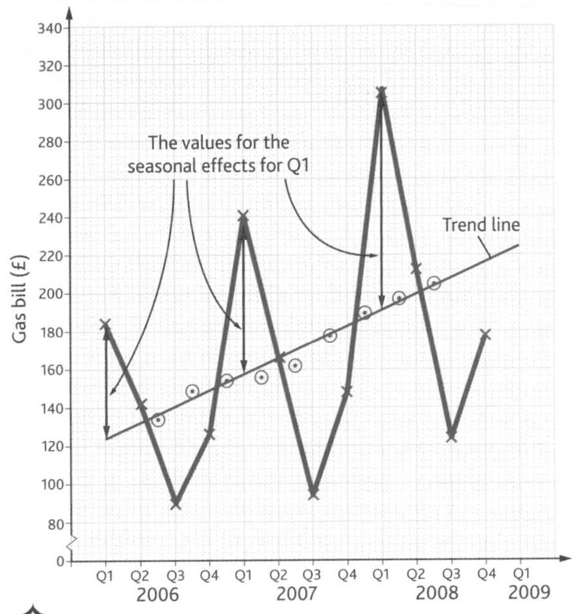

D *Part (e)*

f To estimate Javed's gas bill for Quarter 1 2009, you take the value by reading from the trend line at that point and add on the value of the average seasonal effect.

Q1 2009 estimate $= 226 + 85.333... = 311.333...$

$= £311$ to the nearest full pound.

Z-charts

A **Z-chart** shows several features of time-related data at once. This avoids the need to draw several diagrams to show different aspects of a time related data set; instead all information is contained in this one chart. What information is in a Z-chart?

1. The actual data values for each data point are shown (these form the base of the 'Z').
2. The cumulative values for each data point beginning at the first point on the chart (these form the diagonal upward line on the 'Z').
3. The total of the data for one whole 'period' at each data point (these form the roof of the 'Z'). For example, if sales data in a shop were collected over five days each week:

- the Z-chart would be drawn for one week
- the base of the chart would be the five separate daily values
- the upward line would be the cumulative sales for each day over that week
- the roof would be the total sales for each five-day period finishing on the particular day
- the first value on the roof would be

last Tuesday + **last** Wednesday + **last** Thursday + **last** Friday + **this** Monday

This calculation is a lot like a moving average but without the figures being averaged.

The 'roof' of the Z gives an idea of trend over the period in question, using the previous period as a comparison.

Real world

With the permission of the person who pays the bills at home, try to find out the size of one or all of the gas, electricity and telephone bills over the last two or three years. Which of these bills have noticeable seasonal variations in them? Find appropriate moving averages in order to predict the size of the next bill – one that hasn't arrived yet. How close was your prediction?

If you managed to make predictions for all three, which was the closest to your prediction and which was the furthest out? Can you think of reasons for the accuracy or otherwise of your predictions?

Key terms

Z-chart: a time series graph that also shows cumulative data and the running total for the data over a complete cycle.

AQA Examiner's tip

You will need to know how to draw a Z-chart and what each part of the chart means. In-depth analysis of the gradients of the base and roof of the chart will not be tested.

Example 2

The table shows monthly sales data for a small IT company.

Month	Sales (£1000s)	Total for the year up to ...
January	84	1040
February	82	1064
March	102	1062
April	94	1086
May	100	1086
June	98	1122
July	102	1118
August	88	1132
September	108	1150
October	104	1142
November	106	1160
December	110	

a Complete the total column.

b Find the cumulative frequencies for the sales.

c Complete a 'Z-chart' for this information.

d Write down two patterns in the data.

Solution

a The total for the year up to December is the total of all the values for this particular year (and is the same as the final cumulative frequency). (Other totals include hidden data from the previous year.)

$84 + 82 + 102 + \ldots + 110 = 1178$

b The cumulative frequencies are 84, 166, 268, 362, 462, 560, 662, 750, 858, 962, 1068 and 1178.

c The Z-chart for this data is shown to the right – each of the three sets is joined with straight lines:

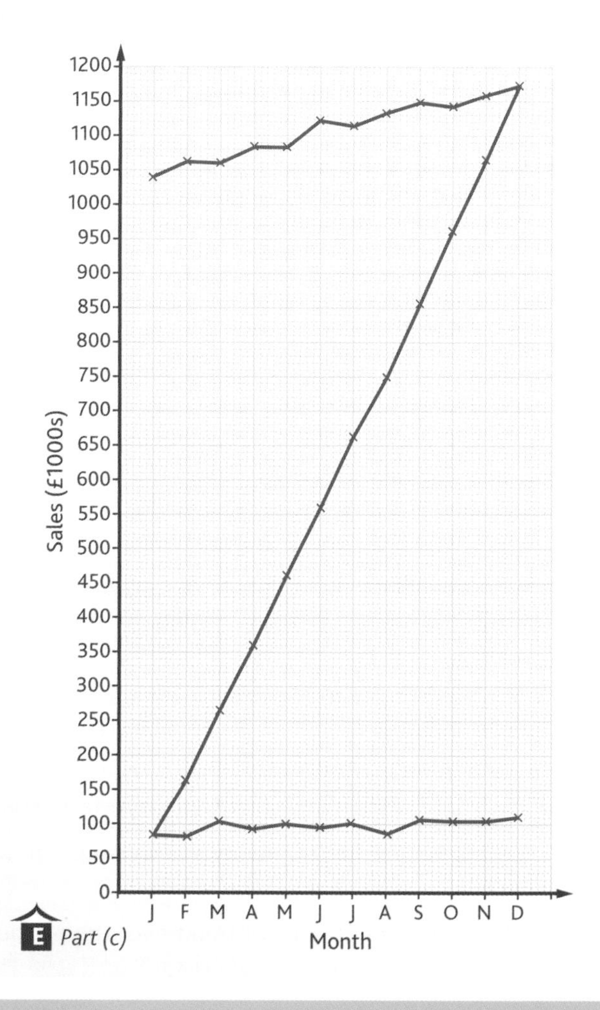

E Part (c)

d The sales throughout the year are fairly steady with dips early in the year and in August.
The trend as shown by the top of the chart indicates an increase in sales over the period in comparison to the year before.

Exercise 9.2

1 What three aspects of time series data are shown on a Z-chart?

2 What is the main benefit of completing a Z-chart?

3 Look at the Z-chart to the right:

a On a copy of the chart, place these three labels with arrows pointing to the appropriate line:

'Current line shows short-term achievement.'

'Moving total line shows long-term improvement.'

'Cumulative line shows medium-term progress.'

b For which of the points A, B or C do each of these statements refer?

'Units sold this month.'

'Units sold over the last 12 months.'

'Units sold since March, the start of the financial year.'

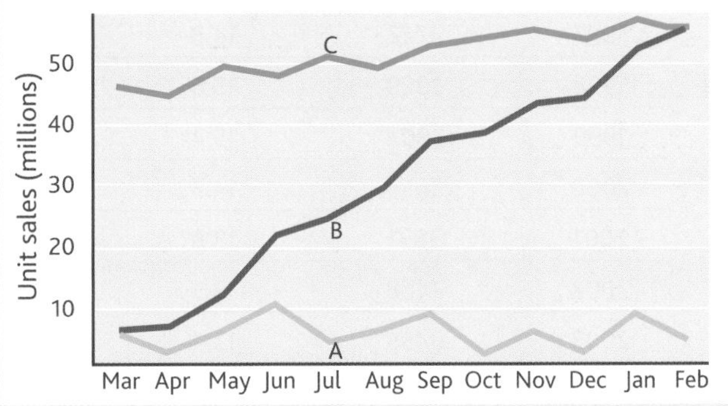

Source: http://syque.com

4 The number of large electrical items sold in a shop each day over one week is shown in the table to the right, along with some other information.

a Copy and complete the table.

b Draw a Z-chart to illustrate the data.

c Write down **two** patterns in the data.

Day	Electrical items sold	Cumulative frequency	Total for last seven days
Monday	23	23	245
Tuesday	21	44	242
Wednesday	28		235
Thursday	26		220
Friday	33		222
Saturday	42		213
Sunday	36		209

5 Look at the time series below.

a Find the average seasonal effect for Monday.

b Use this to make a seasonally adjusted estimate for the sales at the sandwich shop on Monday of Week 4.

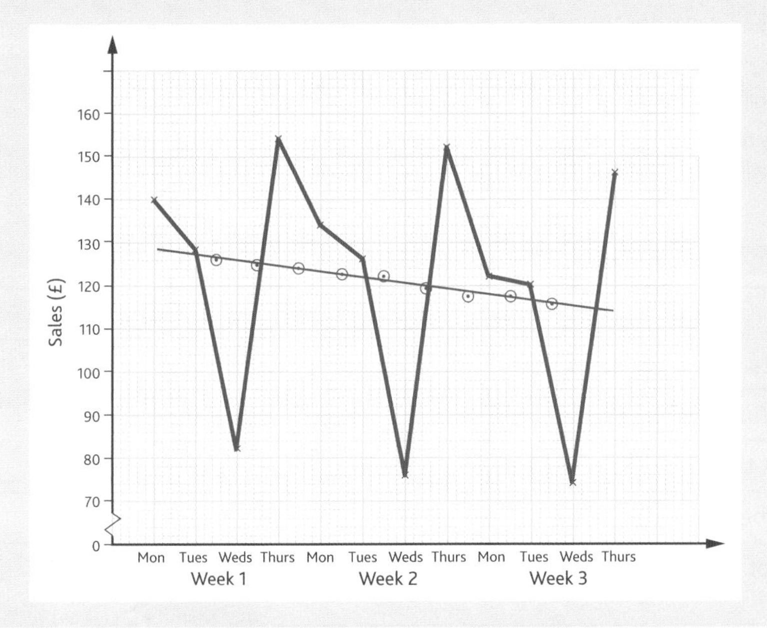

6 The table shows statistics for the number of suicides in the UK from 1996 to 2005.

	Men		Women		Total	
Year	Actual number	Rate per 100 000	Actual number	Rate per 100 000	Actual number	Rate per 100 000
1996	3654	14.6	1239	4.7	4893	9.5
1997	3722	14.8	1259	4.8	4981	9.7
1998	3929	15.6	1225	4.6	5154	10
1999	3904	15.4	1284	4.8	5188	10
2000	3659	14.3	1262	4.7	4921	9.4
2001	3531	13.8	1163	4.3	4694	9.0
2002	3468	13.5	1194	4.4	4662	8.9
2003	3455	13.4	1197	4.4	4652	8.8
2004	3388	13.0	1205	4.5	4593	8.7
2005	3223	12.3	1113	4.1	4336	8.1

Source: ONS, 2007, Mortality statistics, Series DH2 no. 32

a Plot the data for your gender as a time series.

b Calculate and plot four-point moving averages and draw in a trend line.

c Discuss whether there are any seasonal effects present.

d Use the trend line to estimate the figure for 2006.

7 The table shows natural gas production from 2004–2008 for BP.

Units	Q1 2004	Q2 2004	Q3 2004	Q4 2004	Q1 2005	Q2 2005	Q3 2005	Q4 2005
Natural gas production (c)	8600	8425	8275	8714	8745	8661	7841	8458

Units	Q1 2006	Q2 2006	Q3 2006	Q4 2006	Q1 2007	Q2 2007	Q3 2007	Q4 2007
Natural gas production (c)	8713	8624	8086	8256	8502	7859	7879	8123

Source : www.investis.com/bp_acc_ia/

a Display this data as a time series graph.

b Calculate and plot appropriate moving averages for the data.

c Draw a trend line.

d Estimate the average seasonal effect for Quarter 1.

e Use your answer to parts (c) and (d) to make a seasonally adjusted estimate for the value for Q1 2008.

f Write down two features of the data.

Summary

You should:

- know that a time series graph shows data over a period of time
- know that moving averages are used to find trends in data where seasonal effects are present
- know that trend lines are lines of best fit through data or moving average plots
- know that average seasonal effects can be calculated using trend lines and used to predict future values
- know that Z-charts show three aspects of time series data on the same chart: actual values, cumulative frequencies and moving totals.

AQA Examination-style questions

1. The owner of a small cinema changed the film shown every three weeks. He recorded the attendance at the cinema each week for ten weeks. His results are given in the table.

Week number	Attendance	Three-point moving average
1	382	
2	356	336
3	270	
4	394	342
5	362	352
6	300	
7	463	380
8	377	385
9	315	
10	505	

Some of the three-point moving average values for the attendance figures have been calculated.

(a) Calculate the three missing values and write them in the table. *(4 marks)*

(b) Why is it appropriate to calculate three-point moving average values? *(1 mark)*

The attendances have been plotted on the graph below.

(c) Plot the moving average values. *(2 marks)*

(d) (i) Draw a trend line on the graph. *(1 mark)*

(ii) Comment on what your trend line shows. *(1 mark)*

(iii) Give a reason why this trend may not be expected to continue. *(1 mark)*

AQA, 2002

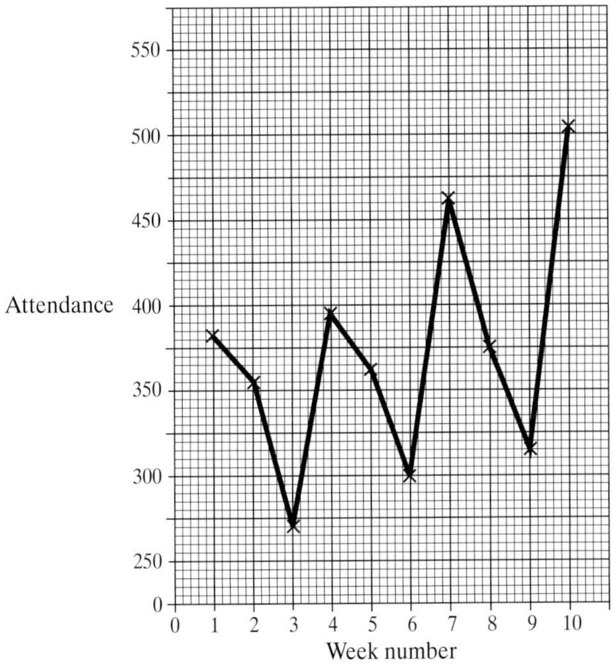

2. The table shows the number of visits to America by UK residents each quarter from 2002 to 2004.

Year	Quarter	Visits (tens of thousands)
2002	Q1	88
	Q2	100
	Q3	118
	Q4	114
2003	Q1	98
	Q2	106
	Q3	124
	Q4	118
2004	Q1	102
	Q2	116
	Q3	134
	Q4	126

Source: Adapted from Social Trends 2005

(a) The data for 2002 have been plotted on the time series graph.

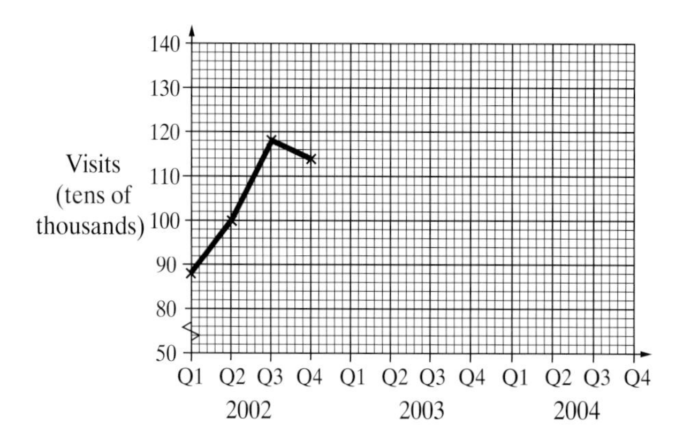

Complete the graph. *(2 marks)*

(b) Describe two different patterns in the data. *(2 marks)*

(c) Calculate the four-point moving averages. *(3 marks)*

(d) (i) Plot the moving averages on your graph. *(2 marks)*

(ii) Draw a trend line. *(1 mark)*

(e) Work out the average seasonal effect for Q1, the first quarter. *(3 marks)*

(f) Use the trend line and your answer in part (e) to estimate the number of visits to America by UK residents in the first quarter of 2005. *(3 marks)*

AQA, 2003

3. The number of sales made by a shop from 1999–2001 are given in the table below.

Period	Sales (1000s)	Three-point moving average
Jan–April 1999	30	
May–August 1999	27	30
Sept–Dec 1999	33	31
Jan–April 2000	33	33
May–August 2000	30	33.3
Sept–Dec 2000	37	34.3
Jan–April 2001	36	
May–August 2001	32	
Sept–Dec 2001	40	

(a) Give **one** reason why three-point moving averages are appropriate. *(1 mark)*

(b) Calculate the value of the next two moving averages and write them in the table. *(3 marks)*

(c) Calculate the average seasonal variation for the periods May–August. *(2 marks)*

(d) Draw the graph below and plot all the three-point moving averages. *(2 marks)*

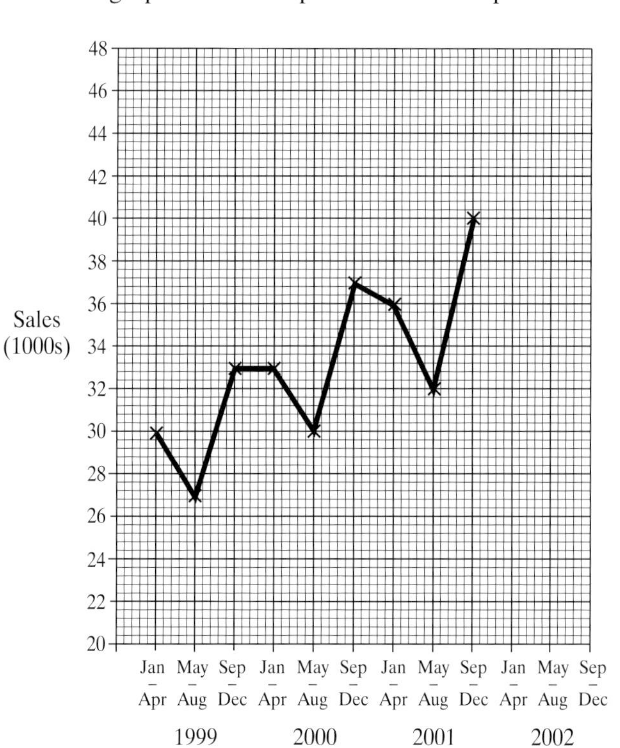

(e) Predict the sales for May–August 2002. *(3 marks)*

AQA, 2003

4. The number of weddings, in thousands, for 14 consecutive quarters are given in the table. Some of the four-point moving averages have been calculated.

Year	Quarter	Number of weddings (thousands)	Four-point moving average (1 d.p.)
1996	1	41.0	
1996	2	91.4	
			79.4
1996	3	129.4	
			79.0
1996	4	55.8	
			77.9
1998	1	39.3	
			77.8
1997	2	87.1	
			77.6
1997	3	128.9	
			77.2
1997	4	54.9	
			76.8
1998	1	37.7	
			75.9
1998	2	85.6	
			76.2
1998	3	125.5	
1998	4	56.0	
1999	1	36.9	
1999	2	83.2	

(a) Calculate the value of the next two four-point moving averages and put them in the table. *(3 marks)*

(b) The original data is plotted on the grid below. Plot all the four-point moving averages on the same grid. *(2 marks)*

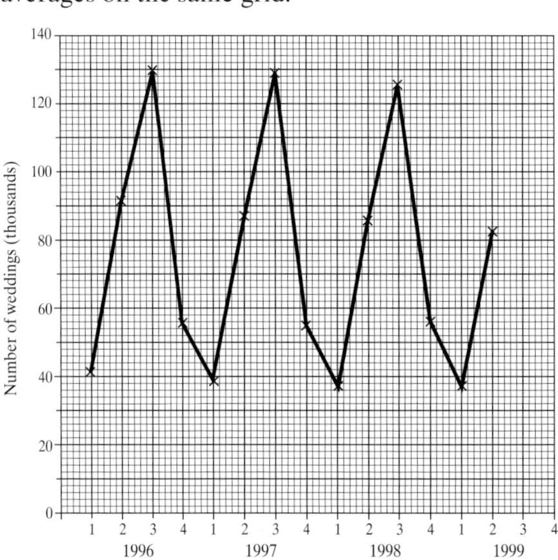

(c) The seasonal variations for quarter 3 are 52 000, 52 000 and 49 000.

Use this information together with a trend line to obtain an estimate for the number of weddings in quarter 3 of 1999. *(4 marks)*

AQA, 2005

10 Further statistical diagrams

In this chapter you will learn:

- to recognise poorly presented diagrams including misuse of scale, areas or volumes
- to check, through calculation, that appropriate values are used in diagrams involving scale, area and volume
- how to draw and interpret choropleth maps
- how to draw and interpret output gap charts
- how to transform from one type of diagram to another.

You should already know:

✓ about all the types of diagram you have met in earlier chapters.

What's this chapter all about?

The first part of this chapter is in many ways a survival guide for looking at graphs and diagrams in newspapers and magazines and on the television and internet. Sadly, many of the people who seem to be responsible for producing the output of diagrams in these media do not seem to understand how many of the diagrams you know and love should be presented. Even more importantly, some people do know how to present such information and will use their skills to produce a diagram that will look convincing and correct but will have significant flaws. A skilled statistician can present the same data in many different ways, which may be attractive to different people at different times – if you can't beat them make sure you join them and avoid being conned!

Later in this chapter we tidy up the course with a few diagrams that didn't really fit anywhere else. But don't miss them as they are not covered in maths and you will need to learn them thoroughly!

Focus on statistics

Sometimes in maths you might have considered the good or bad points of a diagram but much of 10.1 will be in more depth than this and the consideration of the use of area and volume (at higher tier) will be new.

All of 10.2 will be new, except where data is transformed from one type of graph to another, where both happen to be graphs also used in maths.

In other words, there is not much here that you will have met before.

10.1 Misleading and poorly presented diagrams

The world is full of statistical diagrams. Newspapers, the internet and television news are just some of the places where you will meet a multitude of different forms of data presentation. It is important that you are fully aware of the accidental or deliberate occurrences of misleading and poorly presented diagrams that are common in these media. Having this awareness improves your own capabilities as a statistician.

Objectives

In this topic you will learn:

- to recognise poorly presented diagrams including misuse of scale, areas or volumes
- to check, through calculation, that appropriate values are used in diagrams involving scale, area and volume.

Poorly presented diagrams

1 Misuse of scales through:

a Not starting a scale at zero. For example, see the diagram on the right.

The impression given by this diagram is that the March rainfall is extremely high, and way above average. But have a look at the vertical scale, if this was plotted on a graph which was scaled from 0 on the vertical axis, you would see that the difference between average and actual rainfall for March was tiny.

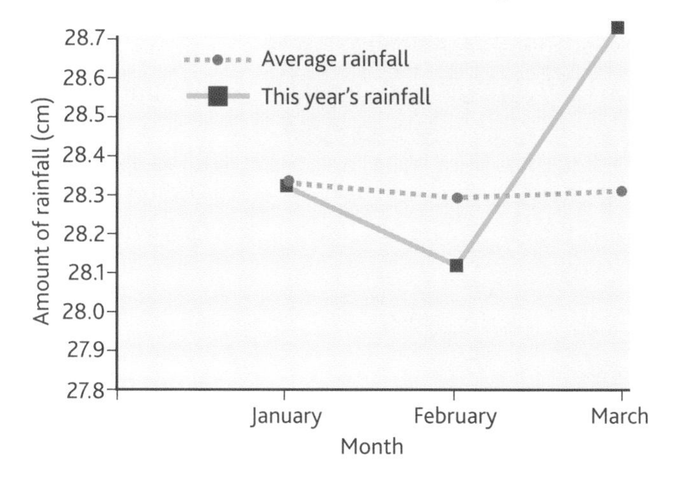

b Having a scale numbered unevenly

i *Vertical scale*

The effect of having a larger gap between 0 and 10 than between 10 and 20, and 20 and 30 changes the apparent difference between the results significantly.

ii *Horizontal scale*

Look at the dates on the horizontal scale. Approximately the same length of scale represents 1938 to 1968 (30 years) as then represents 1968 to 1981 (13 years).

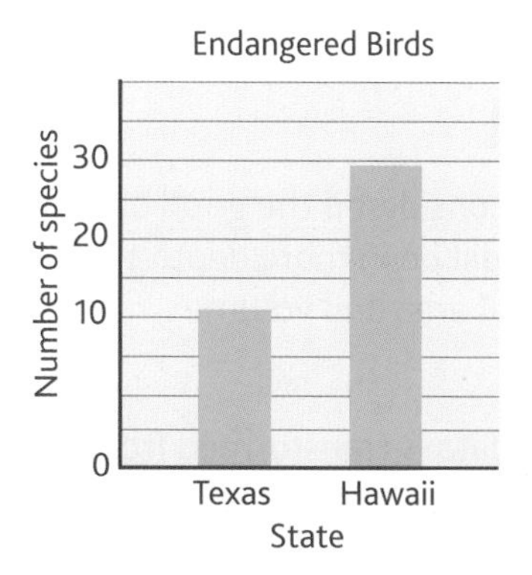

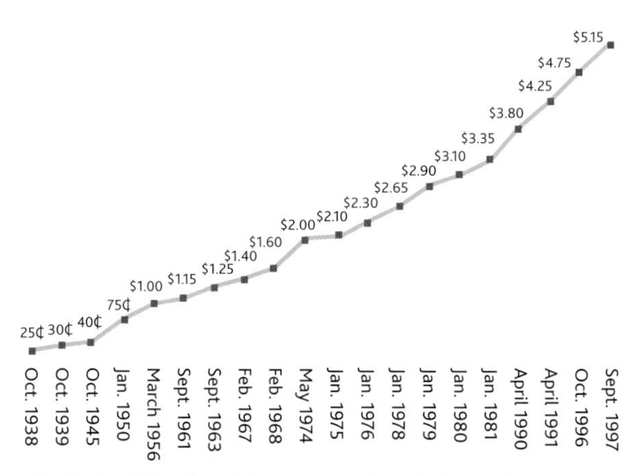

The Federal hourly minimum wage since its inception

2 Missing information

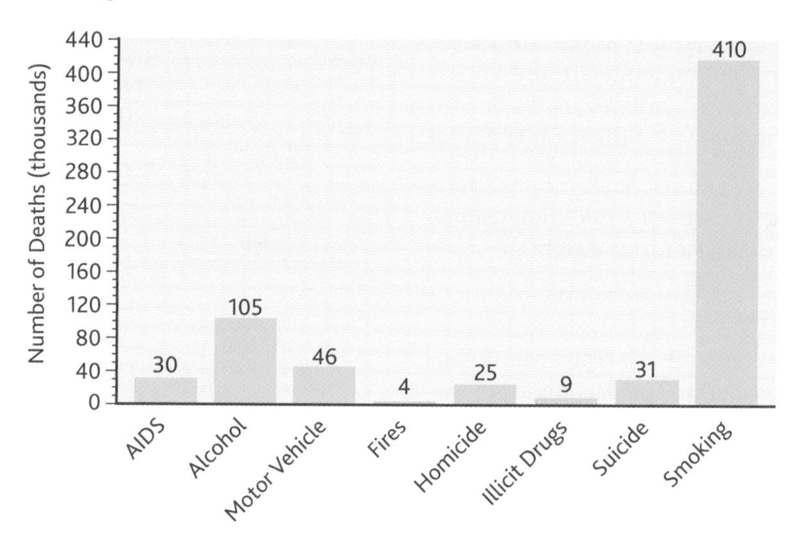

Whilst there is no doubt that smoking is a massive cause of death amongst people, its importance is exaggerated by the fact that other major causes of death, such as heart attacks, are missing.

3 Misuse of area and volume

Newspapers especially are often guilty of using inappropriate diagrams when comparing data. The effect of this is usually to exaggerate differences beyond those that actually exist. For example, see the pie chart on job security.

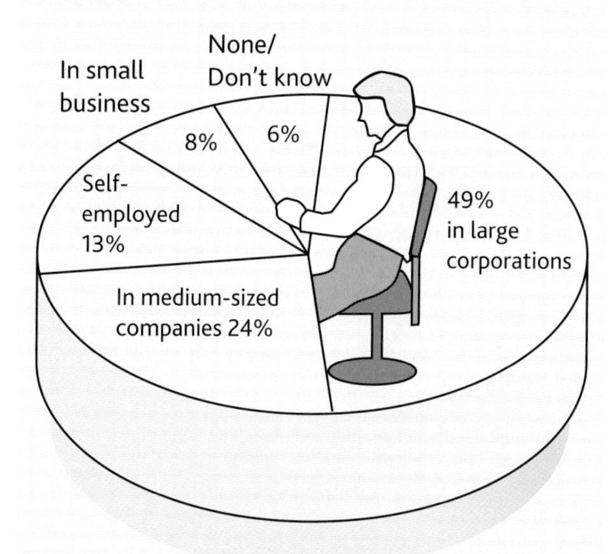

4 Other issues are regularly seen, for example:

- making graphs so complicated that they are difficult to understand
- presenting data as having a correlation, when the order of the plots is arbitrary
- presenting a very small amount of data as representing a general trend.

Be a statistician

You are a statistician and you are going to deliver a short talk to a group of journalists about how to present statistics correctly in their newspapers. Prepare the first part of this presentation – a set of five or so examples of diagrams from newspapers that are misleading or inaccurate, or simply not good statistics in some way. Remember to write down in your script the reasons why you are unhappy with each diagram. Perhaps you could present to the rest of your class, or invite a reporter from a local newspaper!

AQA Examiner's tip

Be as precise as possible on questions asking for errors in diagrams. Students often lose marks for making vague comments like 'the scale is wrong'. Say why the scale is wrong.

Calculating areas and volumes for scales

Recall from your earlier work on histograms that it is important that the area of a rectangle is in proportion to the frequency it represents. This idea is the same for any type of diagram where area, or equivalently volume, is used as a factor in the way a diagram is drawn.

You will only be asked to work in simple situations, and will need to be able to calculate appropriate dimensions for two- or three-dimensional objects. For example, consider the simple data set to the right.

Day	Number of items sold
Friday	12
Saturday	24
Sunday	48

To show this data in a standard bar chart would quite simply require bars of equal width drawn to heights of 10, 20 and 40 on a scaled axis – the heights of the bars are in the ratio 1 : 2 : 4. However, if you decided, for whatever reason, that you wished to use rectangles of varying widths to display the data, the areas of the rectangles would need to be in the ratio 1 : 2 : 4.

Therefore if the rectangle for Friday was a 6 cm high by 2 cm wide rectangle, giving an area of 12 cm^2, the rectangle for Saturday would need an area of 24 cm^2, and for Sunday 48 cm^2. This is easily achieved in the bar chart case, as the widths of the bars are the same, leaving the heights to dictate the areas. However, if the Saturday rectangle was drawn 3 cm wide, instead of 2 cm wide, the height of the rectangle would need to be 8 cm. Given either dimension, x, for the Saturday rectangle, $24 \div x$ would be the other dimension.

If a three-dimensional shape, such as a cuboid, is used, then you would need the volumes to be in the correct ratio.

Therefore, if you decided to use a 2 by 3 by 4 cuboid for Friday, then you have determined that a volume of 24 cm^3 will represent 12 items. So for Saturday to represent 24 items, you would need a symbol with a volume of 48 cm^3, and for Sunday a volume of 96 cm^3. This then keeps the volumes in the correct ratio of 1 : 2 : 4 . Once you have drawn two of the dimensions for the cuboid, the third is fixed by this need to have the correct volume.

Example

To compare sales between offices selling houses, two cuboids are drawn.

Office A sold 50 houses and has a cuboid dawn with dimensions 10 cm by 5 cm by 2 cm.

Office B sold 35 houses.

a What should the volume be for the cuboid drawn to represent Office B?

b If the Office B cuboid is 8 cm wide and 6 cm high, what should the other dimension be?

Solution

a The volume of Office A cuboid is $10 \times 5 \times 2 = 100$ cm^3

So the volume of Office B cuboid should be $\frac{35}{50} \times 100 = 70$ cm^3

b Let the value of the third dimension on this cuboid be x.

Then $6 \times 8 \times x = 70$

so $x = 70 \div 48 = 1.45833\ldots$

$= 1.46$ cm (to 2 d.p.)

Exercise 10.1

1 Draw the misleading graph for the Federal Minimum Wage on correctly scaled axes, labelled with years on the horizontal axis (ignore the months). What do you notice about the data that could not be seen from the misleading graph?

2

a Describe why each of the following graphs are in some way misleading.

b Say what steps you would take to present the data in a more appropriate manner.

i

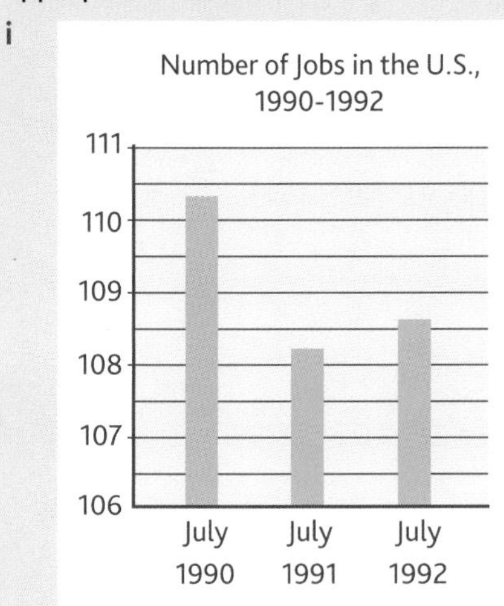

ii

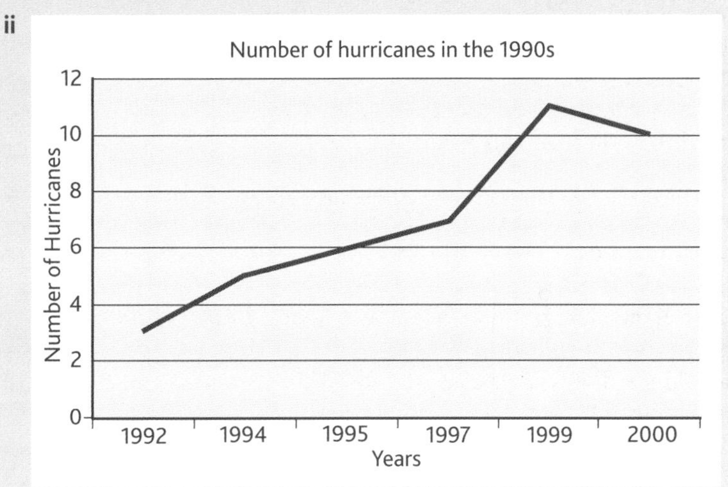

iii

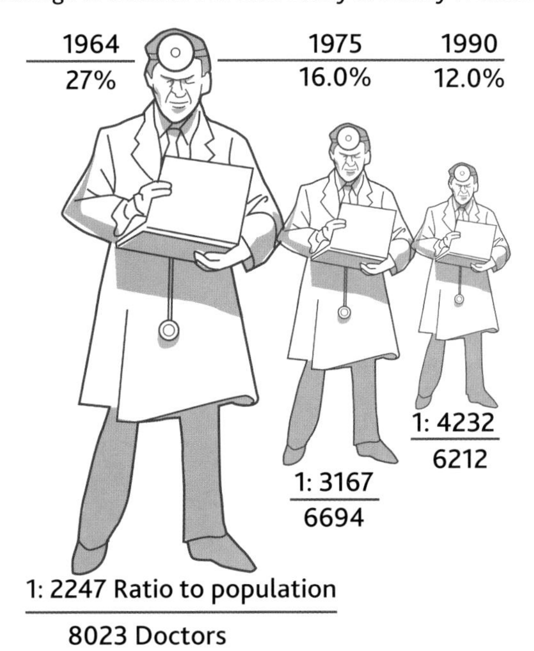

iv

Traffic fatalities

330 320 310 300 290 280 270 260 0

1955 1956

v

Total Expenditures on Health as a Percentage Share of GDP, by OECD Country, 2004

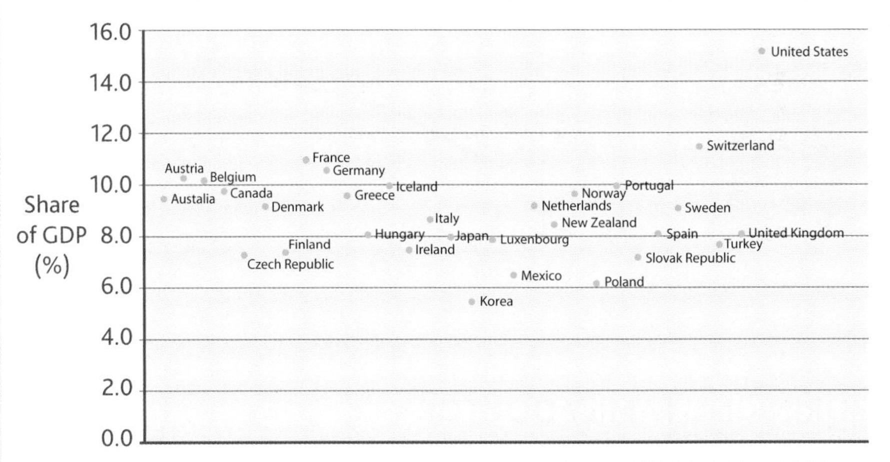

Source: OECD Health Data 2007

Note: For the United States the 2004 data reported here do not match the 2004 data point for the United States in Chart 1 since the OECD uses a slightly different of 'total expenditures on health' than that used in the National Health Expenditure Accounts.

3 John sells 20 televisions in one week at his shop. His assistant Jeff sells 16. To illustrate these sales, various graphs are drawn.

a A chart is drawn showing the sales as square television sets. John's square has a side length of 8 cm. What side length should Jeff's chart have?

b A chart showing the sales as two cubes is drawn. Jeff's cube has a side length of 5 cm. What side length should John's cube have?

4 Tasmin correctly scales two three-dimensional diagrams of a box of vegetables to compare the number of vegetable portions eaten by two classes of students in one week.

Class I has in total eaten 140 portions of vegetables and has a box of volume of 72 cm^3.

Class J has in total eaten 168 portions of vegetables and has a box drawn with length 8 cm and width 3 cm.

Find, to two decimal places, the height of the diagram for Class J.

10.2 Other statistical diagrams

Choropleth maps

A **choropleth map** is a diagram used to show the distribution of items across a geographical area. A choropleth map is sometimes called a shading map. They are often used for population density and other rates such as birth rates and GDP.

This is an example of a typical choropleth map:

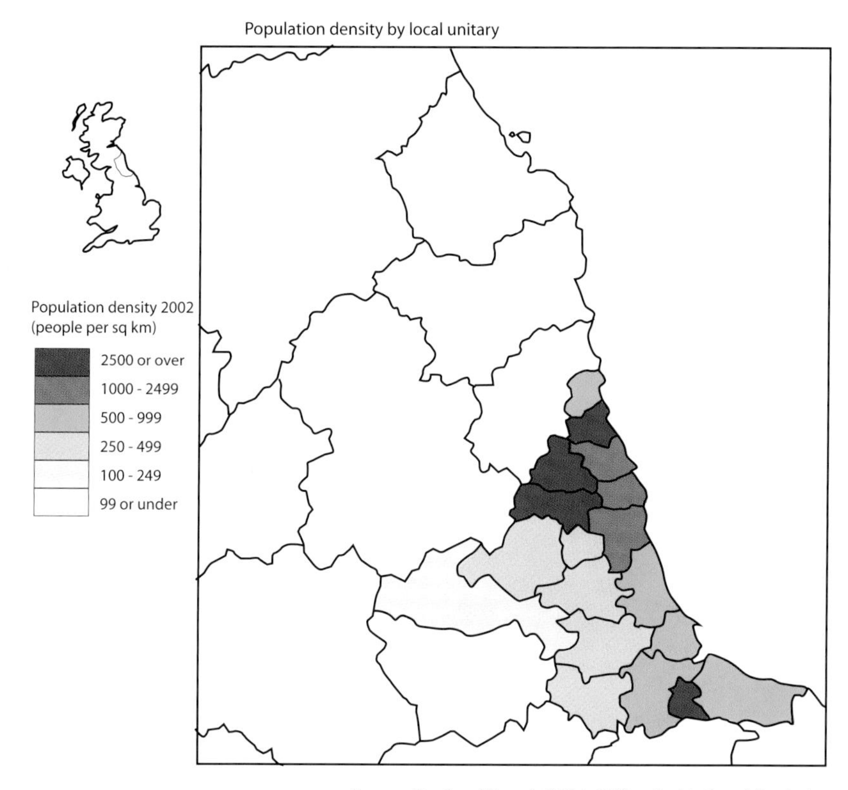

Source: Regional Trends 2004, Office for National Statistics

This map gives an instant visual impression that the most heavily populated areas are congregated around the same coastal area.

To show the development of a choropleth map from scratch, consider this situation. A bag of marbles is emptied on to an enclosed grid of 36 squares. Each dot on Diagram A shows the point at which each of the marbles comes to rest.

The number of marbles coming to rest in each square on the grid is show by Diagram **B**.

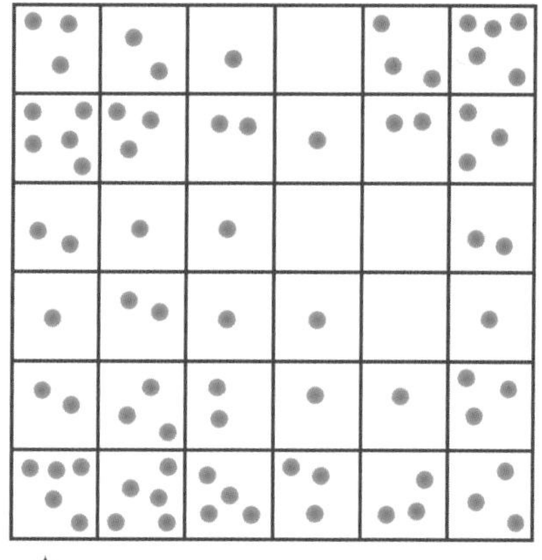

3	2	1	0	3	5
5	3	2	1	2	3
2	1	1	0	0	2
1	2	1	1	0	1
2	3	2	1	1	3
5	5	4	3	3	3

Objectives

In this topic you will learn:

- how to draw and interpret choropleth maps
- how to draw and interpret output gap charts
- how to transform from one type of diagram to another.

Key terms

Choropleth map: a diagram with different colours, or shades of colours, to represent different amounts of items over a given area, usually an area of land.

links

More on the meaning and calculation of rates such as birth rates can be found in Chapter 11.

You now use a key to allocate a colour or degree of shading to each result or group of results. You don't want too many or too few colours or shades on a choropleth map. Between three and six different groups is ideal.

Often one colour is used with different shades of increasing darkness to represent a greater density of item. Whatever you decide to do, a key is essential, otherwise your choropleth map will be meaningless.

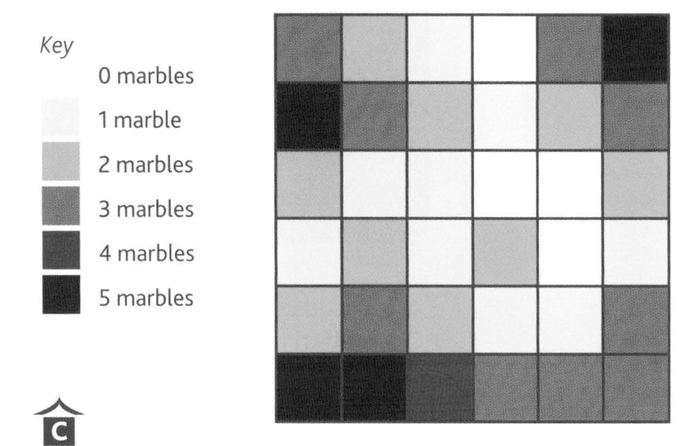

C

Often there are many more data values than this, meaning that the key will be for grouped values.

Many of the actual contexts where you use choropleth maps use continuous data anyway, thus the shading key will have to be for grouped ranges of values.

It is also important to be aware that you can to some extent control the look and effect of a choropleth map by your choice of colour, and the decisions you make regarding the groups you use.

Example 1

The two choropleth maps show the same data regarding numbers of males and females across the USA according to the 2000 Census.

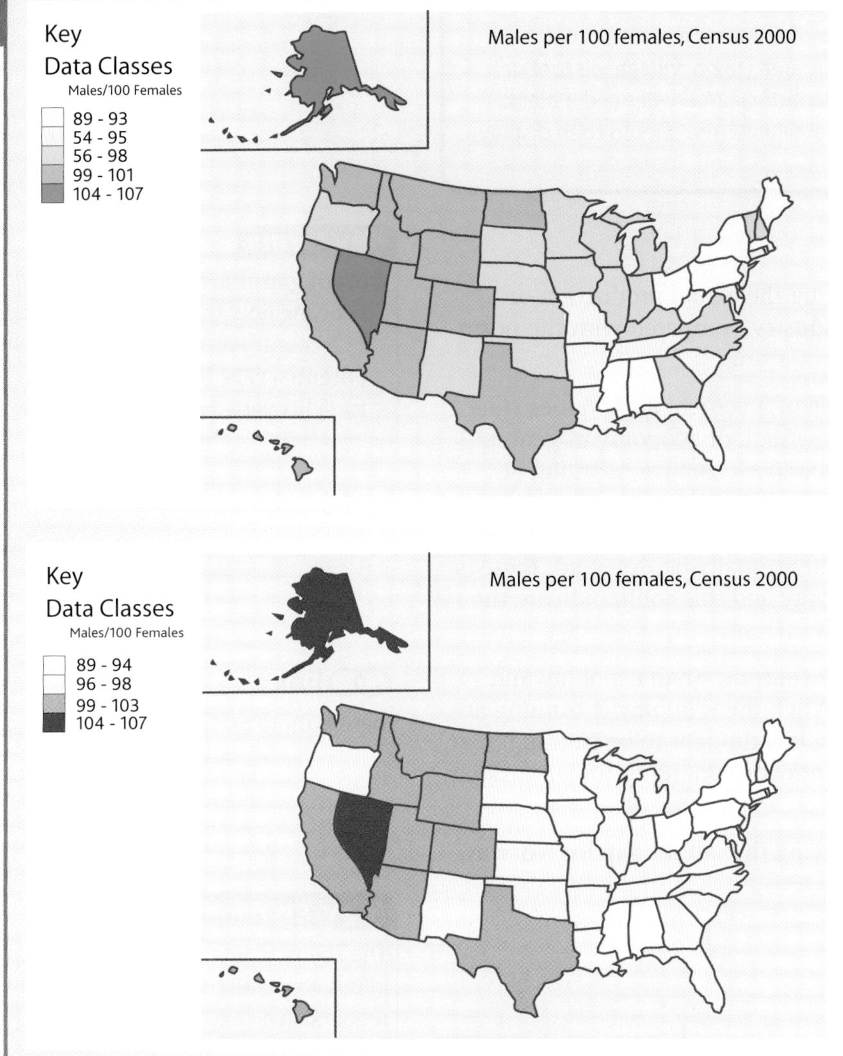

a Comment on the different construction of these choropleth maps.

b Write down three features of the data, as shown by one or both of the maps.

Solution

a The red shading provides a more stark presentation style and could be used (in different contexts) to really make an area stand out. The groupings for the two maps are different. This has had a particular effect on the representation for the north east of the country. Whilst it is only by one shading category, the grouping category of several states has been changed. If you were doing a presentation within one of these states, or working for a group with a vested interest in one of these areas, your choice would be very important.

b The majority of states have a greater proportion of women than men in their population. The states with higher proportions of men are on the western side of the country. No state has fewer than 89 men per 100 women.

Real world

Find some data relating to the areas around where you live. Possible ideas could include birth or death rates, crime figures, percentage pass rates at different qualification levels. Perhaps you could survey information about sizes of families in your area using samples of students in your school. You will need the figures that relate to your area (town, village, district or county) and that for neighbouring equivalent areas. You will need a blank map of the areas for which you have data available. You should then display this data using a choropleth map.

Output gap charts

Key terms

Output gap chart: a chart showing how the real output of a country is compared to the potential GDP for that country, usually over a period of time.

Output gap charts are a way of showing the actual production or output in an economy, or particular industry, compared with the norm for that economy or industry.

They are scaled with zero in the middle of a vertical axis. Values (lines) above zero show that the industry concerned, or economy in general, is performing above the standard level of performance, working at, or close to, full capacity – sometimes this will indicate the likelihood of a boom in an economy. Values (lines) below zero indicate that the performance is below that which would normally be expected; there is spare capacity in the industry or economy and this could indicate the likelihood of a slump.

The measure compares the actual GDP (output) of an economy and the potential GDP (efficient output). When the economy is running an output gap, either positive or negative, it is thought to be running at an inefficient rate, as the economy is either overworking or underworking its resources.

links

There is more on GDP in Chapter 11.

Here is a typical output gap chart showing the output gap for Norway.

AQA Examiner's tip

The ability to interpret the charts in this way is all that will be required in an examination.

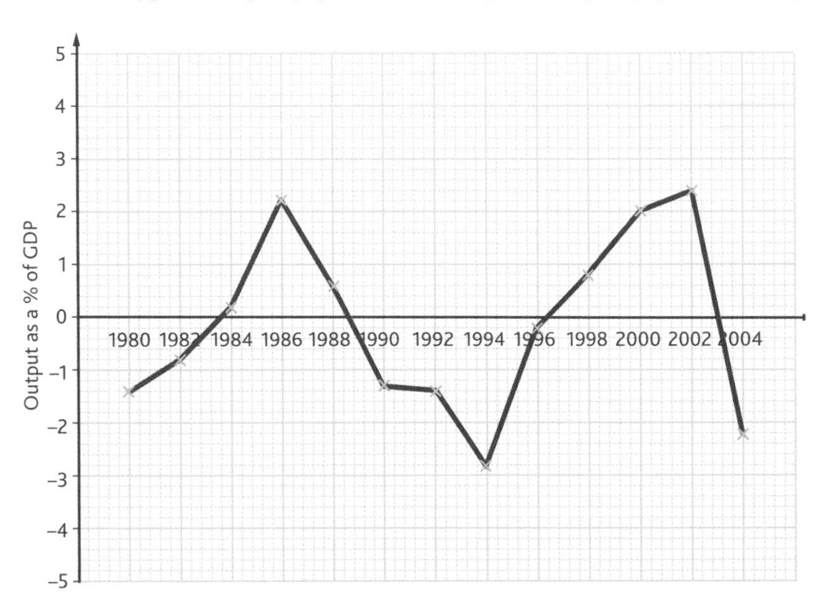

It can be seen that the periods of positive output gap are around 1984 to 1989, and 1996 to 2003.

Transforming from one diagram to another

Throughout this book you have met many types of diagram available to statisticians today. One of the skills you could be asked to demonstrate in the examination is to use your knowledge of the meaning and structure of one statistical diagram to transform the representation into another statistical diagram. It is important, however, that you ask yourself the question as to whether the form of presentation is appropriate.

Example 2

The pie chart shows the main ways in which 60 families say that they have tried to save energy in the last 12 months.

a Draw an appropriate frequency diagram for this data.

b Give one advantage of the frequency diagram over the pie chart.

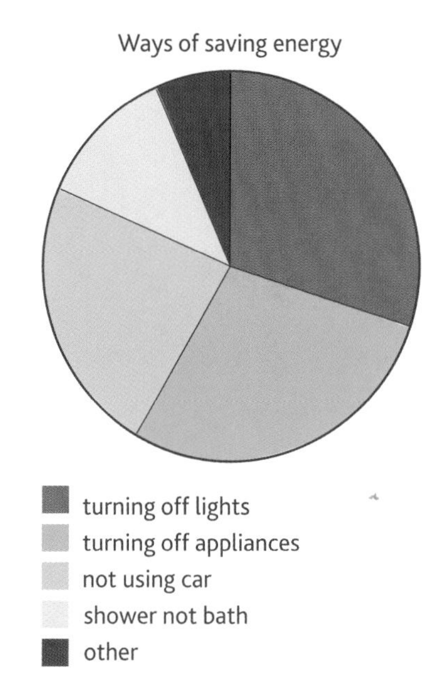

Solution

a Measure the angles for each of the sectors in the pie chart. For example, 'turning off lights' = 108 degrees. As the total frequency is given in the question as 60, you can now work out the number of families choosing each item.

For example, number 'turning off lights' $= \frac{108}{360} \times 60$

$= 18$ families

Similar calculations give the following frequencies:

turning off appliances = 17

not using car = 14

shower not bath = 7

other = 4.

These frequencies can now be used to prepare a labelled frequency diagram – the type of data (qualitative) supports the use of a bar chart as below.

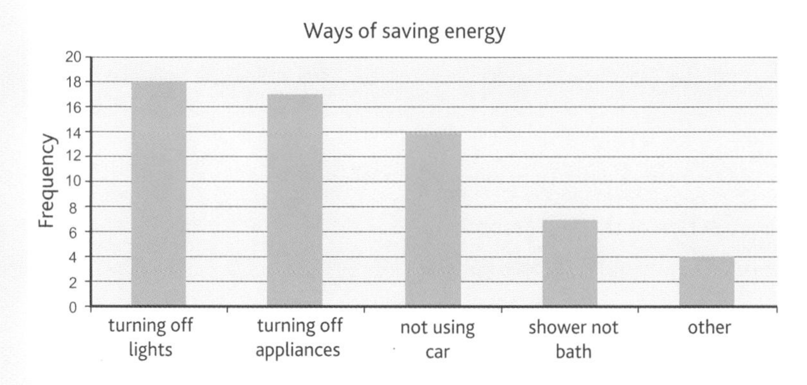

b It is easier to see the actual frequencies using the bar chart compared to the pie chart.

Spot the mistake

Lilith is the human resources manager for a publishing company. She has produced a correct histogram to show the age of her workforce. She now wishes to present something more visually interesting to the company directors. She decides to produce a pictogram from the histogram. Spot her mistake.

Exercise 10.2

1 The choropleth map shows the population density for the countries of the world.

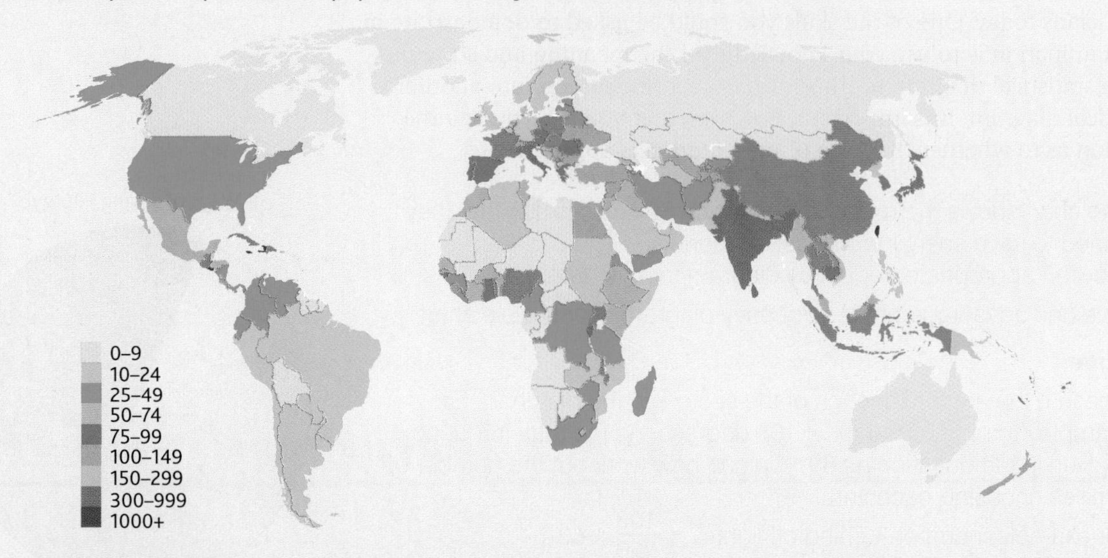

a Write down three features you notice about the population density around the world.

b Comment on the key used.

2 The two choropleth maps show population data for the same area of England.

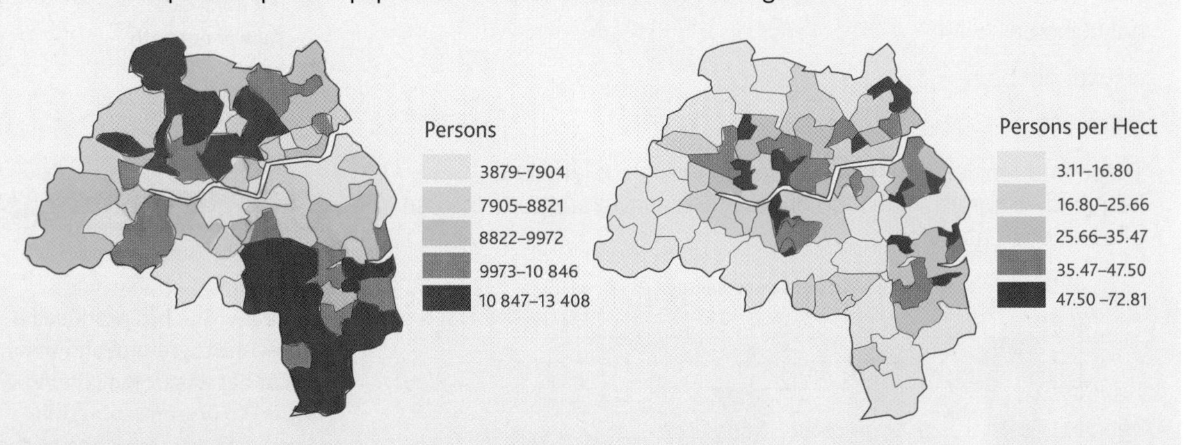

a Comment on the differences in the construction of the two graphs.

b Write down three features of the data revealed by one of the graphs, or by considering the graphs as a pair.

3 The grid shows the number of people in different parts of a beach.

a Use the key given to produce a choropleth map on a blank copy of the grid.

b There are a set of steps from the promenade down to the beach. On your answer to part (a), write 'S' in the rectangle where you think the steps are.

Promenade

23	38	26	21	14	8
22	42	22	21	12	5
18	27	15	13	8	4
14	21	11	10	4	2

Sea

0–9 people
10–19 people
20–29 people
30–39 people

4 During a geography field trip, students count the number of a particular type of small plant in each of several square metres of an area of land. The diagram shows the number of this type of plant in each square metre.

Draw a choropleth map to show this data. You will need to decide on suitable class widths. Remember to use a suitable key.

1	3	2	7	2			
3	5	6	8	6	3	3	1
6	9	11	13	12	7	8	2
4	8	14	15	18	12	7	4
2	6	13	16	18	13	6	3
	4	8	15	15	11	7	3
		5	8	10	8	5	2
			3	6	2	3	1

5 The frequency diagram shows the location of the 45 accidents in the home that occurred in a village over the course of one year.

a Draw a pie chart to represent the data.

b Give one advantage of the pie chart over the frequency diagram.

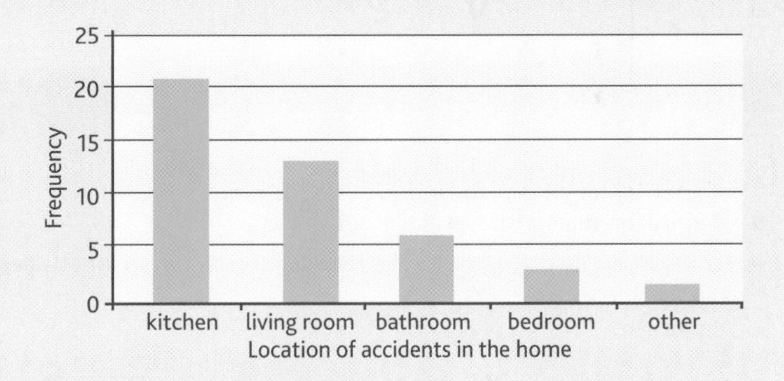

6 The scatter diagram shows the number of customers waiting in a queue at a bank and the waiting time of the person at the back of the queue taken at five-minute intervals over two hours.

a Draw a grouped frequency diagram for the waiting time of these 24 customers. Use four class intervals of equal width.

b Give one advantage of using the grouped frequency diagram over the scatter diagram for displaying information about these waiting times.

c Give one advantage of using the scatter diagram over the grouped frequency diagram for displaying information about these waiting times.

d The manager of the bank wanted information only about the number of people waiting at each of these 5-minute intervals.

i Suggest a suitable diagram that could be used to serve this purpose.

ii Draw your suggested diagram.

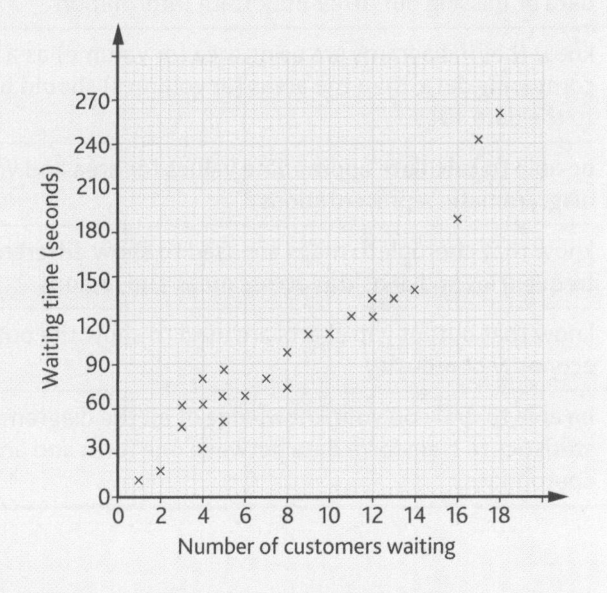

7 The output gap chart shows the output gap for Brownland for the years 1999–2009.

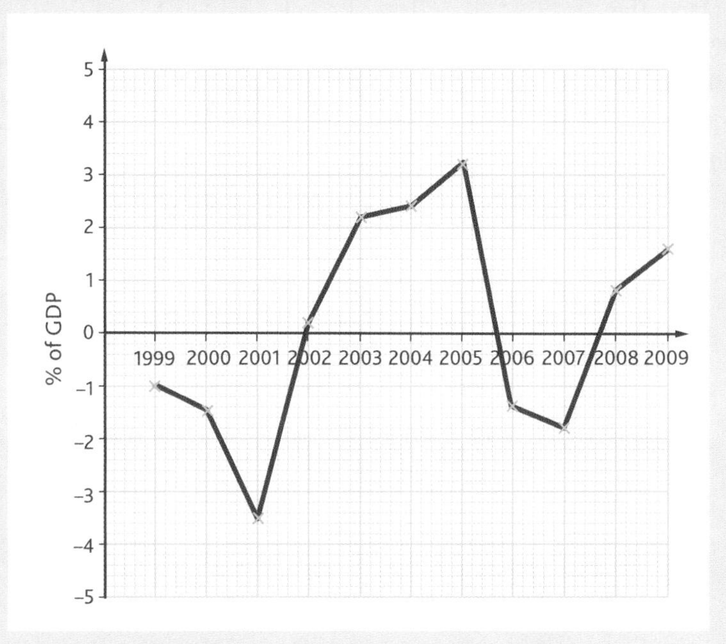

a Identify the years with a positive output gap.

b Identify the years with a negative output gap.

c In which year, and how much was, the largest output gap for this period in Brownland?

Summary

You should:
know that many graphs are misleading due to issues of scale, missing data or missing out other important information
know that if diagrams are using area (or volume) as a way of comparing data, then the areas (or volumes) should be in proportion to the data sizes
be able to calculate appropriate values for area and volume diagrammatic representations
know that choropleth maps are used to show differences in rates, frequencies and densities across areas and regions
know that output gap charts are used to show the output of an economy or industry
be able to draw on your knowledge of all the diagrams met in statistics to transform data between one form and another, if appropriate.

AQA Examination-style questions

1. The diagram shows a rectangular field divided into small square areas. The number in each square shows the number of sheep in that square.

5	7	11	6	3
3	6	7	4	2
0	2	4	3	1
0	0	0	2	2

(a) Use the key provided to produce a shading (choropleth) map on the blank copy of the field below. *(3 marks)*

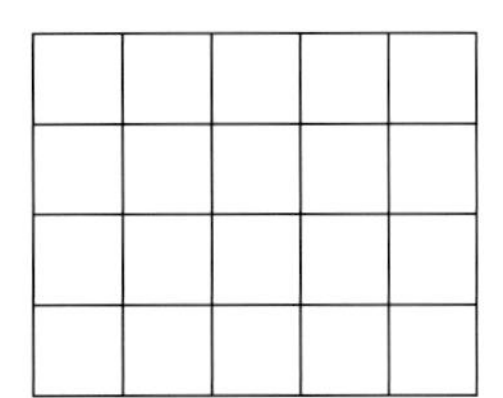

Key

- 0 sheep
- 1–4 sheep
- 5–8 sheep
- 9 or more sheep

(b) A sheepdog is sitting in the field. Mark with a 'D' the likely position of the sheepdog in the field. Give a reason for your answer. *(1 mark)*

AQA, 2007

2. The choropleth map (shading map) shows population changes in ten districts of a large city between 1991 and 2001.

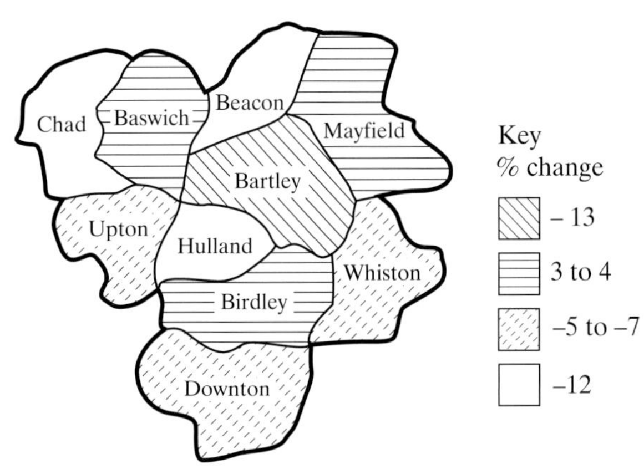

(a) In which district has there been the largest population change? *(1 mark)*

(b) How many districts show a decrease in population of 5% or more? *(1 mark)*

AQA, 2005

3. **(a)** The diagram shows the population of Hong Kong from 1993 to 1999.
Give two reasons why this diagram is misleading.

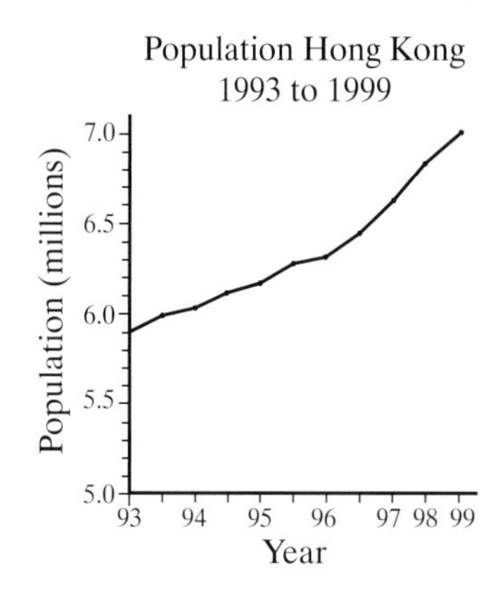

(2 marks)

(b) Explain why this graph does **not** show that drivers aged over 79 are the safest on the roads.

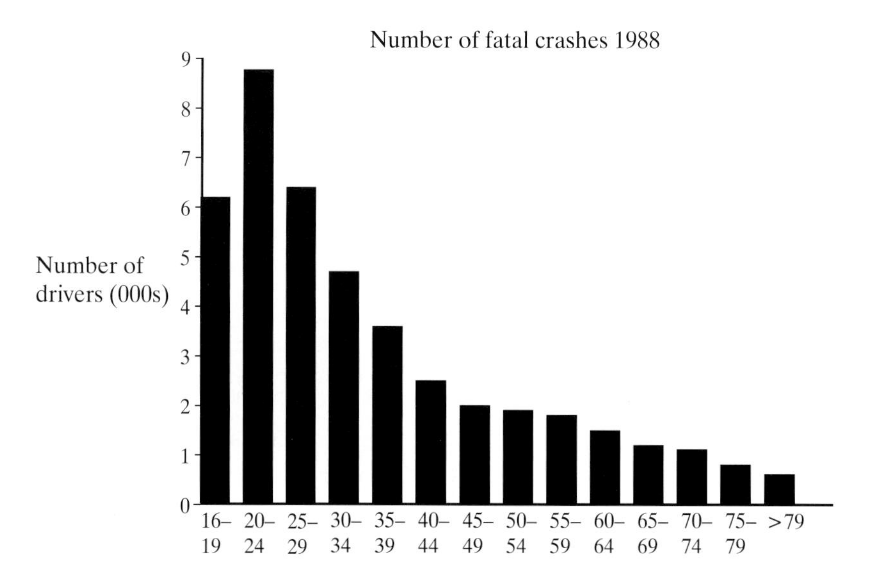

(2 marks)

AQA, 2007

4. The number of pupils in each unit area of a playground is shown below.

2	3	4	8	9	7	8	4
3	4	8	10	12	14	10	7
3	7	9	8	13	15	12	8
0	1	4	7	8	10	6	3
0	0	3	2	4	2	1	0

(a) Complete the choropleth map using the given key.

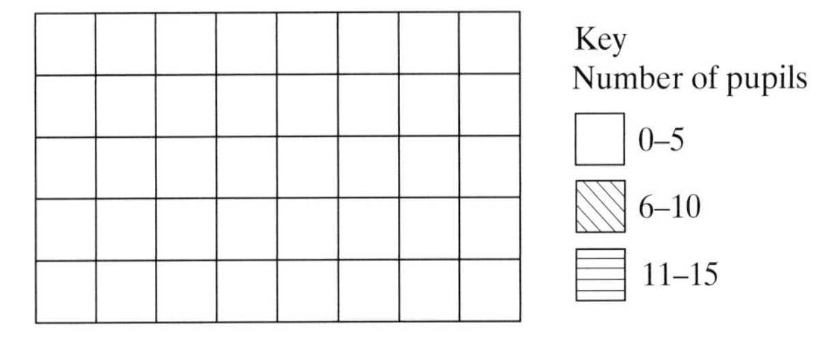

(2 marks)

(b) There is a teacher in the playground. Where do you think the teacher is? Explain your answer. *(1 mark)*

AQA, 2004

5. An estate agent has one house for sale at £100 000 and one at £150 000.
The diagram has been drawn to represent this information.

(a) Explain why the diagram is misleading in representing these house prices. *(1 mark)*

(b) The agent decides to represent the houses using a two-dimensional diagram.

The width representing the £100 000 house is 3 cm. Calculate the width representing the £150 000 house.

(3 marks)

AQA, 2004

6. The diagram shows mobile phone ownership in Hong Kong in 1994 and 1998.

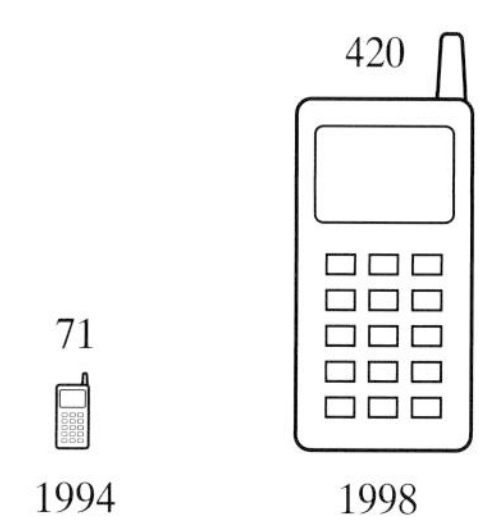

Give one criticism of this diagram.

(1 mark)

AQA, 2007

11 Economic and social statistics

In this chapter you will learn:

- how to calculate and interpret index numbers
- how to calculate and interpret weighted index numbers
- how to calculate and interpret chain base numbers
- about the Retail Price Index (RPI), Consumer Price Index (CPI) and gross domestic product (GDP)
- how to calculate and interpret crude rates
- how to interpret standardised rates and their difference to crude rates
- how to use control charts to check issues of quality assurance.

You should already know:

- ✔ how to calculate with percentages
- ✔ how to rearrange an equation.

What's this chapter all about?

In a nutshell, this chapter is about statistics in the real worlds of economics and demographics (the study of population). Index numbers tend to be used to describe changes in the price of items, so if the price of downloaded music, for example, has gone up, the index is over 100! The chapter also includes the most well known of the national and international measures of wealth and inflation. There is also a section on crude and standardised rates, which usually refer to births and deaths.

Focus on statistics

You will not have met any of this in maths, so it is a rather important chapter. It's also a good place for cunning statistics examiners to get some real meaty questions from!

11.1 Index numbers

Index numbers

An **index number** gives a measure of the change in a value, often a price, of an item or group of items over a period of time. The index number is a percentage given for the new value of an item where the original value or value in the **base year** is 100. To calculate an index number use this formula

$$\text{Index number} = \frac{\textit{current value of item}}{\textit{value of item in base year}} \times 100$$

Alternatively, it is the percentage of its previous value. So 110 means it is 110 per cent of its previous value (or 10 per cent higher) and 96 means it is 96 per cent of its previous value (or 4 per cent lower).

Index numbers above 100 indicate that a value has risen since the base year, whereas index numbers below 100 indicate a fall in values since the base year.

Objectives

In this topic you will learn:

- how to calculate and interpret index numbers
- how to calculate and interpret weighted index numbers
- how to calculate and interpret chain base numbers
- about the Retail Price Index (RPI), Consumer Price Index (CPI) and gross domestic product (GDP).

Key terms

Index number: $\frac{\textit{current value of item}}{\textit{value of item in base year}} \times 100$

Base year: the year that is taken as the time from which all other years are compared, with the base year having value of 100.

AQA Examiner's tip

If ever there was a typical question on index numbers, then Example 1 is it. The first part of the question asks you to find an index number, the second part then asks you to use a given index number. We set a lot of questions like this, so make sure you understand it!

Example 1

The price of a litre of petrol, as on 1 January, is given for four years in the table below:

Year	2005	2006	2007	2008
Price (pence)	96	102	???	124

a Calculate the index number for 2008 using 2005 as base year.

b The index number for 2007 using 2005 as base was 114.

Find, to the nearest penny, the price of a litre of petrol on January 1st 2007.

Solution

a Index number for 2008 using 2005 as base year $= \frac{124}{96} \times 100$

$= 129.1666\ldots$

$= 129.17$ (to 2 d.p.)

(This represents a rise of just over 29% in the price over that time.)

b Given that the index number is 114, you know that $114 = \frac{x}{96} \times 100$ where x is the price for 2007.

Rearranging this equation gives $x = \frac{114 \times 96}{100}$

$= 109.44$

So, to the nearest penny, the price for 2007 was 109p.

Real world

The price of petrol tends to vary quite a lot, and is allegedly dependent upon oil prices. For a period of a few months, keep a record of the price of petrol at the same petrol station once per week. Also use the internet to find out the price of a barrel of crude oil each week. Use index numbers to calculate indices for both petrol and the price of oil each week during the period that you collect data. Do the index numbers follow a similar pattern?

Weighted index numbers

Where the price of an item is made up of different elements, it is possible that these different elements may have differing rates of increase (or reduction) in price. This will matter if the elements are not all of the same importance in making up the price of the item (some are weighted more than others) and you will need to find a **weighted index number.**

A weighted index number requires you to find separate indices for each item, and then multiply each by its weight. These are then added up, and finally divided by the total weighting (often 100 as weights are often given as a percentage);

i.e. weighted index number $= \dfrac{\Sigma\, (\textit{index number} \times \textit{weight})}{\Sigma\, \textit{weights}}$

Key terms

Weighted index number: if an index number is calculated for items made up of many sub items with differing importance (weighting), a weighted index number will take into account the importance of each subitem in the calculation.

Example 2

A company advertises through newspapers and on television. The costs of typical adverts using these two media are shown in the table below for 2008 and 2009.

Year	2008	2009
Cost of a newspaper advert	£120	£125
Cost of a TV advert	£480	£560

Given that this company give a weighting of 75 : 25 on newspapers: TV, find a weighted index number for 2009 using 2008 as base.

Solution

First of all find the separate index numbers for the two elements.

For newspapers, index number $= \dfrac{125}{120} \times 100 = 104.1667$

For TV, index number $= \dfrac{560}{480} \times 100 = 116.667$

Now the weighted index number $= \dfrac{(104.1667 \times 75) + (116.667 \times 25)}{(75 + 25)}$

$= 107.291$

Notice how the much higher weighting of the newspaper has meant that the overall index shows a value much closer to the newspaper's individual value, than the TV's individual value.

Chain base numbers

Index numbers in a given situation tend to refer back to a set base year. If you want to have an idea of changes on a year-by-year basis then you can use **chain base numbers**.

Chain base numbers use the previous year as the base each time, thus chain base numbers immediately tell you whether a price is higher or lower than the preceding year.

Key terms

Chain base numbers: a set of index numbers where the base year changes each time to be taken as the year before the year for which the chain base number is found.

RPI (Retail Price Index): the RPI is a measure of inflation used to determine changes in costs and prices.

Example 3

Consider Example 1 once more. The price of a litre of petrol on 1 January is given for four years.

Year	2005	2006	2007	2008
Price (pence)	96	102	109	124

Calculate the set of chain base numbers for this data.

Solution

There will be three chain base numbers for this data.

The first will be the index for 2006 using 2005 as base $= \frac{102}{96} \times 100 = 106.25$

The second will be the index for 2007 using 2006 as base $= \frac{109}{102} \times 100 = 106.86$

The third will be the index for 2008 using 2007 as base $= \frac{124}{109} \times 100 = 113.76$

This reveals that not only has the price gone up each year compared to the previous ones but the rate of increase is itself increasing as the chain base numbers are getting bigger.

Well-known indices and measures

Your GCSE statistics course requires that you know several measures and indices that are used in the world of economics and business.

RPI (Retail Price Index)

This index is the most familiar measure of inflation in the United Kingdom. It measures increases (or reductions) in the cost of household items. The **RPI** is regularly used by the government to award increases in pensions and other benefits. Unions often use it for bargaining in wage negotiations. Several versions are available – the 'all items' means what it says where everything is taken into account. There are 'all items excluding mortgage payments' and others with different aspects excluded. It is important to be aware which value is being used. The RPI is a weighted index with the weightings designed to reflect the 'average' family.

To the right is an extract of the all items RPI data using January 1987 as a base of 100. The data is monthly for 2007 and the first eight months of 2008. You can see that most months' prices rise but, for example, in July 2007 prices were lower than the previous month.

links

View the full data sets going back to 1947 at **www.statistics.gov.uk/StatBase/tsdataset.asp?vlnk=229&More=N&All=Y**

2007 01	201.6	2008 01	209.8
2007 02	203.1	2008 02	211.4
2007 03	204.4	2008 03	212.1
2007 04	205.4	2008 04	214.0
2007 05	206.2	2008 05	215.1
2007 06	207.3	2008 06	216.8
2007 07	206.1	2008 07	216.5
2007 08	207.3	2008 08	217.2
2007 09	208.0		
2007 10	208.9		
2007 11	209.7		
2007 12	210.9		

CPI (Consumer Price Index)

The **CPI** is a similar measure to the RPI and is also calculated each month.

This index uses a sample of the costs of goods and services that a typical household might buy such as food, energy and travel.

The 'annual rate of inflation' is the popular name given to the yearly change in the prices of this sample of goods and again, like the RPI, this is often used in such areas as wage negotiations.

Given their similarity it is important to know the differences between the CPI and the RPI. The CPI excludes a number of items that are included in the RPI and these are mainly related to housing, for example council tax, mortgage interest payments and insurances. An extremely detailed breakdown of what is included in these indices can be found at **www.statistics.gov.uk/cci/article.asp?id=1059**.

Key terms

CPI (Consumer Price Index): the CPI is an increasingly used measure of inflation similar to but containing some different items from the RPI.

GDP (gross domestic product): the GDP is a measure of the income and output of a country's economy.

GDP (gross domestic product)

The **GDP** is a measure of a country's national income and output based on its economy. It is the value of all the goods and services produced within the country in a given time (usually a year).

The larger the GDP of a country, the 'richer' the country is considered to be, although remember from the output gap charts in the previous chapter that it is not necessarily healthy for an economy to produce beyond its reasonable capabilities. The following is a link to a spreadsheet listing the GDP in millions of dollars for every country in the world over quite a long period of time:

www.ggdc.net/Maddison/Historical_Statistics/horizontal-file_03-2007.xls

A small extract is given below. This is used for one of the questions in Exercise 11.1.

Western Europe	Year 2000	Year 2001	Year 2002	Year 2003
Austria	167 860	169 255	170 886	173 311
Belgium	211 631	213 848	217 062	219 069
Denmark	122 580	123 444	124 015	124 781
Finland	100 952	101 990	104 237	106 749
France	1 263 467	1 289 387	1 305 136	1 315 601
Germany	1 560 098	1 579 443	1 580 379	1 577 423
Italy	1 084 305	1 103 435	1 107 664	1 110 691
Netherlands	343 756	348 661	348 928	348 464
Norway	112 907	115 985	117 268	118 591
Sweden	184 517	186 448	190 133	193 352
Switzerland	163 035	164 733	165 237	164 773
United Kingdom	1 199 910	1 227 529	1 249 236	1 280 625
Total 12 Western Europe	**6 515 016**	**6 624 158**	**6 680 180**	**6 733 431**

Be a statistician

Use this link to investigate values of the CPI for a time period of your choice:
www.statistics.gov.uk/statbase/tsdataset.asp?vlnk=7174& More=N&All=Y
The D7G7 column gives the well known 'annual rate of inflation' figures, first year by year and then month by month. Use calculations and graphs to illustrate patterns in the CPI data as if you were presenting a statistical report to your teacher.

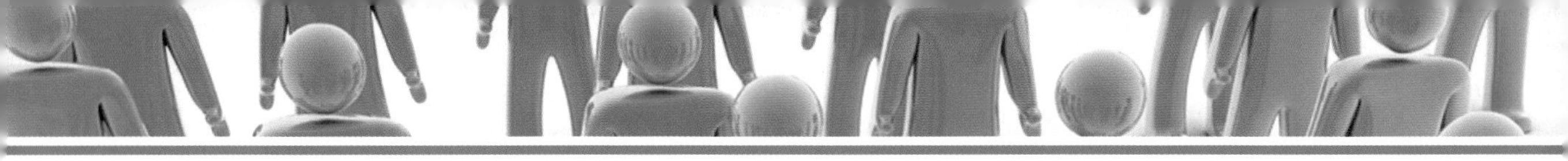

Other measures

You should be aware of other measures that are used internationally to measure similarities and differences between countries. This could include population density or GNP (gross national product) but because other measures are not given in the specification they would need to be defined if they were used in a question.

Exercise 11.1

1 The price of a child's toy over five years is given in the table.

Year	2004	2005	2006	2007	2008
Price (£)	39	35	29	29	32

a Find the index number for the 2005 price using 2004 as base.
b Find the index number for the 2008 price using 2007 as base.
c The index number for 2009 using 2008 as base was 105. Work out the price of the toy in 2009.

2 The index numbers for the number of visitors to a museum are given for five years. Each index number uses the year 2000 as base.

Year	2000	2001	2002	2003	2004
Price (£)	100	104	110	106	120

a In 2000 there were 24 000 visitors.
  i How many visitors were there in 2001?
  ii How many visitors were there in 2004?
b Between which two years was there a drop in the number of visitors? Explain your answer.
c In 2005 there were 31 200 visitors. Find the index number for 2005 using 2000 as base.

3 The cost of various grocery items at intervals over the last 100 years or so is given in the table.

	1914[1]	1950[1]	1975	2000
250 g cheddar cheese	2	3	25	126
500 g margarine	3	5	see part (d)	80
250 g butter (home produced)	3	6	18	82
Half dozen eggs (size 2)	3	11	21	84
125 g loose tea	2	5	11	81
1 kg granulated sugar	2	5	25	55
800 g white sliced bread	1	see part (e)	16	52
1 kg old potatoes	1	1	12	67
1 pint pasteurised milk	1	2	7	34

[1] Prices and weights are given to the nearest decimal equivalents.

Source: Office for National Statistics

Use the table to answer the following questions.
a Using 1914 as base, find the index number for half a dozen eggs in 1975.
b Using 1950 as base, find the index number for 125 g of loose tea in 1975.
c Using index numbers, or otherwise, which grocery item had the biggest percentage increase between 1914 and 2000?
d Using 1950 as base, the index number for 500 g margarine for 1975 is 460. Find the cost of the margarine in 1975.
e Using 1950 as base, the index number for 800 g white sliced bread in 1975 is 800. Find the cost of the bread in 1950.

4 Use the data from the table in Q1 to calculate the set of chain base numbers for the price of the child's toy.

5 Use the data from the table in Q2 to calculate the set of chain base numbers for the number of visitors to the museum.

6 Use the data from the GDP table just before this exercise. Calculate chain base numbers for the UK GDP over the given time period.

7 The chain base numbers for successive years for the cost of a games console is given as 100, 125, 110, 80, 40.

Describe how the price of this games console has changed over this period. Use percentages in your answer.

8 A school budget is 'spent' using 85 per cent for staff costs, 12 per cent for energy and services and 3 per cent on other items. In 2008 the indices for these items using 2007 as base were: staff costs 103, energy and services 118 and other items 106. Calculate the weighted index for the school budget spending for 2008 using 2007 as base.

9 The table shows the weightings of the different items used to calculate the CPI in 2005.

a Large increases in the prices in which items would cause the greatest rise in the CPI?

b Large increases in the prices in which items would cause the least rise in the CPI?

c The index numbers for each of the 12 items are given in the table below. Use all the information to calculate a weighted index number for the CPI 2005.

Allocation of items to CPI divisions in 2005

1	Food and non-alcoholic beverages	10.6
2	Alcohol and tobacco	4.6
3	Clothing and footwear	6.3
4	Housing and household services	10.5
5	Furniture and household goods	6.5
6	Health	2.4
7	Transport	14.8
8	Communication	2.5
9	Recreation and culture	15.1
10	Education	1.7
11	Restaurants and hotels	13.9
12	Miscellaneous goods and services	11.1
Total weighting		100

Item	Index Number	Item	Index Number	Item	Index Number
1	115	5	105	9	115
2	102	6	98	10	103
3	105	7	112	11	96
4	106	8	100	12	110

11.2 Crude and standardised rates

Crude rates

It is very important to know the rates at which certain events are happening. Births, deaths, marriages, unemployment and similar issues affect the way the country will run and information about them is necessary for any government to be able to plan for the future. Rates for all these events are found by comparing the number of instances of each event with the total population in the country.

A **crude rate** is the number of times an event occurs per thousand of the population.

For example, the crude birth rate $= \dfrac{\textit{number of live births recorded}}{\textit{total population}} \times 1000$

Example 1

The village of Blyton has a population of 3500. In 2008 there were seven births recorded. Calculate the crude birth rate for Blyton.

Solution

Crude birth rate $= \dfrac{\textit{number of births recorded}}{\textit{total population}} \times 1000$

$= \dfrac{7}{3500} \times 1000$

$= 2$

In other words, for every 1000 people in the village there were two births recorded.

Standardised rates

Some of these rates, such as death rates, are significantly affected by the age profile of the residents in a town or village. If there are lots of older people in the village you would expect more people to die there than in a village with a lot of younger people. Conversely, you would expect more births, and probably marriages, in a village with a younger population age profile.

Rates are often used to compare conditions for living in different places. Therefore for a fairer comparison to be made, it is important to take the age distribution into account.

Age distributions are taken into account in **standardised rates**.

A standardised rate takes an average of the individual crude rates for each age group, weighted according the numbers in those age groups.

Objectives

In this topic you will learn:

- how to calculate and interpret crude rates
- how to interpret standardised rates and their difference to crude rates.

Key terms

Crude rate: the number of times an event such as a birth or death occurs per thousand of the population.

Standardised rates: the number of times an event occurs per thousand of the population, taking into account the age profile of the population.

Spot the mistake

Edith is trying to decide whether to go and live in a retirement village, or stay in the seaside town where she currently lives. She finds out that the crude death rate at the retirement village is five times higher than in the seaside town. 'I am not going there', she says, 'it must be a really unhealthy place to live.' What mistake has Edith made?

Children playing in a village school

Example 2

Stokeville and Valeport are neighbouring and very similar towns. The age distribution of the two towns is given in the table below in percentages.

Age group	Stokeville	Valeport
0–15	12	19
16–30	18	30
31–45	22	28
46–60	21	16
over 60	27	7

AQA *Examiner's tip*

In your exam you will not be expected to calculate any standardised rates. You will need to know the difference between crude and standardised rates and why you would want to use the standardised rates in certain circumstances.

a Which town is likely to have the higher crude death rate?

b Explain why it would be wiser to use standardised death rates to compare the two towns.

Solution

a Looking at the percentages in the age groups, you can see that Stokeville has a much higher percentage of residents over 60 years old. As a result of this, you would expect a larger number of deaths per 1000 of the population than in Valeport. Therefore Stokeville will have a higher crude death rate.

b Where such large differences in the age distributions occur, the only fair comparison is one that takes age into account, so standardised death rates should be used.

The standardised death rate for Stokeville will be lower than the crude death rate, as it takes this older age distribution into account.

Exercise 11.2

1 In the town of Scotteringham, 450 of the 9000 inhabitants of working age are unemployed. Work out the crude unemployment rate.

2 The table shows the population and the number of births and deaths in Gainstown during 2008.

Measure	Total population at start of 2008	Number of births in 2008	Number of deaths in 2008
Value	40 000	88	64

a Calculate the crude birth rate for Gainstown in 2008.

b Calculate the crude death rate for Gainstown in 2008.

c As there were more births than deaths in the town in 2008, the population will have increased.

Is this statement definitely true, probably true, probably false or definitely false? Explain your answer.

3 The crude birth rate in Babytown for 2009 was 25. There were 375 births in Babytown that year.

a What was the population in Babytown at the beginning of the year?

b Can you tell the population of Babytown at the end of the year? Explain your answer.

4 Look again at the age distributions for Stokeville and Valeport from Example 2.

a Which town is likely to have the higher crude birth rate? Explain your answer.

b Is the standardised birth rate for Valeport likely to be higher or lower than its crude birth rate? Explain your answer.

c Why it is preferable to use a standardised birth rate to compare the birth rates in these two towns?

Age Group	Stokeville	Valeport
0–15	12	19
16–30	18	30
31–45	22	28
46–60	21	16
over 60	27	7

11.3 Quality assurance

In the manufacture and production of items it is important that the items are being produced to meet certain criteria regularly and consistently. For example, a 'size 7' shoe will need to be the right size and more than a slight variation in shoe length will mean that it does not conform to the dimensions required to be a size 7 shoe.

To give another example, a machine filling crisp packets, where the packets are labelled 25 g, must be filling these packets to conform as closely as possible to the required weight. If the machine produces packets consistently containing less than 25 g, the producers run a serious risk of upsetting authorities such as the Trading Standards Department, for short selling the general public. If the machine produces packets that regularly contain more than 25 g, this may well please the buying public and the regulators but it means that the company profits will be reduced as product is being 'wasted' by going above the stated weight.

Quality assurance is where quality is checked by obtaining samples of the products and seeing whether the means, medians or ranges of these samples are within appropriate levels (often called tolerance, action limits, or warning limits) compared to those expected.

The sample measures are displayed on **control charts**. It is then easy to see whether the process is on target or not. If it is not on target then someone must judge whether it is far enough off target to warrant some action being taken, either on the machinery itself, or on the personnel responsible for preparing the products.

Objectives

In this topic you will learn:

how to use control charts to check issues of quality assurance.

Key terms

Quality assurance: quality assurance is where quality is checked by obtaining samples of the products and seeing whether the means, medians or ranges of these samples are within appropriate levels.

Control charts: graphs showing the measures found during quality assurance so that they can be monitored for appropriateness.

Quality control charts

For measures of average (means and medians)

For the graphs that follow, you can assume that representative samples were obtained over regular periods of time to produce the plots.

Notice the design of the chart with the horizontal axis in the middle of the graph, as readings can be above or below the expected mean/ median. The x-axis represents the value of the mean or median that should be produced if the process was completely correct and exact.

1 No action required

This chart shows that production is completely under control. The samples are producing averages very close to the expected value with no pattern in the values being slightly above or slightly below. It is virtually impossible for all plots to fall exactly on the line – this chart is probably about as good as it could get.

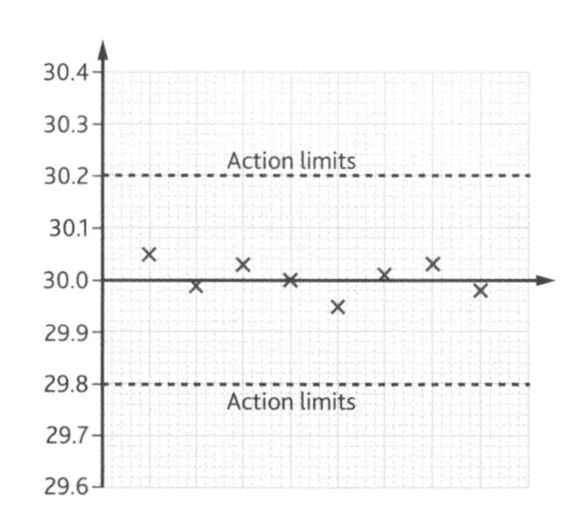

2 Minor action required

a Though the plots do not cross the lower action limit line, it is clear here that the machine or person is producing output that is somewhat below the expected value. A minor adjustment to the machine or piece of advice to the person should enable samples to return to the expected levels.

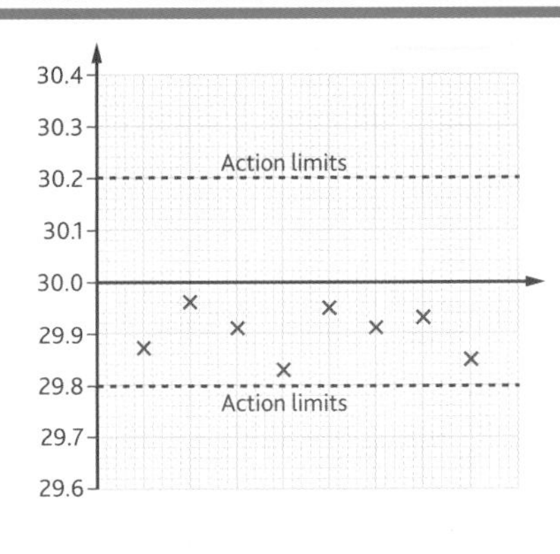

b Most of the samples here seem to perform well but there is one particular sample that has a value way outside the action limits. It would probably be wise to keep an extra close watch here and to take additional samples to verify the performance of the machine or person.

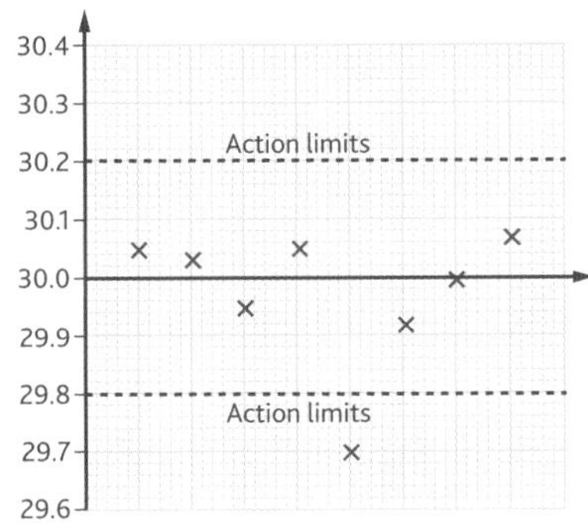

3 Significant action required

a Here the sample values are erratic and frequently over the action limits. This machine must be shut down and repaired. The operator of the machine may need retraining.

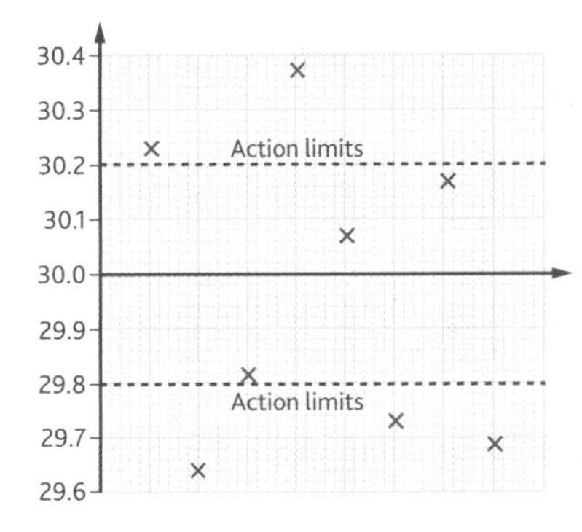

b Here the machine was initially working satisfactorily, but has suddenly begun to produce samples outside the action limits. The machine is out of control by the end of the period, and similar action to that outlined above is required.

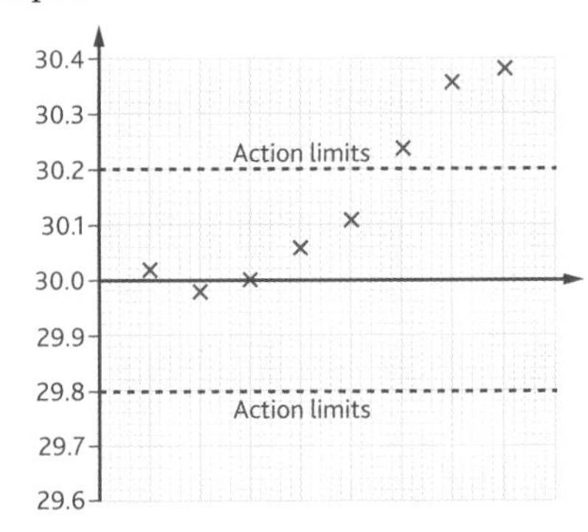

For ranges

Control charts for ranges are simply a way of keeping variation of samples in check. It is perfectly possible for a machine to be producing samples that have an appropriate mean or median (especially the latter) but the overall range in the sample might be quite large, making the majority of the items unfit for the purpose for which they were designed. Notice, on these charts, that as a small range is desirable, there is a single high action limit.

1 No action required

All sample ranges are small and well within the action limit.

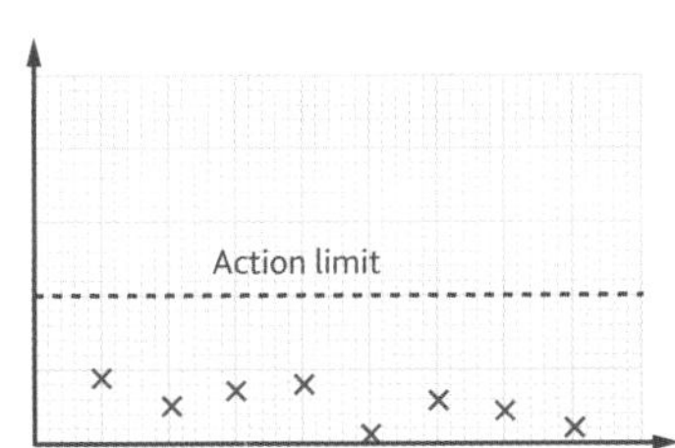

2 Minor action required

a One sample has a range beyond the action limit. This machinery or person needs closely watching, and additional sampling may well be a good idea.

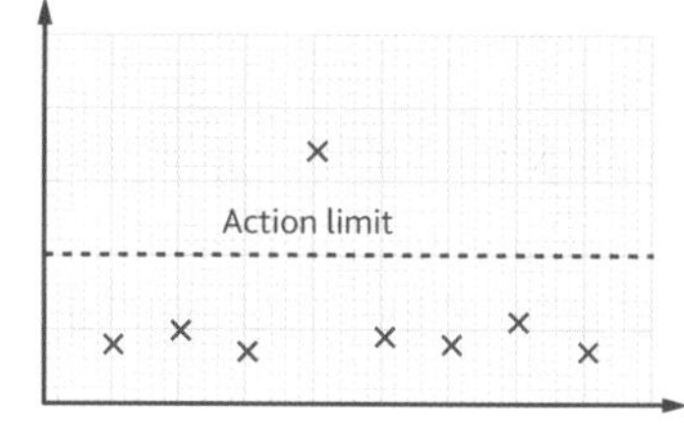

b The range values are not outside the action limit but it would be unwise to ignore the clear trend in the data. It looks very much as if the next sample is going to be above the action limit.

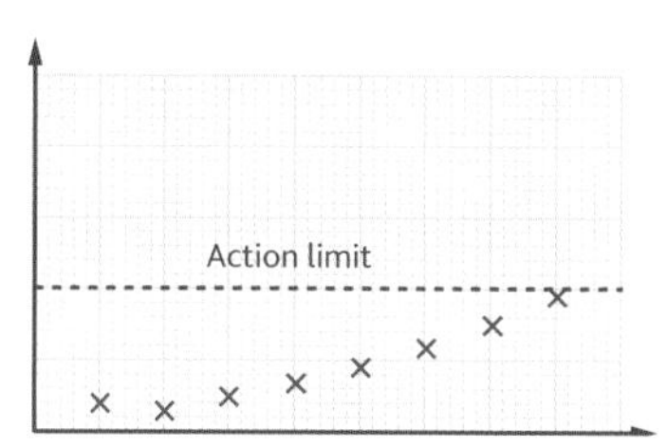

AQA Examiner's tip

Most questions on this topic require you to interpret charts already drawn. You may have to plots some points on occasions but not as many as in the 'Be a statistician' example given here.

3 Significant action required

Many of the samples' ranges are beyond the action limit. Either the machine or the person operating the machine is out of control, with the variation in production likely to cause significant problems. The machine needs to be shut down and repaired, or the person needs to be retrained.

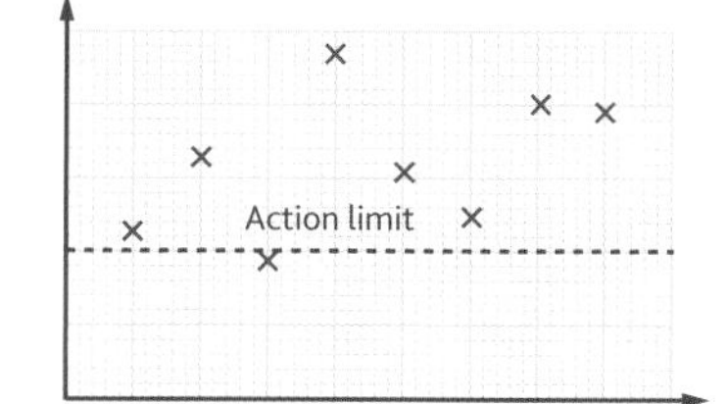

Be a statistician 

Samples of orange squash are taken to test a machine that is filling bottles. The bottles are labelled as containing 1 l. Eight samples of five bottles each are taken, one per hour through the day. Action limits for the mean are mean values more than 5 ml from the 1 l mark. Action limits for the range are range values more than 15 ml.

Use the data provided below to make appropriate calculations, complete appropriate control charts and make any suggestions you see fit for the machinery.

Sample 1: 998 ml, 1001 ml, 1003 ml, 1008 ml, 1012 ml

Sample 2: 996 ml, 1000 ml, 1002 ml, 1010 ml, 1011 ml

Sample 3: 995 ml, 999 ml, 1004 ml, 1009 ml, 1013 ml

Sample 4: 996 ml, 998 ml, 1002 ml, 1007 ml, 1010 ml

Sample 5: 998 ml, 1001 ml, 1005 ml, 1010 ml, 1014 ml

Sample 6: 999 ml, 1004 ml, 1007 ml, 1010 ml, 1016 ml

Sample 7: 996 ml, 1005 ml, 1007 ml, 1012 ml, 1015 ml

Sample 8: 998 ml, 1004 ml, 1009 ml, 1013 ml, 1018 ml

Exercise 11.3

1 Three different machines bag sweets in a factory. The bags are labelled as having a weight of 100 g of sweets. Samples are taken every two hours from the machines and the mean weights of the bags in the samples are calculated. The daily means are then plotted on a control chart. Gemma is the manager who deals with the machinery, and she inspects the control charts at the end of the day. In each of these cases what action, if any, should Gemma take?

a Machine A:

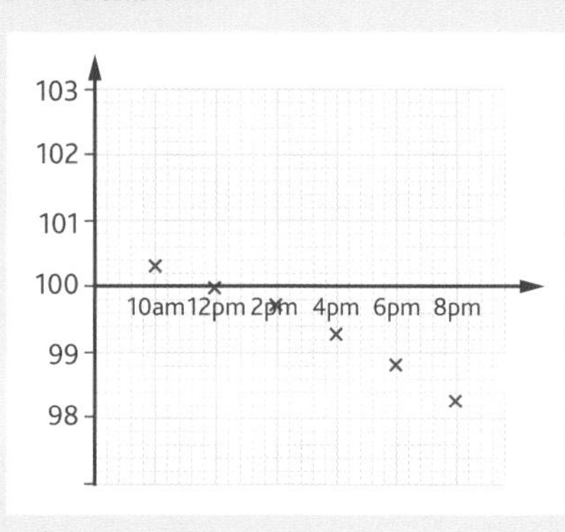

b Machine B:

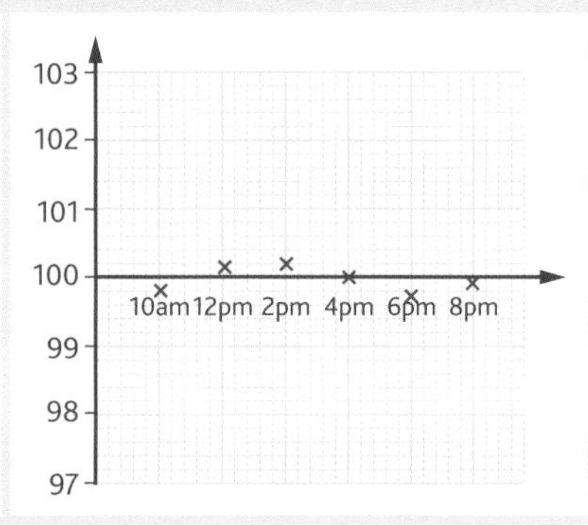

c Machine C:

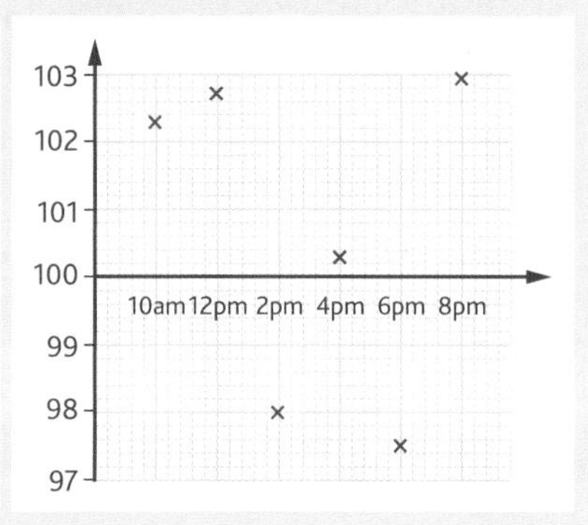

2 Tennis balls are required to bounce to a set height under certain conditions. Murray is responsible for ensuring that the tennis balls that his company produces conform to the rules on bounce. Over the last few years, although control charts have shown that the median of samples is appropriate, a large proportion of balls have been sent back as they do not bounce to the set height. He decides to sample and look at control charts for the range of bounce heights in each sample.

The 3 machines producing the balls gave the following control charts for range. In each case, suggest what action, if any, should be taken.

a Machine A:

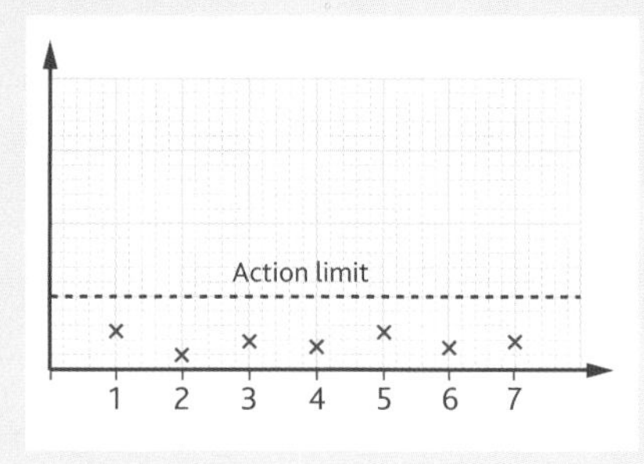

b Machine B:

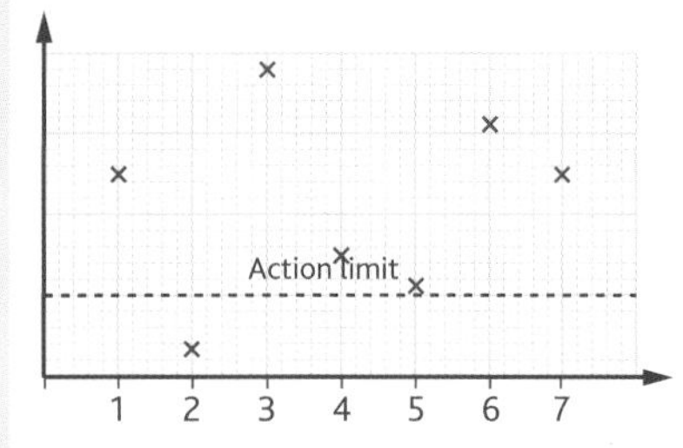

c Machine C:

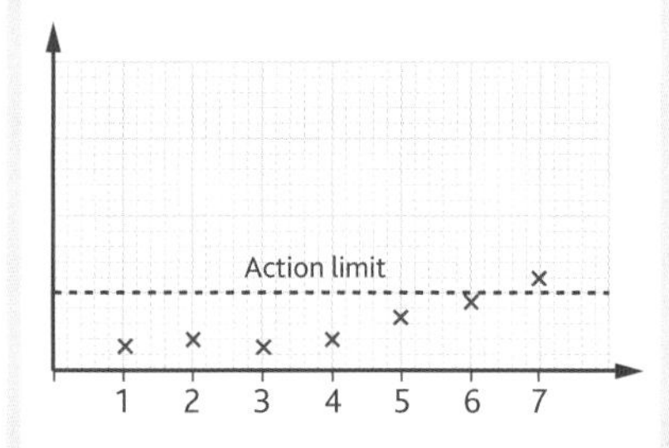

Summary

You should:

- know that index numbers chart changes in data values over time using a base year as value 100
- know that chain base numbers use the previous data value as base each time
- know that weighted index numbers give some items more importance than others in a combined list of indices
- know about the Consumer Price Index (CPI), Retail Price Index (RPI) and gross domestic product (GPD)
- know that crude rates measure items like births and deaths per thousand of the population
- know that standardised rates measure the same but take age distribution of a place into account
- know that control charts for averages or ranges track samples of items for conformity to expected values.

AQA Examination-style questions

1. The table shows some of the prices and price indices of a litre of petrol for the years 2002–2005. The prices are correct to the nearest penny.

Year	Price (pence)	Index
2002	71	100
2003		109
2004	81	114
2005	95	

Take 2002 as the base year.

(a) Find the price of a litre of petrol in 2003. Give your answer to the nearest penny. *(3 marks)*

(b) Find the price index for 2005. *(2 marks)*

AQA, 2007

2. A local newspaper investigates unemployment in the town of Stokeham.
The table gives unemployment data for Stokeham.

Age group	Population in thousands	Number unemployed
16–24	16	1020
25–44	28	1540
45–54	32	1206
55–64	14	680

Calculate the crude unemployment rate for Stokeham. *(3 marks)*

AQA, 2006

3. The table shows the cost indices for renting a shop, using 1997 as the base year.

Year	**1997**	**1998**	**1999**	**2000**	**2001**	**2002**
Cost index	100	115	96	118	110	113

(a) In which years did the rent fall? *(2 marks)*

(b) The annual rent was £6900 in 1998.

(i) Calculate the annual rent in 1997. *(2 marks)*

(ii) Calculate the annual rent in 2002. *(2 marks)*

(c) In which year was the annual rent the highest? *(1 mark)*

(d) Calculate the percentage increase to the annual rent between 2001 and 2002. *(3 marks)*

AQA, 2005

4. **(a)** The village of Scotter has a population of 6000. In 2007 there were 24 births in Scotter.

Calculate the crude birth rate for Scotter in 2007. *(3 marks)*

(b) The village shop sells calendars at Christmas each year. The table shows the price of the calendars for the last three years.

Year	**2005**	**2006**	**2007**
Price	£8	£9	£10

Calculate an index for the price of the calendars in 2007 using 2005 as base year. *(3 marks)*

AQA, 2008

5. A firm produces tins of baked beans. For quality control purposes, a sample of five tins of baked beans is taken every hour and the mass of each tin is measured.

The mean mass and range of masses of each sample is calculated and plotted on separate graphs.

The graphs below show the mean mass and range of masses of the first seven samples. The eighth sample has tins of baked beans of the following masses:

1.072 kg 0.998 kg 1.024 kg 1.037 kg 1.046 kg

The mean mass of this sample is 1.0354 kg.

(a) Calculate the range of masses of this sample. *(1 mark)*

(b) Plot the values of the eighth sample on the appropriate graphs.

The target mean is 1 kg.

The acceptable range is 0.08 kg. *(2 marks)*

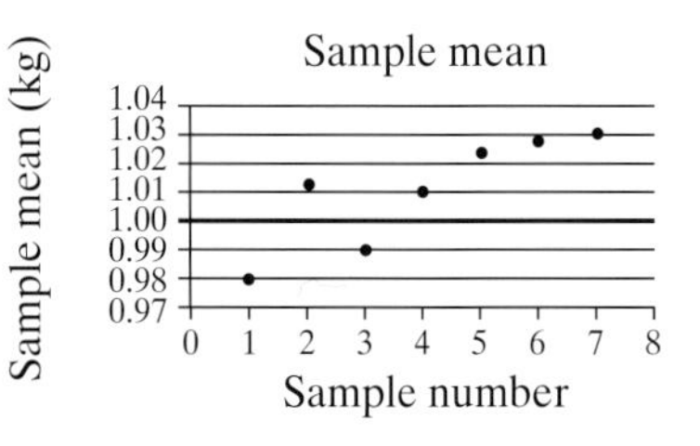

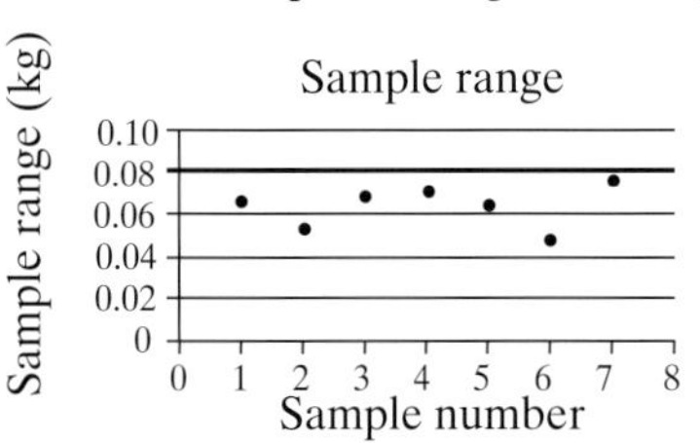

(c) Make one comment on each graph in relation to the production process. *(3 marks)*

AQA, 2003

6. The table gives the population distribution by age in town Q and the number of deaths for each age group.

Age group	Population in thousands	Number of deaths
0 to under 17	25	463
17 to under 30	22	201
30 to under 45	23	257
45 to under 65	18	329
65 and over	14	589

(a) Calculate the crude death rate for town Q. *(2 marks)*

(b) Why are standardised death rates better measures than crude death rates? *(1 mark)*

AQA, 2003

7. The table shows the breakdown of the Retail Price Index at July 2004.

	Group	Weight	Index (1987 = 100)
1	Food	111	152.9
2	Catering	49	231.8
3	Leisure goods	46	99.1
4	Leisure services	70	251.3
5	Housing	209	263.8
6	Fuel and light	28	139.4
7	Household goods	71	145.3
8	Household services	59	180.4
9	Clothing and footwear	51	99.0
10	Personal goods and services	42	200.0
11	Motoring expenditure	146	184.5
12	Fares and other travel costs	21	215.6
13	Alcoholic drink	68	204.3
14	Tobacco	29	313.7

(Based on the Retail Price Index, Monthly Digest of Statistics, July 2004)

The ‘all groups’ index for July 2004 is 198.4

(a) (i) Calculate an 'all groups' index, excluding expenditure on both alcoholic drink and tobacco. Give your answer to 1 d.p. *(4 marks)*

(ii) Explain why the 'all groups' index has decreased. *(2 marks)*

(b) The table shows the annual cost of Rashid's car insurance for the past five years.

Year	**2002**	**2003**	**2004**	**2005**	**2006**
Annual cost (£)	500	546	670	640	625

(i) Use the chain base method to calculate index numbers for the years 2003 to 2006 inclusive.

Give your answers to one decimal place. *(4 marks)*

(ii) Describe what these chain base index numbers show. *(2 marks)*

AQA, 2007

12 Probability

In this chapter you will learn:

- how to use probability as a measure of how likely an event is
- to link words (such as 'impossible') and probabilities
- to use a probability scale
- to recognise equally likely events
- to recognise when all possible events are listed
- how to draw and use sample space diagrams
- how to recognise mutually exclusive events
- how to recognise independent events and find probabilities of combined events
- to use the general multiplication law (the AND rule) and when it can be used
- how to draw and use tree diagrams in simple situations
- how to use tree diagrams for more complex situations
- how to use relative frequencies
- to appreciate that the more trials carried out, the more reliable the estimates of probabilities become
- to use diagrams to find probabilities (such as Venn diagrams and tables)
- how to find conditional probabilities from diagrams and tables
- how to calculate expected frequencies
- how to apply probabilities to generate simulations
- how to apply probabilities to estimate the size of a population (capture/recapture).

You should already know:

- ✔ how to add, subtract and multiply fractions and decimals
- ✔ how to reduce fractions to their lowest terms
- ✔ how many cards there are in a pack of cards
- ✔ the definitions of different types of number, such as square, prime, odd and even.

What's this chapter all about?

This chapter delivers the last exciting instalment of learning in this book. It will probably cover all the probability you will ever need to know . . .

Probability is a topic that has been around for quite a while. Blaise Pascal, a French mathematician, did a lot of work on probability. He was trying to make money using his work on gambling but it probably took him a long time to learn that gambling does not usually pay – look at how many people buy lottery tickets every week these days; they spend a fortune, but very few of them ever win!

You will see how probability can be used in a couple of ways that will be new to you – enjoy!

Focus on statistics

Lots of the probability work is in maths, but some of the diagrams, such as Venn diagrams, will be new to you.

In **12.5** you will learn about Venn diagrams.

In **12.6** you will simulate real life using probability.

In **12.7** you will learn how probability can be used to estimate the size of a population – it beats counting them all!

12.1 How likely...

Predictions about the future are not usually guaranteed to be accurate and reliable. For example, when you go out, it is helpful to have some idea as to whether or not it is likely to rain. This will determine whether you will take a coat, or an umbrella, or whether you are going to be prepared to get wet. You do not usually know that it is certain to rain unless, of course, it is raining when you set off.

When you talk about how likely 'something' is to happen, you will use words such as unlikely, 'probably', 'slim chance', 'fat chance', 'no way', 'impossible', 'certain', 'even chance' along with many other similar expressions. The 'something' is called an **event**.

You need to be able to rank expressions for probabilities in order.

Impossible — Unlikely — Evens — Likely — Certain

Objectives

In this topic you will learn:

- how to use probability as a measure of how likely an event is
- to link words (such as 'impossible') and probabilities
- to use a probability scale
- to recognise equally likely events
- to recognise when all possible events are listed.

Key terms

Event: something that takes place, such as rolling a die, tossing a coin, taking a card from a pack.

Example 1

Lucy was asked to give examples of events with the following likelihoods:

- **a** impossible
- **b** very unlikely
- **c** unlikely
- **d** evens
- **e** likely
- **f** very likely
- **g** certain.

Solution

- **a** Lucy said that it was impossible for her to win the lottery this week, as she had forgotten to buy her lottery ticket.
- **b** Lucy said that it was very unlikely that she would be struck by lightning.
- **c** Lucy said that it was unlikely that the car would break down next week, as it had just been serviced.
- **d** Lucy said that there was an evens chance that Birmingham City would win the toss for the kick off at the start of their next football match.
- **e** The weather forecast had been accurate all week and Lucy had just seen that the forecast had predicted rain for tomorrow. Lucy said that it was likely to rain tomorrow.
- **f** Lucy said that Uncle Des was very likely to see the Blues next football match, as he had bought a season ticket.
- **g** Lucy said that it was certain to be Sunday tomorrow, as today was Saturday.

These words are not very specific about how likely an event is. In order to be more specific about how likely an event is to occur, you will need to use numbers in place of these words. The number is referred to as a **probability**.

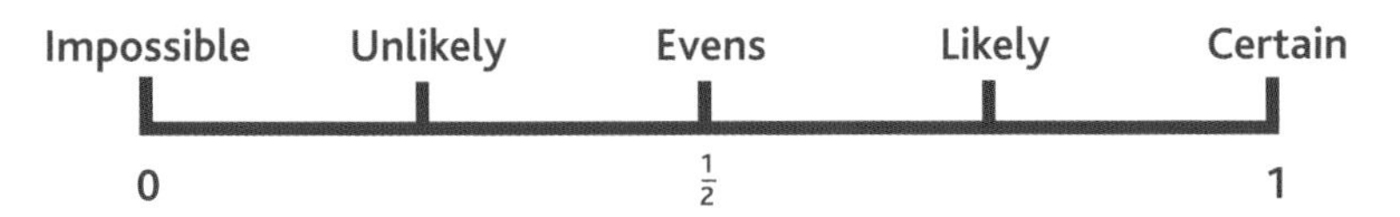

Values closer to 1 indicate 'more likely'; values closer to 0 indicate 'less likely'. 'Even chance' or 50-50 events – events that are as likely to occur as they are not to occur – have a probability of $\frac{1}{2}$. You must use only fractions, decimals or percentages to express these probabilities. Fractions and decimals are usually best, as percentages can be difficult to work with later.

Key terms

Probability: a measure of how likely an event is to occur.

An event that is impossible has a probability of 0.

An event that is certain to occur has a probability of 1.

All other events have probabilities between 0 and 1.

Outcome: the result of an event such as getting a 6 when a die is rolled, or a 'head' when a coin is tossed.

Equally likely: outcomes that are equally likely all have the same probability.

It all adds up to 1

When a coin is tossed it can land either with either 'head' or 'tail' facing up. The result 'head' or 'tail' is called the **outcome**. If the coin is a fair coin, then both outcomes are **equally likely**. As one or other of them is certain to occur, the probabilities must add to give 1. Each outcome has a probability of $\frac{1}{2}$. Instead of writing 'the probability of getting a head is $\frac{1}{2}$' as a sentence, it can be shortened to $P(H) = \frac{1}{2}$.

Finding probabilities

When you have equally likely outcomes, you can use the following formula to find probabilities:

$$P\ (event) = \frac{number\ of\ ways\ event\ can\ occur}{total\ number\ of\ possible\ outcomes}$$

Example 2

A fair die is rolled. Find the probability of getting:

a a 4

b an even number

c a number less than 3

d a prime number.

Solution

The possible outcomes are 1, 2, 3, 4, 5, 6. The question states that the die is a fair die. This means that all outcomes are equally likely to occur.

a $P(4) = \frac{1}{6}$ This is because $\frac{\textit{there is only one way to get a 4 on a die}}{\textit{there are 6 possible outcomes}}$

b $P(\text{Even}) = \frac{\textit{number of ways 'even' can occur}}{\textit{total number of possible outcomes}} = \frac{3}{6} = \frac{1}{2}$

c 'Less than 3' means '1 or 2'. $P(\text{less than } 3) = \frac{2}{6} = \frac{1}{3}$

d Prime numbers on a die are 2, 3, 5. $P(\text{Prime}) = \frac{3}{6} = \frac{1}{2}$

Exhaustive

When all the possible outcomes of an event are listed, it is said to be exhaustive. The sum of all the probabilities is 1. (Remember, 'sum' means you have to add the numbers up.) The probability that an event A does not occur is written P(A'). For any event A, P(A') = 1 – P(A)

In the example above $P(4) = \frac{1}{6}$

So, $P(\text{not a } 4) = P(4') = 1 - \frac{1}{6} = \frac{5}{6}$

AQA Examiner's tip

In Example 2 the fractions in (b), (c) and (d) have been cancelled down. Unless you are specifically asked to do so, you do not need to cancel fractions down to score full marks.

Exercise 12.1

1 Match the probability of each of these events with the labels A to E on the probability scale below.

a Getting a club when a card is drawn at random from a pack of cards.

b Getting a 6 when a die is rolled.

c Getting a 'head' when a coin is thrown.

d The sun rising tomorrow.

e Snow falling in the Sahara desert tomorrow.

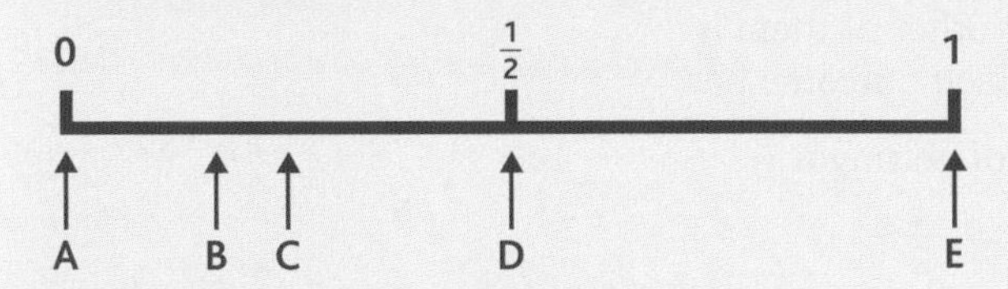

2 A card is drawn from a full pack of playing cards. What is the probability that it is:

a from a red suit

b a card with the number 7 on it

c a picture card (jack, queen, king)

d the king of diamonds

e the ace of spades

f the 4 of clubs

g not the 4 of clubs

h from a green suit?

3 The probability that it will rain tomorrow is 0.2. What is the probability that it will not rain tomorrow?

4 The probability that Sam gets up as soon as his alarm goes in the morning is 0.1. Does this mean that he generally gets up as soon as his alarm goes? Explain your answer.

5 Phil, a snail racing trainer from Kent, always trains his snails to be champions. The probability that his fastest snail, Herbert, wins any race he is entered in is 0.8. What is the probability that Herbert will not win his next race?

12.2 Working with probability

Sample space diagrams

You need to be able to find probabilities of events that might involve more than one thing going on, such as a coin being thrown and a die being rolled. In these cases it usually helps to draw diagrams to show all the possible outcomes. These are called **sample space diagrams**. A sample space diagram is an example of a two-way table.

Objectives

In this topic you will learn:

- how to draw and use sample space diagrams
- how to recognise mutually exclusive events.

Key terms

Sample space diagram: a diagram that show all the possible outcomes in a table.

Mutually exclusive: events that are mutually exclusive cannot occur at the same time; if two events A and B are mutually exclusive, then P(A or B) = P (A) + P(B)

Example 1

A fair coin is thrown and an ordinary die is rolled at the same time. Draw a sample space diagram to show all the possible outcomes.

Solution

		Die					
		1	2	3	4	5	6
Coin	Head	H1	H2	H3	H4	H5	H6
	Tail	T1	T2	T3	T4	T5	T6

Sample space diagrams can then be used to find probabilities using the formula given in topic **12.1**.

links

You learned how to draw two-way tables in Chapter 3.

Mutually exclusive events

These are events that cannot happen at the same time. For example, if a card is taken from a pack of cards it cannot be both a king and an ace.

When two events are **mutually exclusive** you can find the probability of one **or** the other happening by adding together the probability of each event occurring. Thus if two events A and B are mutually exclusive, then P(A or B) = P(A) + P(B)

This is sometimes called the 'OR' rule, as it always includes one event OR another occurring.

It is possible that you need to find the probability of more than two mutually exclusive events occurring. Then P(A or B or C or . . .) = P(A) + P(B) + P(C) + . . . Remember to check that your answer is always between 0 and 1.

Spot the mistake

Will rolled an ordinary, fair die. He said, 'The probability that I will get an even number is $\frac{1}{2}$ and the probability I will get a prime number is also $\frac{1}{2}$. This means that the probability of getting an even number or a prime number when I roll my die is $\frac{1}{2} + \frac{1}{2} = 1$.' What mistake has Will made?

Example 2

Two dice are rolled together. The two numbers showing are added together.

a Draw a sample space diagram showing all the possible outcomes.

b Use your diagram to find the probability of getting a total of:

i 6

ii more than 8

iii less than 3 or more than 10.

AQA Examiner's tip

In this example, the scores on the dice were added. It is possible to add, multiply or subtract the scores on two dice. Remember to read the questions carefully.

Solution

a

Die 1 (columns), Die 2 (rows)

	1	2	3	4	5	6
1	2	3	4	5	6	7
2	3	4	5	6	7	8
3	4	5	6	7	8	9
4	5	6	7	8	9	10
5	6	7	8	9	10	11
6	7	8	9	10	11	12

b i $\frac{5}{36}$ as there are 5 values of '6' and 36 values in the table

ii $\frac{10}{36} = \frac{5}{18}$ as there are 10 values that are 'more than 8'

iii A number cannot be less than 3 and more than 10 at the same time. These events are mutually exclusive.

P(less than 3) $= \frac{1}{36}$ and P(more than 10) $= \frac{3}{36}$

P('less than 3' OR 'more than 10') = P(less than 3) + P(more than 10) $= \frac{1}{36} + \frac{3}{36} = \frac{4}{36} = \frac{1}{9}$

Of course this can be done by counting the four values that are either above 10 or below 3, which is much easier, but this example shows the principles behind the OR rule.

Exercise 12.2

1 Two fair dice are rolled. The difference between the scores is recorded.

a Draw a sample space diagram to show all the possible scores.

b Use your diagram to write down the probability of getting:

i 3

ii 1

iii 6

iv a number less than 3

v 1 or 3.

2 The probability that Colin listens to Radio Cumbria is 0.1, and the probability that he listens to Radio 2 is 0.3.

a Explain why the probability that he listens to Radio 1 cannot be 0.7.

b Explain why the probability that he listens to Radio Cumbria or Radio 2 is likely to be 0.4.

3 Are the following pairs of events mutually exclusive? Explain your answer.

a 'Getting a head' and 'getting a tail' when a coin is thrown.

b 'Getting a 6' and 'getting a 3' when a die is rolled.

c 'Getting a heart' and 'getting a king' when a card is chosen from a pack of cards.

d 'Getting a 3' and 'getting a 5' when a card is chosen from a pack of cards.

e 'Getting a red counter' and 'getting a blue counter' when a counter is taken from a bag of counters.

4 Sally has a box of chocolates containing 5 with nuts, 4 caramels and 6 with crème centres. She takes a chocolate out of the box without looking. Write down the probability she chooses:

a a caramel

b a caramel or a crème

c a chocolate that is not caramel.

12.3 Independent events

Two events are said to be **independent** if the outcome of one event does not affect the outcome of the other.

In the example above, a fair coin is thrown and an ordinary die is rolled at the same time. If the coin lands 'heads' or 'tails' it will not affect the die throw. The two events are independent. You can use a sample space diagram to find the probability of events which are combined in this way.

When two events A and B are independent P(A and B) = P(A) × P(B)
This is often called the 'AND' rule as you are usually calculating the probability of one event AND another event both occurring.

The multiplication law is easily extended. If events A, B, and C are all independent events, then P(A and B and C) = P(A) × P(B) × P(C)

Objectives

In this topic you will learn:

- how to recognise independent events and find probabilities of combined events
- to use the general multiplication law (the AND rule) and when it can be used.

Key terms

Independent: two events are independent if the outcome of the second is not affected by the outcome of the first.

AQA Examiner's tip

Always check your answers carefully. Make sure you do not make errors in decimal place value. Remember that 0.2 × 0.2 is 0.04, and not 0.4.

Spot the mistake

The probability that Seamus forgets to do his homework is 0.3 in any subject. He works out that the probability he forgets to do his homework in history and geography is 0.9. What mistake has Seamus made?

Example 1

The probability that Jim chooses to go to Spain on holiday is 0.3.
The probability Jane chooses to go to France on holiday is 0.6.

a Write down the probability that Jim chooses to go to Spain and Jane chooses to go to France.

b Write down the probability that Jim chooses to go to Spain and Jane chooses not to go to France.

c What assumption did you make in order to work out your answers to (a) and (b)?

Solution

a P(Spain and France) = P(Spain) × P(France)
= 0.3 × 0.6 = 0.18

b P(Spain and not France) = P(Spain) × P(Not France)
= 0.3 × 0.4 = 0.12

c You assumed that Jane's choice and Jim's choice are both independent.

Exercise 12.3

1 One ordinary fair die is rolled and a card is chosen at random from a pack. What is the probability that:

a an odd number is rolled and a red card is drawn

b a six is rolled and a king is drawn

c a square number is rolled and the king of spades is drawn?

2 Janet and John enjoy reading books. The probability that John chooses to read a comedy book from the library is 0.7. The probability that Janet chooses to buy a thriller when she goes to the book shop is 0.4.

a Work out the probability that:

i Janet chooses a thriller and John chooses a comedy.

ii Janet chooses a thriller and John chooses not to get a comedy.

b What did you assume when you calculated your answers to (a)?

3 The probability that Ian goes to town on the train is 0.2. The probability Ian's sister goes to town by bus is 0.4. Explain whether the two events are likely to be independent.

4 Phil has a strong sense of humour. The probability he will tell his teacher a joke in any lesson is 0.6. Phil has two lessons on a Friday afternoon. Work out the probability that:

a Phil will tell a joke in both lessons.

b Phil will not tell a joke on Friday afternoon.

5 The probability that Penny will forget her keys on any day is $\frac{1}{10}$.

a What is the probability that she will forget her keys on Monday and Tuesday?

b What is the probability she will forget her keys on Monday and Tuesday and will remember them on Wednesday, Thursday and Friday?

6 The captain of a football team always called 'Heads' when the coin was thrown at the start of his matches. What is the probability that he called correctly on 3 consecutive matches?

7 A red, a green and a yellow counter are each numbered on both sides as follows:

	Top	Bottom
Red	1	6
Green	2	5
Yellow	3	4

All the counters are placed in a 'shaker', shaken and then turned out on to a table. Each counter is equally likely to show the top or bottom number.

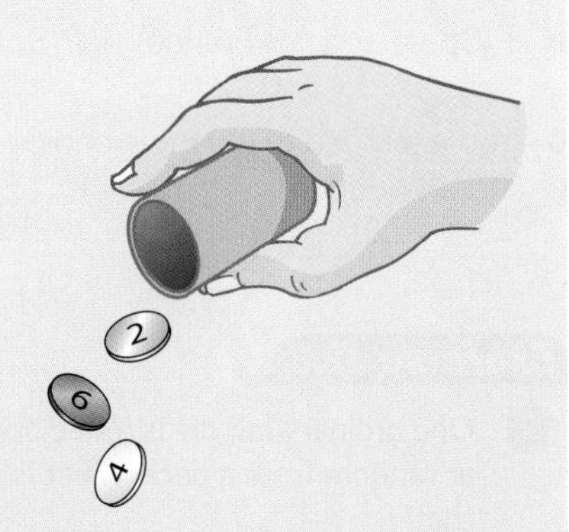

a What is the probability that when the counters are shaken and thrown:

i the red counter shows six

ii the red counter shows six and the green counter shows two

iii the red counter shows six, the green counter shows two and the yellow counter shows four?

12.4 Tree diagrams

Objectives

In this topic you will learn:

- how to draw and use tree diagrams in simple situations
- how to use tree diagrams for more complex situations.

Tree diagrams are usually drawn to help you see all the possible outcomes. Having a tree diagram can be useful when you need to calculate probabilities of combined events. This tree diagram shows the probabilities and outcomes when two fair coins are thrown.

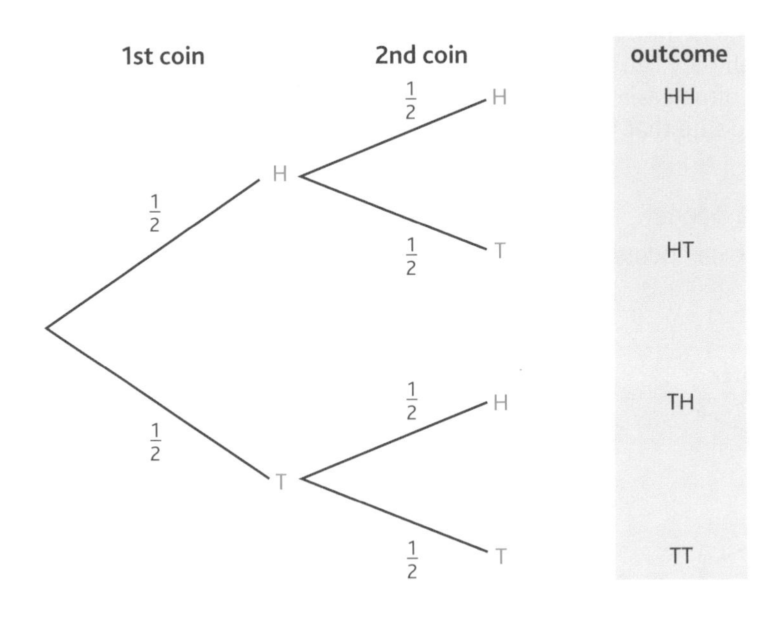

The blue lines are called 'branches'. Each branch is labelled with the outcome at its end. Each branch has the probability of its outcome labelled halfway down it.

The column at the right hand side of the tree diagram shows the combined event outcome. Notice that there are two ways to get the combination of one head and one tail and that is HT or TH. Always check that the probabilities on each 'pair' of branches add up to 1.

Here $\frac{1}{2} + \frac{1}{2} = 1$

In order to calculate the probabilities of the outcomes in the right hand column, you will use the AND rule.

P(HH) = P(Head on the first coin AND Head on the second)

$= \frac{1}{2} \times \frac{1}{2} = \frac{1}{4}$

In the same way P(HT) $= \frac{1}{2} \times \frac{1}{2} = \frac{1}{4}$

and P(TH) $= \frac{1}{2} \times \frac{1}{2} = \frac{1}{4}$

and P(TT) $= \frac{1}{2} \times \frac{1}{2} = \frac{1}{4}$

We can now use the OR rule to find the probability of one each of head and tail when a coin is thrown.

P(One of each) = P(HT OR TH) $= \frac{1}{4} + \frac{1}{4} = \frac{1}{2}$

AQA Examiner's tip

When you are asked to draw and label a tree diagram in your exam, marks will be given for correctly labelling each branch with its outcome and for correctly adding to each branch the probability of its outcome.

With replacement

Rachel's problem

Rachel has a bag of five red counters and ten blue counters. She takes a counter out of the bag, records the colour and then replaces the counter back into the bag. She then takes another counter out of the bag, and again records the colour. What is the probability she takes one of each colour?

Problems such as this are called 'with replacement' problems, as things are replaced after they are taken out. Although they can be solved using just the AND rule and the OR rule, it is often easier to use a tree diagram. Tree diagrams also help you to make sure that you do the correct calculation to solve the problem.

Example 1

a Solve Rachel's problem above using a tree diagram.

b Find the probability she takes at least one blue counter.

Solution

a Draw a tree diagram first.

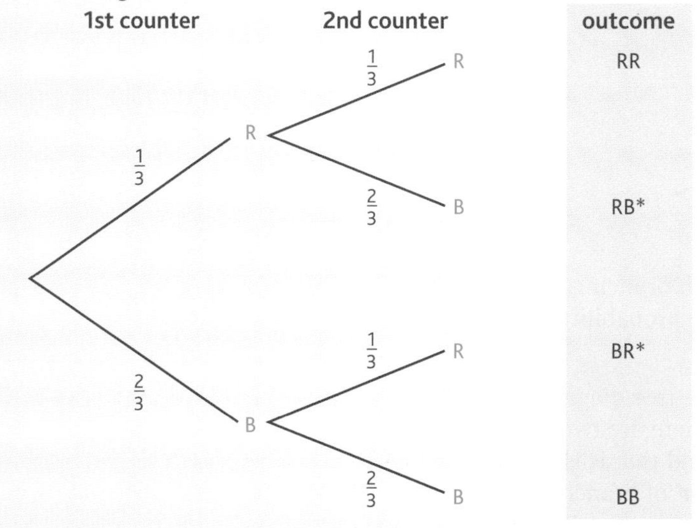

Check that pairs of branches add up to 1: $\frac{1}{3} + \frac{2}{3} = 1$

There are two ways to obtain one of each colour. Either red AND then blue, or blue AND then red. These are starred on the tree diagram.

P(RB) = P(R AND B) = $\frac{1}{3} \times \frac{2}{3} = \frac{2}{9}$

P(BR) = P(B AND R) = $\frac{2}{3} \times \frac{1}{3} = \frac{2}{9}$

Now you use the OR rule.

P(one of each) = P(RB OR BR) = P(RB) + P(BR) = $\frac{2}{9} + \frac{2}{9} = \frac{4}{9}$

b The easiest way to answer this is to calculate the probability that she takes only red ones, then subtract this from 1.

P(both red) = P(RR) = $\frac{1}{3} \times \frac{1}{3} = \frac{1}{9}$

P(at least one blue) = 1 – P(RR) = $1 - \frac{1}{9} = \frac{8}{9}$

The other way to obtain this answer is to find P(RB), P(BR) and P(BB), then add them together. This method involves three multiplications and one addition of three fractions. As there is more to this calculation, there is more that can go wrong.

In this problem it was much easier to use $\frac{1}{3}$ and $\frac{2}{3}$, rather than $\frac{5}{15}$ and $\frac{10}{15}$. If fractions with denominator 15 had been used the numbers would have been much larger. You do not need to worry too much about this in an examination as you are allowed to use a calculator throughout and many calculators will work out the fractions for you. You should still write down which fractions you are using and show the calculation you are trying to do, even if you do use your calculator, in order to ensure that you gain as many marks as possible in your exam.

AQA Examiner's tip

Some candidates show the probability calculations next to their tree diagram. This is allowed but you need to show which calculations you are choosing in the working-out space above the answer line on the exam paper, in order to secure the marks for that particular question.

Without replacement

At the higher tier you are expected to be able to answer questions that involve problems where the outcome of the first event changes the probability of the second event.

Example 2

Calculate the answer to Rachel's problem if she does not put the first counter back.

Solution

a

1st counter		2nd counter		outcome
$\frac{5}{15}$	R	$\frac{4}{14}$	R	RR
		$\frac{10}{14}$	B	RB
$\frac{10}{15}$	B	$\frac{5}{14}$	R	BR
		$\frac{9}{14}$	B	BB

Note that this time the probabilities have not been 'cancelled' down. This helps you check that you are putting the correct probabilities on the tree diagram. As before, check that the probabilities on pairs of branches add up to give 1.

$$P(RB) = P(R \text{ AND } B) = \frac{5}{15} \times \frac{10}{14} = \frac{50}{210}$$

$$P(BR) = P(B \text{ AND } R) = \frac{10}{15} \times \frac{5}{14} = \frac{50}{210}$$

Now you use the OR rule.

$$P(\text{one of each}) = P(RB \text{ OR } BR) = P(RB) + P(BR) = \frac{50}{210} + \frac{50}{210} = \frac{100}{210}$$

If the question asks you to give your answer in its simplest form, then you will need to cancel this down. $\frac{100}{210} = \frac{10}{21}$

b The method is the same as before, however, the fractions are different.

$$P(\text{both red}) = P(RR) = \frac{5}{15} \times \frac{4}{14} = \frac{20}{210} = \frac{2}{21}$$

$$P(\text{at least one blue}) = 1 - P(RR) = 1 - \frac{2}{21} = \frac{19}{21}$$

In this example, there were just two possibilities and only two stages. In your exam you could be given a problem involving up to three possibilities and up to three stages. The following example shows how complex these problems can be. You are very unlikely to be given a problem quite as complex as this one in your exam.

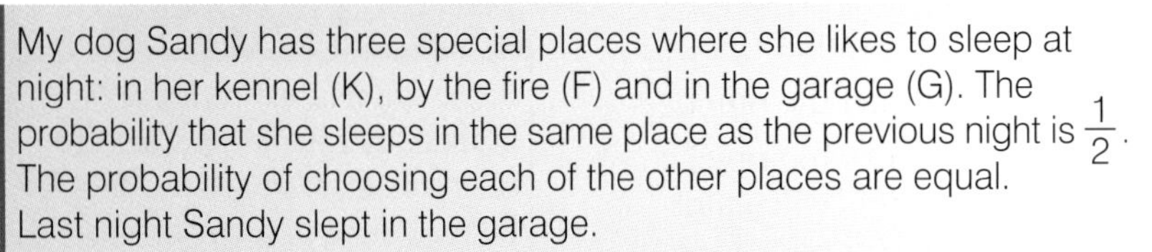

Example 3

My dog Sandy has three special places where she likes to sleep at night: in her kennel (K), by the fire (F) and in the garage (G). The probability that she sleeps in the same place as the previous night is $\frac{1}{2}$.
The probability of choosing each of the other places are equal.
Last night Sandy slept in the garage.

a What is the probability that tonight Sandy will sleep:

i in the garage **ii** in the kennel **iii** by the fire?

b Draw a tree diagram and show the probabilities for tonight and tomorrow night.

Sandy slept in the garage on Monday.

c Use your tree diagram to calculate the probability that on the three nights Sandy will sleep in the garage:

i once only **ii** twice only **iii** three times.

d What is the probability that Sandy will sleep in a different place on each of these three nights?

Solution

a **i** $\frac{1}{2}$ **ii** $\frac{1}{4}$ **iii** $\frac{1}{4}$

b

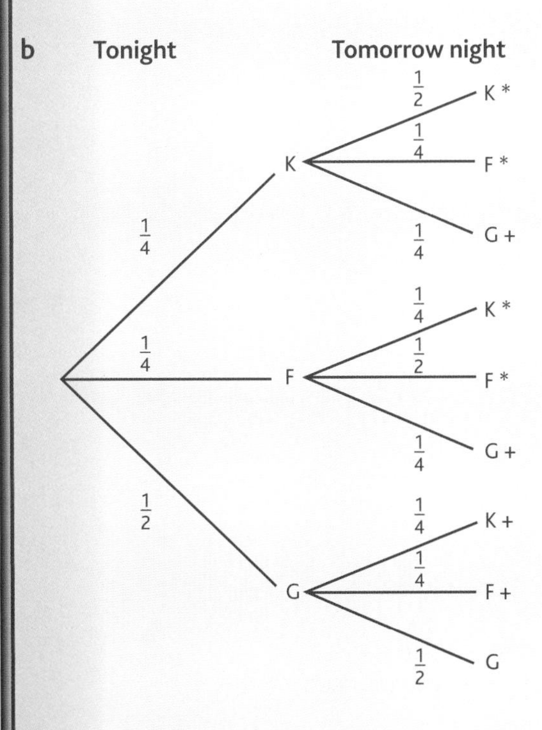

c Use the tree diagram. You must read questions carefully. In this case, the two nights in the tree diagram are Tuesday and Wednesday nights. You already know that Sandy slept in the garage on Monday, so that is one of the three nights.

i Need outcomes marked * in tree diagram

P(K and G') + P(F and G')

$= \frac{1}{4} \times \frac{3}{4} + \frac{1}{4} \times \frac{3}{4} = \frac{3}{16} + \frac{3}{16} = \frac{6}{16} = \frac{3}{8}$

ii Need outcomes marked + in tree diagram

P(K and G) + P(F and G) + P(G and K) + P(G and F)

$= \frac{1}{4} \times \frac{1}{4} + \frac{1}{4} \times \frac{1}{4} + \frac{1}{2} \times \frac{1}{4} + \frac{1}{2} \times \frac{1}{4}$

$= \frac{1}{16} + \frac{1}{16} + \frac{1}{8} + \frac{1}{8} = \frac{6}{16} = \frac{3}{8}$

iii P(G and G) $= \frac{1}{2} \times \frac{1}{2} = \frac{1}{4}$

d P(3 different places) = P(F and K) + P(K and F)

$= (\frac{1}{4} \times \frac{1}{4}) + (\frac{1}{4} \times \frac{1}{4}) = \frac{2}{16} = \frac{1}{8}$

These are examples of 'conditional' probabilities, as the probabilities are conditional on where Sandy slept the previous night.

Exercise 12.4

1 Stephen and Di are members of a squash club. They each play a game one Saturday. They do not play each other. The probability that Stephen wins his game is 0.8. The probability that Di wins her game is 0.7.

a Copy and complete this tree diagram.

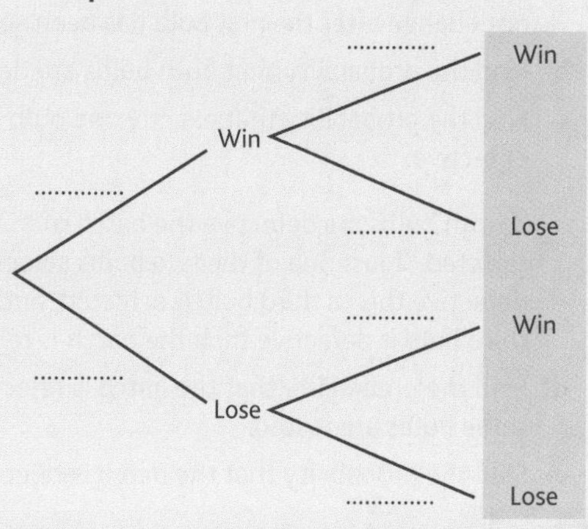

Use the tree diagram to find the probability that:

b both win

c both lose

d at least one of them wins.

2 The probability that Will passes his driving test is 0.7. His mother will pay for him to take up to two attempts to pass before he has to pay for himself.

a Copy and complete the tree diagram.

b Calculate the probability that he passes after two attempts.

c Calculate the probability that he has to pay for himself.

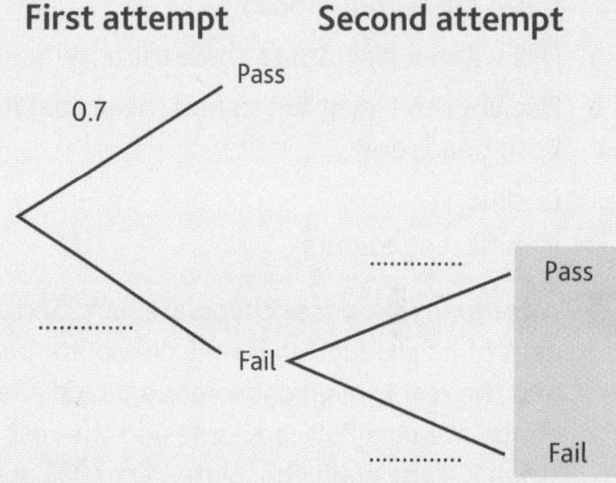

3 Mustafa is playing a game. He needs to roll sixes. He rolls a die twice. Copy and complete the tree diagram.
Remember that 6' means 'not 6'.

Use your tree diagram to find the probability that John rolls:

a two sixes

b exactly one 6

c at least one 6.

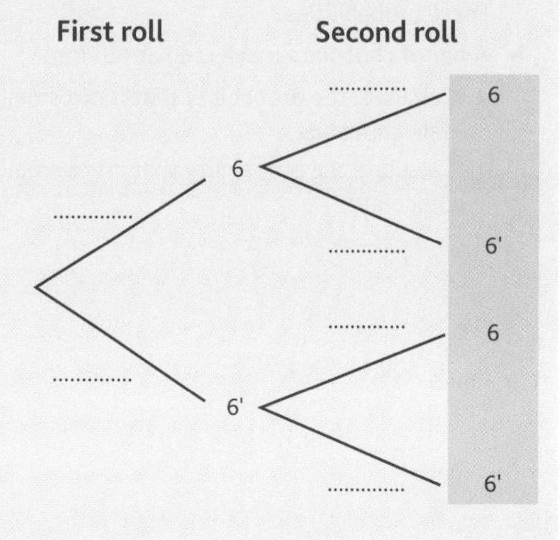

4 In a large batch of light bulbs, 5% are defective. Two bulbs are selected at random from the batch and tested.

a Draw a tree diagram to represent these two selections. (Note that the batch of light bulbs is large, so you can assume that the probability will not change after the first bulb has been selected.)

b Find the probability that both bulbs are defective.

c Find the probability that exactly one bulb is defective.

If both bulbs are defective the batch of bulbs is rejected. If just one of the two bulbs selected is defective then a third bulb is selected, and if the third bulb is defective then the batch is rejected.

d Find the probability that the batch is rejected after three bulbs are tested.

e Find the probability that the batch is rejected.

5 A bag contains 8 red beads and 4 blue beads. A bead is taken out of the bag without looking. It is not replaced in the bag. A second bead is then taken out without looking.

a Draw a tree diagram to show these selections.

b Use your tree diagram to find the probability that both beads are:

i blue

ii different colours.

6 A company produces chocolate bars. Sixty per cent of its production is milk chocolate bars and the rest is plain chocolate bars. Of the milk chocolate bars 75% are large and the rest are small. Of the plain chocolate bars 70% are large and the rest are small.

a Draw a clearly labelled tree diagram to represent these probabilities.

b A bar of chocolate is selected at random.

i Calculate the probability that it is a small bar of milk chocolate.

ii Calculate the probability that it is a small bar of chocolate.

7 A green bag contains eight 1p coins and five 2p coins. A yellow bag contains nine one penny coins and six two pence coins. A coin is selected at random from the green bag and placed in the yellow bag. A second coin is then selected at random from the yellow bag and placed in the green bag.

a Copy and complete the tree diagram illustrating the two selections.

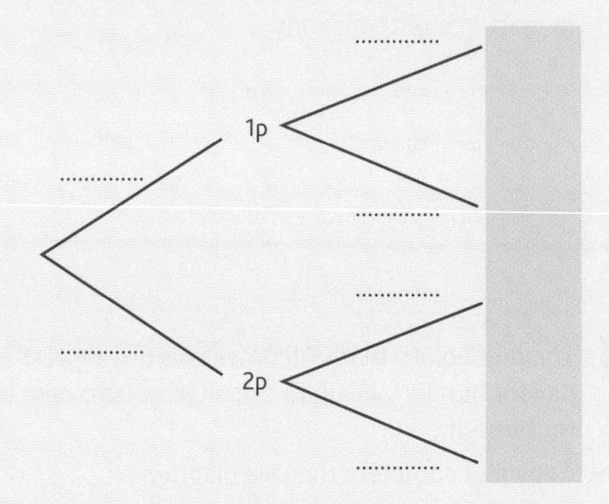

b Calculate the probability that the sum of money in each bag is unchanged after the two transfers.

For a challenge, a third coin is selected at random from the green bag and placed in the yellow bag.

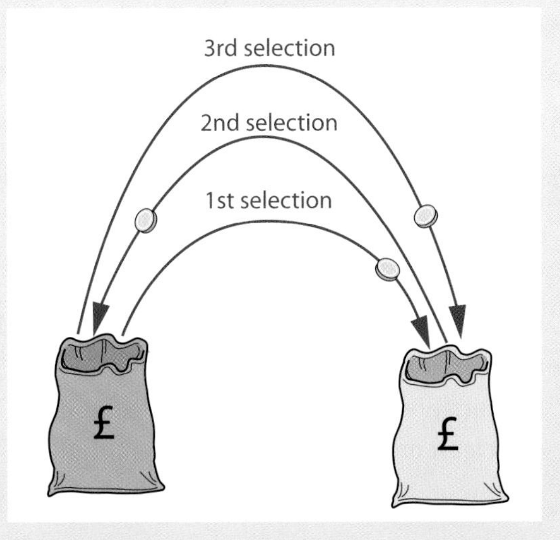

c Calculate the probability that the sum of money in each bag is unchanged after the three transfers.

12.5 Experimental probabilities

So far the probabilities you have considered have all been theoretical probabilities. You have used equally likely events to find probabilities. Sometimes it is not possible to use theoretical probabilities. For example, Mark has a die. It is weighted. This means that the scores are not all equally likely to occur. He rolls the die 50 times and records the results he gets.

	1	2	3	4	5	6
Frequency	3	7	8	6	9	17

He then rewrites these to find the **relative frequency** for each number. The relative frequency is the frequency with which he gets each score on the die relative to the total number of rolls.

	1	2	3	4	5	6
Relative frequency	$\frac{3}{50}$	$\frac{7}{50}$	$\frac{8}{50}$	$\frac{6}{50}$	$\frac{9}{50}$	$\frac{17}{50}$

Relative frequencies are often used to estimate probabilities. They are often called **experimental probabilities** or **empirical probabilities** as they are found by carrying out an experiment. The number of times the experiment is repeated is often called the number of trials. The larger the number of trials, the more reliable the relative frequencies are as estimates of the probabilities.

Objectives

In this topic you will learn:

- how to use relative frequencies
- to appreciate that the more trials carried out, the more reliable the estimates of probabilities become
- to use diagrams to find probabilities (such as Venn diagrams and tables)
- how to find conditional probabilities from diagrams and tables
- how to calculate expected frequencies.

Key terms

Relative frequency: the frequency with which one outcome of an experiment occurs relative to the total number of times the experiment is carried out.

Experimental probabilities: the name given to a relative frequency when it is used as an estimate for a probability.

Empirical probabilities: another term used for experimental probabilities.

Be a statistician

Use relative frequencies to estimate the probability of a drawing pin landing 'point up' by carrying out the following experiment. Drop a drawing pin 10 times. Count how many times it lands point up. Keep a record of your results in a table. Fran did this experiment and started her table like this:

Number of trials	Running total	Point up	Running total of point up	Relative frequency
10	10	3	3	0.3
10	20	6	9	0.45

Continue your table down so you have a total of 10 rows of trials. Put your own data into the table. When Fran had completed her table she plotted the data onto axes like the ones to the right:

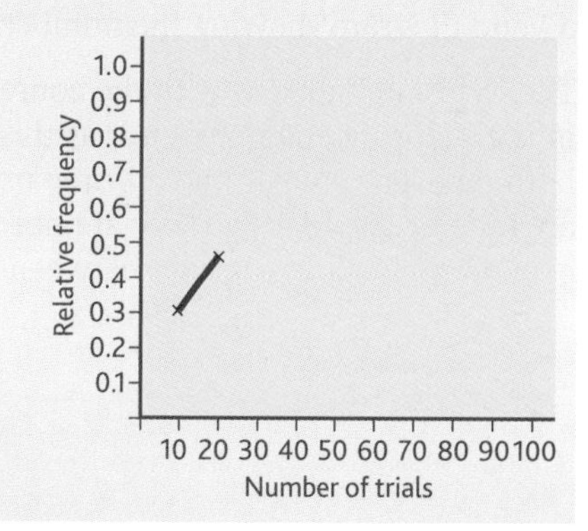

Draw a graph like Fran's to show your own data. Looking at your graph, you should notice that the changes between the relative frequencies become smaller as the number of trials increases. Compare your graph with the graphs of other students in your class.

Sometimes you will be able to use data that either you or someone else has recorded. Usually this will be in a table. It may be recorded as frequencies or percentages. This can then be used to find relative frequencies.

Example 1

Jenny asked her statistics class whether they could roll their tongue. She recorded her data in a two-way table as follows:

	Can roll tongue		
	Yes	No	Totals
Male	10	4	14
Female	9	6	15
Totals	19	10	29

a Use the table to estimate the probability that a child chosen at random from Jenny's class is:

i female

ii a female who can roll her tongue

iii a male who cannot roll his tongue.

b A person is chosen at random from Jenny's class. Given that the person is male, what is the probability that he can roll his tongue?

Solution

a i P(female) $= \frac{15}{29}$

ii P(female and can roll tongue) $= \frac{9}{29}$

iii P(male and cannot roll tongue) $= \frac{4}{29}$

b Here you are considering only the 14 males. Of these males, 10 can roll their tongue. P(can roll tongue given they are male) $= \frac{10}{14}$

This is an example of a 'conditional' probability, the condition here being that they are male. The clue to this is in the use of the word 'given'.

AQA *Examiner's tip*

Many students write frequencies when they are asked for relative frequencies in exam questions. Remember to write them as a fraction or decimal.

Venn diagrams

These show how data can be grouped. Loops are used to represent each group. Sometimes the loops overlap. Items that are within an overlap belong to all the groups that overlap. This is an example of a Venn diagram showing the numbers from 1 to 9 inclusive.

The loops are labelled with the contents. Note that the only number in both loops is the number 2, as this is the only number which is both an even number and a prime number. The numbers 1 and 9 are outside the loops as they are not prime numbers, nor are they even numbers, but they are between 1 and 9.

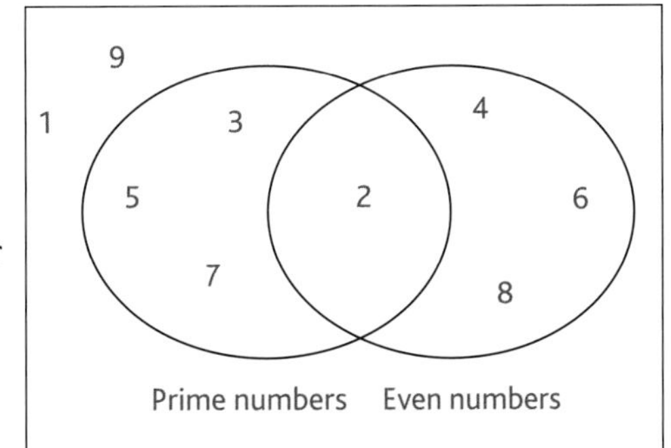

Probabilities from Venn diagrams

The next example shows how Venn diagrams can be used to find probabilities.

Example 2

The Venn diagram on the right shows the options subjects that students in a year 10 form group have chosen to study at GCSE.

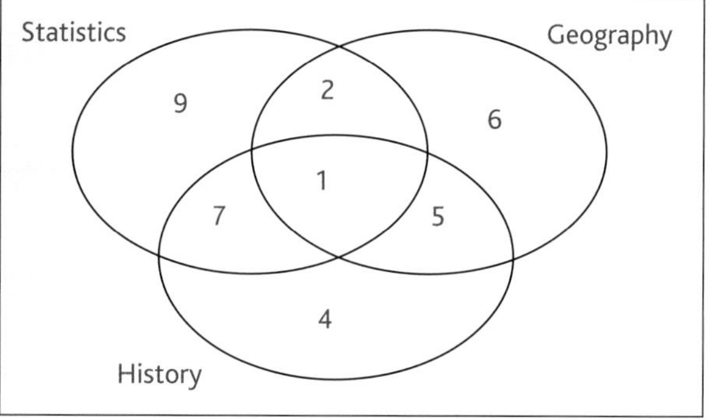

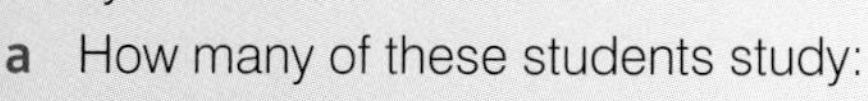

a How many of these students study:

- **i** statistics
- **ii** geography
- **iii** history
- **iv** both history and geography
- **v** geography and history but not statistics?

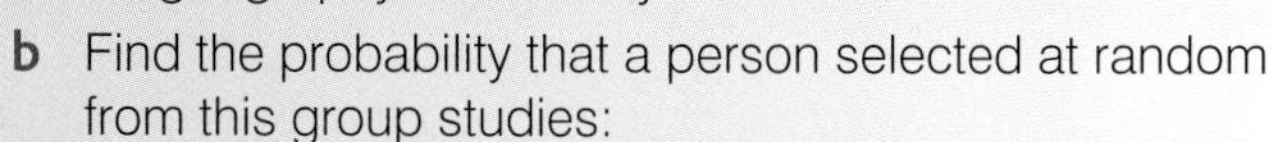

b Find the probability that a person selected at random from this group studies:

- **i** history
- **ii** statistics or geography or both, but not history.

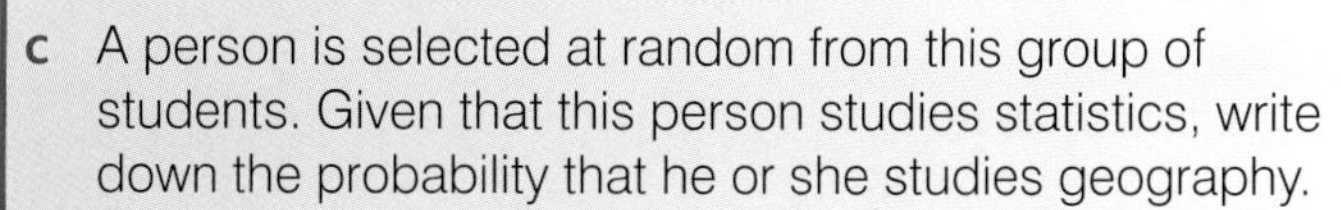

c A person is selected at random from this group of students. Given that this person studies statistics, write down the probability that he or she studies geography.

Solution

The Venn diagram shows that there is one person studying all three subjects. There are $2 + 1 = 3$ studying statistics and geography, but 1 of these studies history as well.

a **i** $9 + 2 + 1 + 7 = 19$ (these are all the students in the statistics loop)

ii $6 + 2 + 1 + 5 = 14$ (these are all the students in the geography loop)

iii $7 + 1 + 4 + 5 = 17$

iv $1 + 5 = 6$

v 5

b **i** $\dfrac{(7 + 1 + 4 + 5)}{34} = \dfrac{17}{34} = \dfrac{1}{2}$

ii $\dfrac{(9 + 2 + 6)}{34} = \dfrac{17}{34} = \dfrac{1}{2}$

c There are 19 students studying statistics. Of these students three study geography. P(geography given they do statistics) $= \dfrac{3}{19}$

Expected frequencies

The number of times you would expect something to happen is called an expected frequency. You can work out expected frequencies using probabilities. To work out an expected frequency you multiply the number of trials by the probability of getting the outcome you are interested in.

Example 3

Kerry was playing a game that involved rolling a die. She thought that the die was biased. She rolled the die 30 times and got the following results.

Score	1	2	3	4	5	6
Frequency	4	6	4	5	4	7

Was Kerry correct?

Solution

If the die was unbiased, then the probability of getting any single score is $\frac{1}{6}$. The expected frequencies are $30 \times \frac{1}{6} = 5$

So, on average, Kerry should expect to get each score five times. As Kerry has actually got each score about five times the die is unlikely to be biased – based on this evidence. Kerry could carry out her experiment again and would be unlikely to get exactly the same results as she has here.

Exercise 12.5

1 A survey of the ages of cars was carried out in a local car park. The cars were parked in rows. The table below shows the data collected.

Row	Number of cars in the row	Number of cars more than 4 years old	Relative frequency of cars more than 4 years old
A	10	4	0.4
B	20	6	
C	15	3	
D	10	2	
E	10	0	
F	15	3	
G	20	4	
H	10	5	
I	20	7	
J	10	4	

a Complete the relative frequency column.

b Calculate the best estimate of the probability of a car, selected at random from the car park, being more than four years old.

2 Malcolm wanted to estimate the probability of getting a six with his new die. He rolled it six times. On two of the rolls he got a six. He called his die 'Lucky' and told his friends that the probability of getting a six with 'Lucky' was $\frac{1}{3}$.

a Was Malcolm correct with his estimate of the probability? Explain your answer.

b What could Malcolm do to improve his estimate of the probability of getting a 6?

3 Barry the PE teacher thought that all his students loved PE. He surveyed all the students in one of his classes, and got the following results.

	Male	Female	Totals
Love PE	12	11	
Not so keen on PE	3	4	
Totals			

a Copy and complete the two-way table for Barry's class

b A student is chosen at random from Barry's class. What is the probability that the student:

i is male
ii is female
iii loves PE
iv is female and loves PE
v loves PE given that he is a male student?

4 In Laddam Languages School they proudly boast success in teaching students languages. They say that if any student is chosen at random, the probability that the student is good at speaking French is 0.9. There are 180 students in Year 10. How many of these students would you expect to be good at speaking French?

5 Dee Jay carried out a survey of musical tastes. Her results are shown in the Venn diagram below.

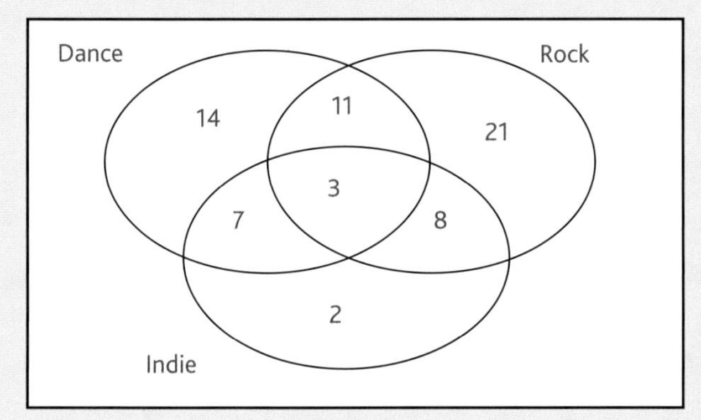

a Write down the number in Dee's survey who liked:

i dance
ii both dance and rock
iii indie but not rock
iv both dance and indie.

b Dee chose a person at random from her survey. What is the probability that the person she chose likes:

i dance but not indie

ii rock

iii all three types of music?

c Dee chose a person at random from her survey. Given that the person likes dance music, what is the probability that the person she chose likes:

i rock

ii indie but not rock?

6 This fair spinner is spun 250 times.

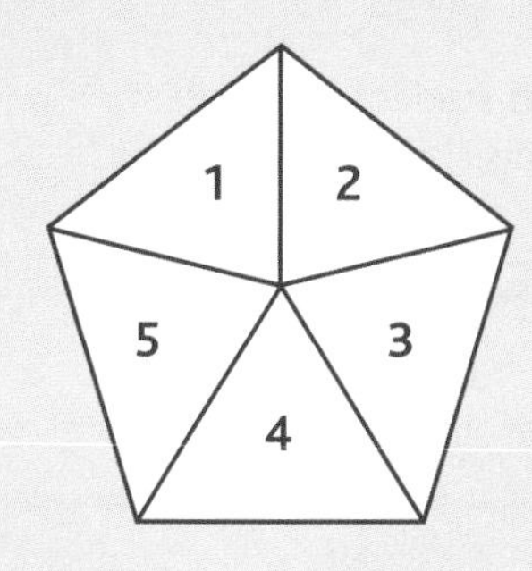

Francis records which number it lands on each time it is spun.

a Which one of the following tables is most likely to show her results?

Table A

score	frequency
1	95
2	35
3	48
4	38
5	34

Table B

score	frequency
1	52
2	48
3	53
4	48
5	49

Table C

score	frequency
1	49
2	48
3	62
4	17
5	74

b Explain your answer to part (a).

c The spinner is spun 200 times. How many times would you expect it to land on 4?

12.6 Using probabilities – simulation

Probabilities are used to set up models where it may be very expensive, not practical, or impossible to carry out experiments.

Random numbers are used in simulations.

Example

A doctor's surgery studied the ages of patients visiting the surgery one week. It found that $\frac{1}{3}$ of its patients were under the age of 40, and $\frac{1}{6}$ were between the ages of 40 and 65. The rest were all over 65. Use a die to generate random numbers to simulate the ages of the next 10 people visiting the surgery.

Solution

Generate

First you need to decide what method is appropriate for generating random numbers. As the ages are divided into sixths, a die will be perfectly acceptable.

Allocate

Next you need to allocate the random numbers to ages. Numbers on a die are 1, 2, 3, 4, 5, 6. You need two of these numbers for patients under 40, as $\frac{1}{3} = \frac{2}{6}$; one of these numbers for the age group 40 to 65 and the remaining numbers for the over 65s.

Numbers 1 and 2 = under 40 (U40)

Number 3 = between 40 and 65 (40–65)

Numbers 4, 5 and 6 = over 65 (65+)

Simulate

Roll the die 10 times, record the scores and match the numbers to your allocation. Shaun did this and got the numbers:

Score on die	1	2	5	3	6	1	5	4	4	1
Age	U40	U40	65+	40–65	65+	U40	65+	65+	65+	U40

You would not expect to get the same results every time you repeat this simulation. Just as if you sat in the doctor's waiting room for 10 minutes, a different period of 10 minutes would bring in a different set of people.

Objectives

In this topic you will learn:

how to apply probabilities to generate simulations.

links

See the sampling chapter (Chapter 2) for methods of obtaining random numbers.

AQA *Examiner's tip*

Remember to read the question carefully, it will tell you what the probabilities or proportions are that you should use. Make sure that you then allocate the numbers carefully, without any repeats or any missed out.

Exercise 12.6

1 A minibus takes a circular route through some villages. At any stop it is likely to collect up to five people. It is also likely to drop off up to five people. It starts with five people on board.

a The bus can carry up to 15 passengers. Set up and run a simulation to see how many stops the bus is likely to make before it becomes full.

b The bus company decides to use a bigger minibus for this route. Run your simulation several times. What is the minimum number of seats the bus company is likely to need?

2 A local authority is collecting data about traffic leaving town on a main road. It is going to carry out some long-term road works and plans to put up traffic lights for safety. It finds that in any period of 10 seconds it is possible that 0, 1, 2 or 3 cars may pass the place where workers are going to dig. The results of the survey are shown in the table.

Number of cars	0	1	2	3
Frequency	2	4	3	1

The local authority sets the traffic lights to red for 90 seconds. Set up and run a simulation to find how many cars are likely to be in a queue at the end of the 90 seconds.

3 A company puts free toy cars in children's meals. There are five cars to collect for a full set. It is known that the toys are in the following proportions:

red: 1 blue: 2 green: 3 white: 2 yellow: 2.

Set up a simulation using the numbers 0 to 9. Use either your calculator's random number function or the table of random numbers at the back of this book to generate single digit numbers from 0 to 9. How many meals would you expect to need to buy before you have collected a full set of all five cars?

links

You will find one of these simulations that a Year 11 student has already produced on the Nelson Thornes website.

Real world

Collect some data about the cars passing the front of your school. You will need to design a data collection sheet before you start. Make sure that you stand safely away from the road when you do this. You will probably want to work with a partner.

Count how many cars go past in the same direction in a period of time. You might decide to collect the same kind of data as the local authority did in Question 2 in the exercise. Decide how you will generate random numbers. Allocate them to your data and simulate the cars passing your school.

You might decide to write a short program to run the simulation on a computer.

12.7 The capture/recapture technique for estimating the size of a population

Objectives

In this topic you will learn:

how to apply probabilities to estimate the size of a population (capture/recapture).

Probability is applied to estimate the size of a population. For example, if the number of fish, x, in a lake is not known, then a sample of, say, 20 can be caught – harmlessly and humanely of course! These can then be marked in some way (for example, with a non-toxic dye) and released back into the population. These fish need to be allowed to mix with the population. The probability of a fish that is caught at random being marked is $\frac{20}{x}$.

A second sample is now caught. This sample does not have to be the same size as the first sample. It might catch, say, 15. You count the number of fish which are marked in this second sample. It may be 4. From your second sample you know the relative frequency of marked fish is $\frac{4}{15}$.

Assuming the relative frequency to be a good estimate of the probability of catching a marked fish, then you know:

$$\frac{20}{x} \approx \frac{4}{15}$$

You can rearrange this to find x

$15 \times 20 \approx 4x$

$300 \approx 4x$

$75 \approx x$

Your estimate of the size of the population of fish is 75. This is an estimate because you are using sample data to estimate the size of a population. If you repeated the experiment you would not necessarily expect to get the same outcome.

To summarise, the steps are:

- capture a sample of n items from the population
- mark these n items
- return the items to the place you captured them from
- allow time for the marked items to mix completely with unmarked items
- recapture m of the items
- suppose x of these items carry the mark you earlier gave to the n items
- then, N, the population size can be estimated using $\frac{x}{m} = \frac{n}{N}$ which when rearranged gives $N = \frac{nm}{x}$

  You are using this formula:

  $$P(\text{Marked}) = \frac{\text{number in first sample}}{\text{number in population}} \approx \frac{\text{number marked in second sample}}{\text{number in second sample}}$$

Which can be rearranged to give:

$$\textit{number in population} \approx \frac{\textit{number in first sample} \times \textit{number in second sample}}{\textit{number marked in second sample}}$$

Clearly, the larger the sample, the better the estimate.

You need to be aware of three important features of this process:

- The first sample when returned to the population must have completely mixed up with the rest of the population before the second sample is taken. With relation to animal experiments, this may take many days or even longer.
- The process of catching and marking the items must not make them behave differently from the unmarked items, otherwise this will lead to an artificial proportion of marked items in the second sample. For example, causing unnecessary trauma to fish during the capture of the first sample will affect their behaviour.
- The process can only be valid if there are not likely to be short-term major fluctuations in the size of the population that may occur between the times of the first and second samples.

Be a statistician

You will need:

- a non-transparent jar, bag or box
- a set of at least 100 beads, or similar, which are small enough so that it is not at all clear how many are in the container just by feel
- a marker pen.

Task:

With a friend you are going to estimate the number of beads using the capture recapture technique. One of you needs to put all the beads in the chosen container without the other seeing the number.

- Take a sample of 20 beads out of the bag (without replacement).
- Mark each of these beads clearly, using the marker pen.
- Now replace these 20 beads back into the container.
- Shake the container well to mix the marked beads with the unmarked beads.
- Now take another sample and note the proportion of the sample that consists of marked beads.
- Use this value to estimate the population size.
- How close were you?
- Are there any reasons why your estimate is not a good one – any flaws in your experimental design?

Example

Ailsa is estimating the number of birds on a Scottish island. She captures 50 birds and rings them. A few days later she returns to the island and captures a sample of 32. She finds that 8 of these are already ringed. Estimate the number of birds on the island.

Solution

Let n be the number of birds on the island. Then

$$\frac{50}{n} \approx \frac{8}{32}$$

$$50 \times 32 \approx 8 \times n$$

$$\frac{1600}{8} \approx n$$

$$\text{So } n \approx 200$$

There are about 200 birds on the island.

AQA Examiner's tip

In your exam you are likely to be asked how to improve the estimate of the population size. You need to make sure that your answer is in context. For example, an answer such as 'catch more birds' would be appropriate for this example.

Spot the mistake

David, Simon and Ibrahim are studying GCSE statistics. Their teacher has given them a large box, which has a large number of pieces of paper in it. The teacher has asked them to use capture/recapture to estimate the number of pieces of paper in the box. They mark 60 pieces of paper with a coloured felt-tipped pen, put them back into the box and give the box a good shake to mix the paper up. Simon takes out 50 pieces of paper. Fifteen of these are marked with the pen. David then takes out 20 pieces of paper. Twelve of these are marked with the pen.

Simon works out

$60 \times \frac{50}{15} = 200$, and states confidently there are about 200 pieces of paper in the box.

David works out

$60 \times \frac{20}{12} = 100$, and states with confidence that there are 100 pieces of paper in the box.

Who has the best estimate and what mistake have they made?

Ibrahim says that a better estimate can be obtained by combining the data from the two samples. Find Ibrahim's estimate.

Exercise 12.7

1 During a census of the seal population around an island, 100 seals were caught and marked with a coloured dye. These were then returned to the sea. The following day 100 seals were caught and five of them were found to be marked from the previous day. Estimate the number of seals around the island giving a reason for your answer.

2 A group of fresh-water biologists wants to estimate the number of fish in a lake. They catch and mark 50 fish. They return the 50 fish to the lake. They then catch 80 fish. Of these 80 fish, 6 are marked.

a Use this informatin to calculate an estimate of the total number of fish in the lake.

b What could the biologists do to improve this estimate?

3 Bertie knows that the buses in his town are allocated at random to particular routes at the beginning of the day, but he does not know how many buses the company own. He puts a sticker on 10 buses, and then one week later spends some time watching buses go past his house. He notices that out of 32 buses going past his house only two have the sticker on.

a Use this information to estimate the number of buses owned by the company.

b Give three reasons why this method of estimating the number of buses owned by the company is probably flawed.

4 Rod and Annette are estimating the number of fish in two different lakes.

a Rod catches 27 fish and marks them with a harmless dye. He then returns them to the lake. One week later he catches 40 fish and sees that 12 of them are carrying the mark. Estimate the number of fish in this lake.

b Annette catches some fish and also marks them with a harmless dye. She then returns them to the lake. One week later she catches 50 fish and sees that four of them are carrying the marks. From this she estimates there are 375 fish in the lake. How many fish did she mark in total?

Summary

You should:
know how to use probability as a measure of likelihood
be able to draw sample space diagrams
be able to find probabilities of combined events
be able to use tree diagrams to find probabilities
be able to use relative frequencies as estimates of probabilities
know how to find probabilities from tables and diagrams
know how to calculate expected frequencies
know how to find probabilities of combined events using conditional probabilities
know how to set up and run a simulation
know how to use capture/recapture to estimate the size of a population.

AQA Examination-style questions

1. A bag contains:
7×1p coins
6×2p coins
4×5p coins
3×10p coins.

(a) One coin is selected at random. What is the probability that it is worth more than 4 pence? *(1 mark)*

(b) The coin is replaced. A second coin is selected at random. It is worth more than 4 pence. What is the probability that it is a 10p coin? *(2 marks)*

AQA, 2005

2. All students at a school study French and Spanish. The probability that a student is good at French is 0.8. If a student is good at French, then the probability that he/she is good at Spanish is 0.3.

(a) Copy and complete the tree diagram to show the probabilities when a student is selected at random. *(4 marks)*

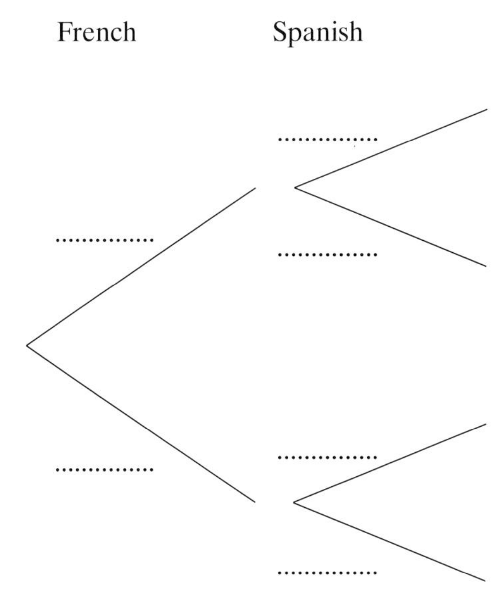

(b) Calculate the probability that a student selected at random is good at French and Spanish. *(2 marks)*

AQA, 2005

3. One hundred and fifty students were asked which daily newspaper(s) they read. The results are shown in the diagram.

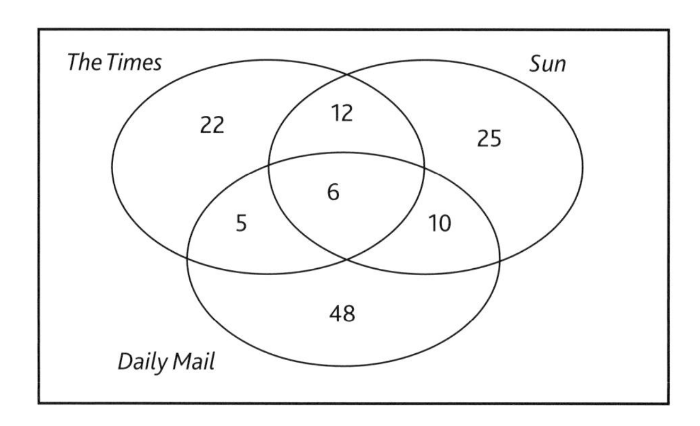

(a) Find the probability that a student chosen at random reads:

(i) *The Times* *(1 mark)*

(ii) only one of the papers *(2 marks)*

(iii) none of the three papers *(3 marks)*

(iv) the *Sun* or *The Times* or both but not the *Daily Mail*. *(2 marks)*

(b) One of the students, Jean, reads the *Daily Mail*. Find the probability that she also reads *The Times*. *(3 marks)*

AQA, 2008

4. A survey of 24 students was carried out about the number of students who wear glasses and wear earrings. The diagram shows some of the information from the survey. The section labelled X has not been completed.

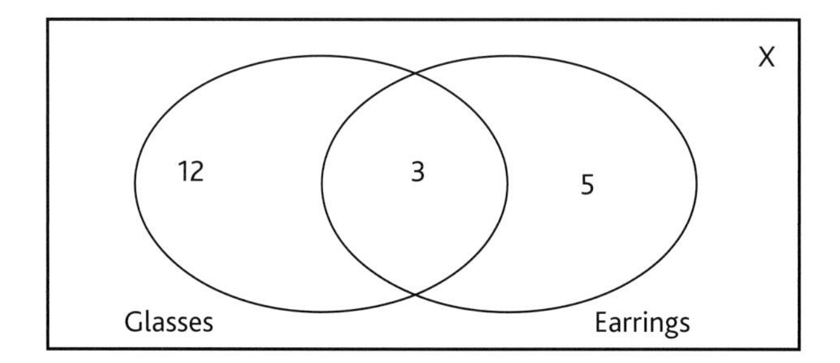

(a) Work out the number that should go in the section labelled X. *(2 marks)*

(b) What can you say about the students in the section labelled X? *(1 mark)*

(c) One student is chosen at random. What is the probability that the student:

(i) wears earrings, but does not wear glasses *(1 mark)*

(ii) wears earrings and wears glasses? *(1 mark)*

(d) A student chosen at random wears earrings. What is the probability that this student also wears glasses? *(2 marks)*

AQA, 2006

5. When motorists call a particular road breakdown service, they are put into one of three categories by the operator at the switchboard. These categories are Emergency (E), Urgent (U) and Non-Urgent (N).

Emergency and Non-Urgent categories are equally likely.

The Urgent category is four times more likely than the Emergency category.

The breakdown services wishes to carry out a simulation of 20 calls to their switchboard using a fair die.

(a) Describe how they could allocate the numbers 1, 2, 3, 4, 5 and 6 on the die to a particular category of call. *(3 marks)*

(b) Using your answer to part (a) list the type of call simulated by the following numbers on the die using the letters E, U or N.

5	1	4	6	4	4	2	3	2	6
1	1	3	4	2	6	6	5	4	5

(2 marks)

AQA, 2007

6. **(a)** A fair six-sided die is rolled once.

(i) What is the probability of obtaining a 4? *(1 mark)*

(ii) What is the probability of obtaining a number which is *not* 4? *(1 mark)*

Alan and Ben are playing a game in which they take it in turns to roll a fair six-sided die. The winner is the first person to roll a 4. Alan is to roll first. The tree diagram below shows the branches for Alan's first turn and for Ben's first turn.

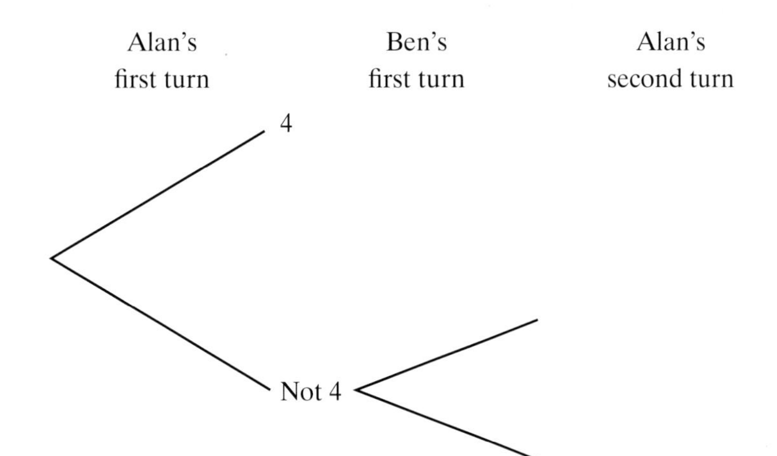

(b) Write on the tree diagram the two possible outcomes for Ben's first turn. *(1 mark)*

(c) Write on the tree diagram the probabilities for Alan's first turn and the probabilities for Ben's first turn. *(2 marks)*

(d) Calculate the probability that Ben wins the game on his first turn. *(2 marks)*

(e) On the tree diagram draw and label the branches for Alan's second turn. *(2 marks)*

(f) Calculate the probability that Alan wins the game on his second turn. *(4 marks)*

AQA, 2002

7. A marble is rolled down the board and scores 4, 7 or 1.

The probability of scoring 1 is 0.5.

The probability of scoring 4 is 0.3.

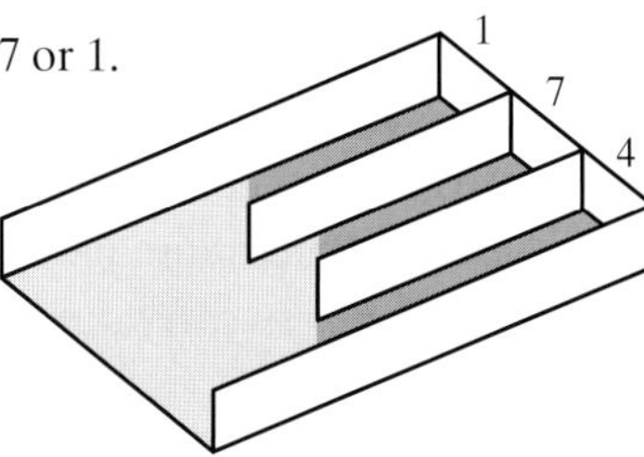

(a) Calculate the probability that a player scores 7. *(1 mark)*

A player rolls the marble down the board 60 times.

(b) Calculate the expected number of times that 4 is scored. *(2 marks)*

Two marbles are rolled down the board and the scores are added.

(c) (i) Calculate the probability of scoring a total of 14. *(2 marks)*

(ii) Calculate the probability of scoring a total that is an odd number. *(3 marks)*

AQA, 1999

8. Alfie has two unbiased five-sided spinners: one red and one blue.

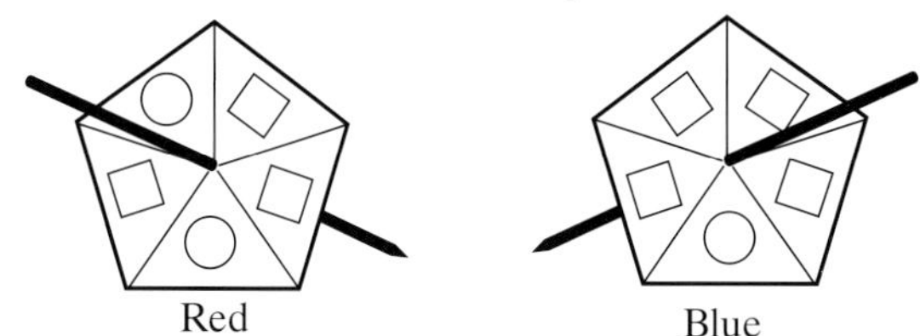

On the red spinner there are three squares and two circles.

On the blue spinner there are four squares and one circle.

Alfie spins one of the spinners. It lands on a circle.

(a) Alfie said the probability of the spinner landing on a circle was 0.2.
Was the spinner red or blue? Give a reason for your answer. *(1 mark)*

(b) Alfie spins each spinner once.
Copy and complete the tree diagram to show the probabilities when each spinner is spun. *(3 marks)*

Red spinner　　Blue spinner

....　Square　...............　Square
　　　　　　　　...............　Circle

....　Circle　...............　Square
　　　　　　　　...............　Circle

(c) What is the probability that both spinners land on squares? *(2 marks)*

AQA, 2006

9. Rashid has a savings box containing 20 coins.

Six of these are 10p coins. The remainder are 50p coins.

Rashid selects at random, one coin at a time from the box. He does not replace the coin.

(a) What is the probability that the first coin he selects will be a 10p coin? *(1 mark)*

(b) If the first coin he selects is a 10p coin, what is the probability that the second coin he selects will also be a 10p coin? *(1 mark)*

(c) Complete the tree diagram below to show all the possible ways in which Rashid could select the first two coins.

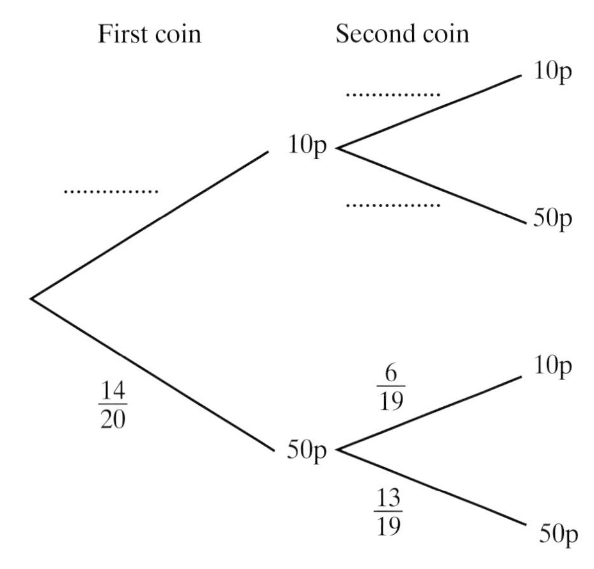

(d) Use the tree diagram, or otherwise, to calculate the probability that the first two coins selected are:

(i) both 10p coins *(2 marks)*

(ii) both of the same value *(3 marks)*

(iii) both different values. *(2 marks)*

(e) If the first six coins Rashid selects are all 10p coins, what is the probability that the next coin selected is:

(i) a 10p coin? *(1 mark)*

(ii) a 50p coin? *(1 mark)*

AQA, 2003

10. Records for a local library show for each book whether it is in the fiction, non-fiction or classics category and whether it is a hardback or softback version.

When the library closed on Wednesday last week 2700 books were out on loan.

Of the books on loan 72 per cent were in the fiction category.

Of the 620 hardback books on loan 55 per cent were in the non-fiction category and 25 per cent in the classics category.

In total 176 classic books were on loan.

(a) Complete the table, entering the number of books on loan in each case.

Category \ Version	Hardback	Softback	Total
Fiction			
Non-fiction			
Classics			176
Totals	620		2700

(4 marks)

(b) A library record for a book on loan is chosen at random. Use the table to calculate the probability that the book is:

(i) non-fiction and a softback version *(1 mark)*

(ii) non-fiction or a hardback version *(2 marks)*

(iii) fiction, given that it is a softback version. *(2 marks)*

(c) How many of the first 200 books taken out on loan on the following day would you expect to be hardback classics? *(2 marks)*

AQA, 2006

11. Students who do their statistics homework have a probability of 0.8 of passing their statistics exam.

Students who do not do their statistics homework have a probability of 0.4 of passing their statistics exam.

In a class, 75 per cent of the students do their statistics homework.

(a) Complete the probability tree diagram.

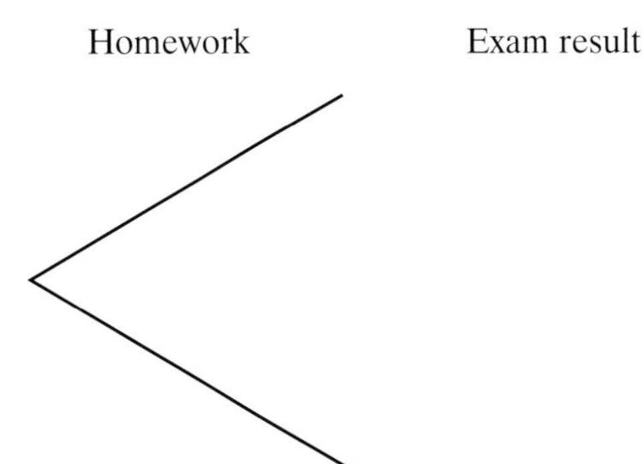

A student is chosen at random from this class.

(b) Calculate the probability that the student passes the statistics examination. *(3 marks)*

In a county, 2800 students passed their statistics examination.

(c) How many of these students would you expect to have done their statistics homework? *(3 marks)*

AQA, 2002

12. In a game of paper, scissors and stone, Chris and Steve place a hand behind their backs. They display their hands, at the same time, as one of the three symbols shown.

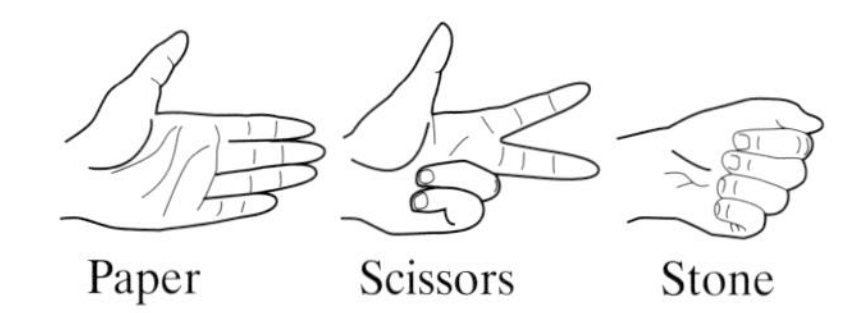

Their choices of symbol are independent.

Chris selects paper with a probability of $\frac{1}{2}$ and scissors with a probability of $\frac{1}{3}$.
Steve selects paper with a probability of $\frac{3}{5}$ and scissors with a probability of $\frac{3}{10}$.

(a) Complete the tree diagram.

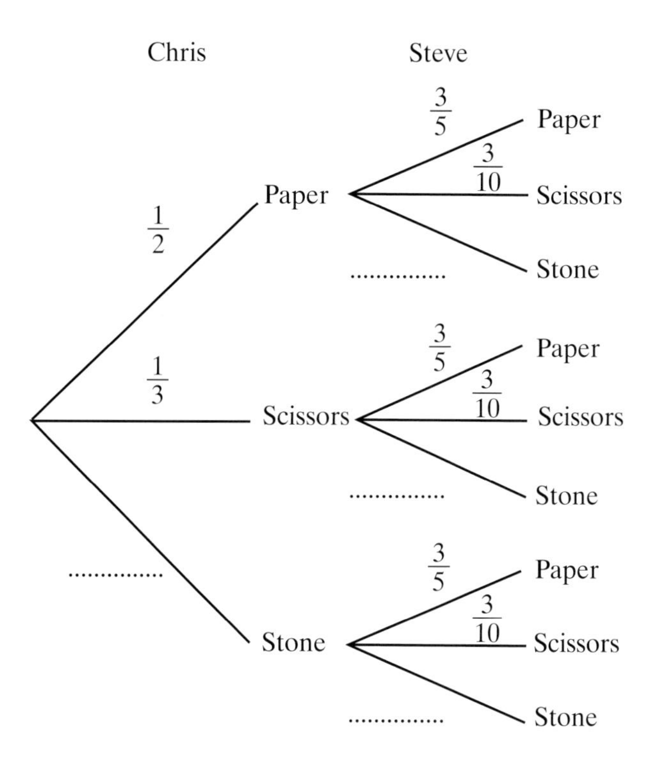

(b) A game is played between Chris and Steve.
Paper beats stone (by wrapping it).
Scissors beat paper (by cutting it).
Stone beats scissors (by blunting them).
The game is a draw if both players display the same symbol.

(i) Show that the probability of a draw is $\frac{5}{12}$. *(3 marks)*

(ii) Calculate the probability that Chris wins. *(3 marks)*

(c) Three games are played.
Calculate the probability that exactly one game is drawn. *(3 marks)*

AQA, 2004

13. **(a)** Kalvinda wants to estimate the area of a wood. He draws a square surrounding the wood on a map. The map is drawn below. The area represented by the square is 25 km^2. He plots 100 points at random. Seventy of these points are in the wood.

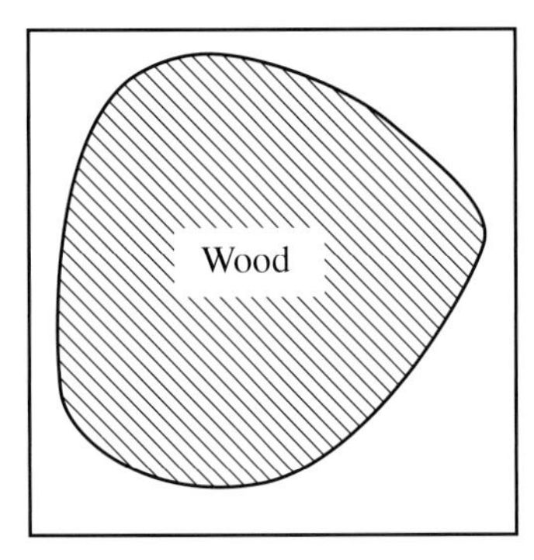

Use these results to estimate the area of the wood. *(2 marks)*

(b) Sarah is going to plot two points at random on the map. Use the estimate obtained by Kalvinda to calculate the probability that exactly one of the points is in the wood. *(3 marks)*

(c) John is going to plot four points at random. He wants to calculate the probability that exactly two points are in the wood.

(i) List the different orders in which the four points could be selected.
One is done for you.
1. In, In, Out, Out *(1 mark)*

(ii) Calculate the probability that exactly two points are in the wood. *(3 points)*

AQA, 2003

14. The table gives the number of pairs of shoes sold by size and width fitting by a local shop.

		Width fitting				
		C	**D**	**E**	**F**	**Total**
Shoe size	**5**	3	5	3	2	13
	6	4	7	8	3	22
	7	2	4	5	3	14
	8	1	2	3	1	7
	Total	10	18	19	9	56

(a) What is the probability that a person selected at random buys a pair of shoes of size 5, width D? *(1 mark)*

(b) What is the probability that a person selected at random buys a pair of size 5 shoes? *(1 mark)*

(c) What is the probability that a person selected at random buys shoes of width D, given that they brought shoes of size 5? *(2 marks)*

(d) Two people are selected at random. What is the probability that they both brought shoe size 5? *(3 marks)*

AQA, 2003

13 The controlled assessment

In this chapter you will learn:

to plan and carry out the investigation for your controlled assessment.

You should already know:

- ✔ about the data handling cycle from Chapter 1
- ✔ how to write a hypothesis
- ✔ how to plan an investigation
- ✔ about representation of data
- ✔ about calculation of summary statistics.

What's this chapter all about?

Now that you are reaching the end of the book it's time to start thinking about exams (and assessment). The good thing about this course is that you can get some marks before you even enter the exam room. To do this you will need to complete a 'controlled assessment'. That is, you carry out a statistical investigation and then you follow it up with a written assessment. This is practical statistics and that's why the assessment is called 'statistics in practice'! The controlled assessment is untiered. This means that you will do the same assessment as all the other candidates, no matter whether you are entered for the foundation tier or the higher tier. This chapter is here to help you make sure that you know what you have to do, so you can score more marks.

Basically, it is your chance to show what you can do and you will need to use pretty much all of the work in this book, starting with the data-handling cycle from Chapter 1. That cycle is crucial for getting around here, so don't lose it! Just remember that the last part of the data handling cycle – the evaluation bit – is assessed in the written assessment and not in the investigation, so you don't need to worry about that until later ...

Good luck!

Focus on statistics

All of this is new as there is no controlled assessment or coursework in GCSE maths. Remember, this is your opportunity to show how much you have learnt doing this course.

13.1 The investigation

This is the first part of the controlled assessment. This part is worth 20 marks, and you are expected to spend no more than 8 hours on the investigation. You will be given a task sheet, which will have a simple statement on it. You are expected to use this as a starting point. The first thing you will need to do is to make sure that you understand any terms that are used. You will need to think of a hypothesis, or possibly two hypotheses that relate to the statement you are given. You are encouraged to discuss this statement with your classmates and your teacher, who will be able to give some advice.

You will need to think about the following questions:

- What data will you need to collect to address your hypothesis?
- Is the data available?
- Where will you collect the data from?

Data selection

You are not expected to collect primary data but you are expected to state where your data comes from. The exam board will give you websites and occasionally data sets for you to work with but you can choose your own if you wish.

You will need to decide on the best method to sample your data and what makes it the best method to use. You will also need to think about what you are going to do with your data – how you will use it to address your hypothesis. This will come into your decision making about what data to collect.

Remember, you should:

- make your hypothesis clear
- keep things as simple as possible
- state any variables that are involved.

You might want to consider a pilot survey before you start on your main study, but this should not extend the time your entire investigation takes. This will allow you to check that your variables and the data you collect will allow you to address your hypothesis.

Once you have collected your data you will need to summarise it. Your summary will probably be in the form of a table.

links

Use what you have learnt in Chapter 4 on data representation.

Diagrams

Ensure that your data representation is appropriate. Think about why the method you have chosen is better than other methods that you could have used. This will be useful in the written part of the assessment.

Calculations

Ensure that your calculations are appropriate. What do you want them to show and why are the methods you have chosen better than other methods you could have chosen? You will want to find some measures

AQA Examiner's tip

Do not write about why you have chosen diagrams and calculations in the investigation part. There are questions about this in the written assessment.

of spread and some measures of average in order to make comparisons. It might be appropriate to find out about correlation.

Interpretation

You will need to write about what your representations and calculations tell you. You will need to refer back to your hypothesis. Do you think your hypothesis was correct or not, or is the evidence that you have found not sufficient to answer this question?

You might wish to develop or extend your work.

When you have completed your work your teacher will collect it in and mark it.

> **AQA Examiner's tip**
>
> Look at the criteria used to mark the task. You do not need to do every type of graph and calculation you can think of. This would take far too long and would leave you nothing to talk about in the written assessment. You may be asked to suggest another diagram or calculation that you did not use in your investigation and give reasons why.

Sample Investigation

'Who does best in the Olympics?'

You are expected to use your knowledge of statistics to write a suitable hypothesis for this question. You should then investigate the hypothesis and draw a conclusion.

Starting point

You will need to decide exactly who the 'who' refers to and how do you assess the word 'best'. There are lots of possibilities.

The 'who' could be:

- males and females
- taller (or shorter) people
- people from which country
- countries with a higher GDP
- countries with a larger population
- countries from further away
- countries with a warmer (or colder) climate
- countries with more (or less) competitors
- countries which did better last time.

There are probably a lot of other possibilities you may choose to consider.

The 'best' could be:

- most medals
- most golds
- most records (Olympic, personal bests)
- most medals per competitor
- most medals for the size of country (area of land, population, population density).

There are probably a lot of other possibilities you may choose to consider but remember that the task should take no more than 8 hours or so to complete, so concentrate on one or two issues.

Sources

Where are you going to collect your data?

There are a lot of websites to explore. For example:

- **www.cia.gov**
- **en.beijing2008.cn**
- **www.olympic.org**
- **www.london2012.com**

This list is just for starters. Remember too that some websites or data will be suggested by AQA.

Hypothesis

Now write your hypothesis. You could consider a pilot survey. How much data do you need? What data will you collect? What sort of analysis will you use? If you are comparing then you will need to use averages and measures of spread – but you will need to decide which ones to use.

links

See Chapter 5 for methods on calculating averages.

What representation will you use? For example, you could consider box plots or histograms. You will need to decide what you want to use your representation for, and what is appropriate for your data.

If you are looking for a correlation then you will need scatter diagrams or correlation coefficients.

You will need to interpret your diagrams and calculations. What do they tell you about your hypothesis? And what is it about your calculations and diagrams that tells you this is the case? In other words, have you found supporting evidence for your hypothesis or not?

At the end of this chapter, you will be asked to tackle the sample investigation outlined above. You should work through this investigation carefully:

- think about the decisions you make as you go
- collect the data
- represent and summarise the data with statistical calculations
- interpret your calculations and diagrams
- write a report for someone else to read, explaining what you have done.

links

See Chapter 8 to remind yourself about scatter diagrams and correlation.

But before working on this assignment, read the next topic, which is all about the second part of the controlled assessment – the written assessment.

13.2 The written assessment

This is the second part of the controlled assessment. This part is also worth 20 marks, and is 45 minutes long. You will do this under exam conditions and at a time chosen by your teacher. At the start of the written assessment you will have your completed investigation returned to you so that you can refer to it in your assessment, but it will not have been marked at this stage. Both the investigation and written assessment will be collected in at the end.

Section A

This consists of a number of questions relating to your own investigation and requires you to evaluate your own findings. It will ask about some of the decisions you made while doing your task, such as why you used the diagrams you did, what other diagrams you could have used and why the ones you chose were best. You could be asked about the sampling method you chose, or the calculations you did. You could be asked about any developments or extensions you could have done. A lot of this is about you evaluating your work.

Section B

At the start of this section you will either be supplied with additional data on a related topic or will be asked to expand your original investigation. The written assessment will always total 20 marks but there may be slight variations in the allocation of marks to each section, according to the nature of the task in a given session.

You are not allowed to bring any notes with you into the assessment and must refer only to your completed investigation during the examination conditions.

You will only be allowed to make one attempt only at a written assessment and redrafting is not permitted.

Sample written assessment

A sample written assessment is provided at the end of this chapter for you to try. You should complete the investigation first and then perhaps a few days later you should try the written assessment. Don't leave too long a gap as it's best to tackle the written assessment while the work you did in your investigation is still fresh in your mind. You should try to do this work under exam conditions and before you start you should ensure you have a copy of your investigation in front of you.

Summary

You should:

be able to plan a line of enquiry

know what is expected in your controlled assessment.

AQA Examination-style questions

Sample Investigation

'Who does best in the Olympics?'

You should use your knowledge of statistical methodology to write an appropriate hypothesis for this question.

You should then investigate the hypothesis and draw a conclusion.

Sample written assessment *(20 marks)*

SECTION A

1. Explain why two of the diagrams you used in your work were appropriate for representing your data.

Suggest another way you could have represented your data. State why it would be appropriate to use this method. *(6 marks)*

2. What were the limitations in the data you used? How could you have developed your task further? *(4 marks)*

SECTION B

You are asked to follow up your investigation by analysing data from the annual sports day at a big secondary comprehensive school in a large city.

3. Explain how you would go about selecting and collecting appropriate primary data.

You should include the sampling method and what data you would want to collect. Explain why you would do it this way. *(5 marks)*

4. What calculations would you use to analyse the data?

Explain why you would use the calculations you have chosen and what you would expect to gain from using them. *(5 marks)*

Answers to revision questions

1 Chapter 1

Exercise 1.1

Throughout this exercise the data needed is entirely dependent on the hypothesis suggested in part (a) of each question. As there is a very large variety of hypotheses, which might be at least as valid as those suggested here, a complete set of correct answers cannot be given. However, the answers that follow suggest possible responses and could be used to promote discussion.

1

a Our products are more popular than those of our competitor.

b Either quantity of items sold, or sales figures from both companies.

c How large the companies are, number of employees, prices of products, quality of the products and how long they last, customer satisfaction. There are other valid factors.

2

a My team is getting better.

b Number of goals scored, number of matches won, position in league for this year and last year.

c How many new players the team has, age of players, attendance of spectators, quality of opponents. Whether they lose to the same teams in consecutive seasons.

d It may be difficult getting details about the players themselves and the records of each match may be difficult to obtain. If Sam's team is a 'local' team the historical data may not exist.

3

a More waste is recycled this year than last year.

b Quantity of waste going to landfill, quantity being recycled.

c Method of collection of household waste, customer satisfaction. Whether other areas are improving in the same way. Fly-tipping.

d They may not be able to find reliable data about fly-tipping. Other areas may keep different records. They may have changed the way they collect refuse (they may have subcontracted some of the recycling). Different areas may recycle different waste products in different ways (for example some recycle cardboard by collection – others do not).

e Look at what other waste products could be recycled. Look at changes in collections to see whether more frequent collections encourage more recycling.

4 Mary could ask her customers and her employees. She could ask people who do not use her salon to see if they would use it if appointments were available. She could ask people who live in a different area of town if they would use her salon if it were closer to where they live. She could even consider the number of salons that other (similar) towns have.

Spot the mistake

Herbie has forgotten that although shoe sizes are available in half sizes, there are no other sizes in between these. That is, you cannot get a size between 5 and 5 ½, for example. Thus the variable is not actually continuous.

Exercise 1.2

1 Primary data from her own company, which she can collect, secondary data from the competitor as it is unlikely that she will be able to obtain primary data from them. For customer satisfaction she could use primary data.

2 Sam could use primary data if he starts recording it now and for future performance. If he has just posed this question and wishes to compare current performance with previous performance then he will need to use secondary data as he may not have data already recorded.

3 The local authority will be able to use primary data for its own recycling but will have to use secondary data from other authorities for comparison as a neighbouring authority will collect its own data. If it uses other companies to help with refuse then it might have to use secondary data.

4

Qualitative	Quantitative		Categorical
	Discrete	Continuous	
Sweetness of a cup of tea Colour of a car	Number of pips in an orange Cost of a newspaper Number of doors on a car Mark in a SATs exam	Speed of a car Weight of an orange Length of a pencil Amount of petrol in a car	Level in a SATs exam Size of an egg in an egg box

5

a Area (or length, width, thickness, weight)

b Cost (or number of perforations around the edge)

c Colour (or texture)

d Class, for example first class, second class or airmail (or country of origin)

6

a C Colour of the counters

b B Number of players

Exercise 1.3

1 All the cars he sells.

2

a Energy-save light bulbs last longer than ordinary ones.
b All the light bulbs that Brightlight make.
c A sample should be used. As they need to measure the length of time the bulbs last until they stop working, then if they use a census (testing all of them) they would not have any left to sell!

3

a Students prefer healthy dinners.
b All the students at the school (not just those that have school dinners).
c Census. As the school is small they can ask everyone.

4 There would be far too much data for her to be able to use if she used census data.

5 A census uses the whole population; a sample uses part of the population.

Exercise 1.4

1 Example:

This is a short questionnaire about part time jobs. Please answer all questions by ticking the appropriate box. All information you provide will be treated confidentially. Thank you.

a Are you male ☐ female ☐ ?
b Do you have a part time job? Yes ☐ No ☐

If you answered 'No', then thank you – you have completed the questionnaire.

If you answered 'Yes', then please continue to answer the questions below.

c How much do you get paid per hour?

less than £4.50 ☐

£4.50 or more, but less than £5.00 ☐

£5.00 or more, but less than £5.50 ☐

£5.50 or more ☐

d How many hours did you work last week?

Less than 5 ☐

5 or more, but less than 10 ☐

10 or more, but less than 20 ☐

20 or more ☐

2

a Leading ('anyone who likes . . .'). Only gives two options when there are many others. Not everyone likes football.
b If you like football please write the name of your favourite team in the space . . .
In this case there are far too many options to ask a closed question.

3 The main reason is to test the questions, although testing the sampling procedures or method of data collection are other reasons.

4 This is a short questionnaire about your stay here. Please answer all questions by ticking the appropriate box. All information you provide will be treated confidentially. Thank you.

a Are you male ☐ female ☐ ?
b Was your stay for Business ☐ Pleasure ☐ Other ☐ ?
c How long was your stay?
1 night ☐
2 or 3 nights ☐
4 or 5 nights ☐
6 nights or more ☐
d What is your age?
Below 30 ☐
30 or more, but less than 40 ☐
40 or over ☐

5 A pilot survey would allow him to get an idea of what people would give as their answers. This would help him avoid options for which everyone ticks the same box when he designs his questionnaire.

6

a i A closed question has options to choose answers from.
ii People are more likely to give their age if they choose from age ranges.
b Beckie could have offered an incentive to people replying – free tickets to the restaurant in a draw. (Beckie could send out reminders, or in this case knock on doors, to ask for them back, as the village is small and she knows that everyone was given a questionnaire.)
c Beckie should have done a pilot survey to test the questions, or have got someone else to read them for her.
d The question is not relevant, asking about pubs when her survey is about a restaurant, some people eat out but may not have a meal.

7

a i Overlaps (£1500 is in two options, as are £2000, £3500 and £6000 – these are all the same problem).
ii There is no option for people who do not go abroad. (There needs to be a 'less than £1000' option, which could allow them an answer.)
b i Offer an incentive (for example, a prize draw).
ii Make it easy to answer. Another possible answer here is to make sure it is freepost for returning (as it is a postal survey).

8 Use a random response method. The teacher may choose to toss a coin, if it lands 'heads' the student ticks the 'yes' box; if it lands 'tails' the student answers the question 'Have you smoked any cigarettes in the last week?' with separate 'yes' and 'no' tick boxes. About half the class should get a 'head' and tick the 'yes' box; the rest will answer the question, so she will be able to work out roughly how many smoke.

9 i Continuous opinion scale:

Agree ______________________________ Disagree

ii Discrete opinion scale:

Agree	☐
Slightly agree	☐
Neither agree nor disagree	☐
Slightly disagree	☐
Disagree	☐

Exercise 1.5

1

a

Age in completed years	less than 15	15 to 24	25 to 34	35 to 44	45 and over
Male tally					
Female tally					

b Change the age classes to be:

Age in completed years	less than 13	13 to 19	20 to 34	35 to 44	45 and over
Male tally					
Female tally					

The teenagers would all be in the 13 to 19 class.

c A data logger would just record the number of people entering, not their gender or age.

2 Mr T Rout should divide the fish into two groups when they are very small/young. Feed one group with the new type of food and one group with the usual food (this is the 'control' group – he cannot leave them unfed!). He can then sample from each group as the fish grow but he does need to make sure that they are not put back with the rest once they have been weighed!

3 Sow identical seed into two seed trays, one with each type of compost. Keep the two trays under identical conditions so that they get the same amount of light, warmth and water. If one seed tray has someone talk to it then so must the other one! (Talking produces carbon dioxide and warm air! It is important that both trays have the same conditions.) Once the seeds start to germinate, count the number of seedlings in each tray every day.

4 The amount of fertiliser is the explanatory variable, and the number of flowers is the response variable.

5

a Matched pairs. Ideally you would use twins. However, the company would probably have to use pairs of people who are as similar as possible, same IQ, age, gender. Test their memory before being given the supplements. Randomly allocate one from each pair to take the supplement. Test them again after two weeks of taking the supplement. Compare the results.

b Types of things in the memory test, IQ, age, time of day, day of the week, job that people do, other medication, lots of other possibilities . . .

2 Chapter 2

Spot the mistake

Niles is asking a biased sample. There is a strong chance that people who take media studies will have different viewing habits to those of the average 15-year-old.

Exercise 2.1

1

a Not true – variations can occur in data even when the data is collected carefully. (Think about rolling a die – do you always get the same results?)

b True – a census means you have to ask or test the whole population.

c Not true – a sample will usually be at most 10 per cent of the population and often far less.

2 Given that tasting the cake makes it unsuitable for sale, this sample size can be very small. It would soon be obvious if there were problems with the cakes. Suggest perhaps three to five cakes at intervals through the day.

3 Sample size is too small and this is really not representative of the whole population as the parents of his friends could be of similar age.

It is too long since 2007; people might not know or remember.

He should record the data immediately, or else he could make a recording error or forget the responses.

Exercise 2.2

1 Number the albums 00–76 or 01–77.

Obtain a few two-digit random numbers using a calculator.

If the first number is over 77, choose a second number, repeating this process as necessary.

Match the chosen number to the list.

2

a Stratified sampling to reflect the proportions of full- and part-time workers

b 20

3

a Musicians may not play much sport or might be too busy to practise! He is also only sampling one type of person.

b Number the pupils from 001 to 150 (for example). Obtain a selection of 3 digit random numbers (probably about 20). Many of the random numbers will be above 150, so use the first 6 at or below 150, ignoring repeats, and match these to the list of pupils.

c It is too small. He should aim for 15–30 ideally as this should not be too difficult to get.

4 Ellie is wrong, because as soon as the starting point for the systematic sampling is chosen, the majority of items cannot appear in the sample, and therefore do not all have an equal chance of being chosen, as is required for a genuine random sample.

5

a Example of calculation: total number of animals looked after = 1237

number of one week cats = $\frac{231}{1237} \times 50 = 9.34$

Values rounded to 2 d.p. are in this table:

	One week	Two weeks
Cats	9.34	4.37
Dogs	16.41	13.06
Rabbits	4.16	2.67

Size of sample for each category are therefore:

	One week	Two weeks
Cats	9	4
Dogs	17	13
Rabbits	4	3

Notice that the value 16.41 is rounded up to preserve the sample size of 50 as required. Rounding 16.41 causes the least error because this value is closest to the 0.5 boundary.

b There is an assumption here that each single animal was from a different customer (house), which is quite unlikely.

6

a Eating habits of males and females may well be quite different

b $0.42 \times 50 = 21$

c Number the boys. Calculate the interval that matches a sample size of 21. Obtain a random starting point. Count on to get the remaining sample members.

7 Choose a year group at random.

Choose a tutor group at random.

If necessary quota sample from within this tutor group (or ask them all as a cluster sample).

3 Chapter 3

Spot the mistake

Paul has used uneven class interval lengths for no good reason. However, if he were to use some narrower groups it would surely make sense to do this where there is more frequency in the middle of the data, not at the extremes. Shelley has used only two class intervals so virtually all the information has been lost.

Exercise 3.1

1

Number of goals	Tally	Frequency
0	||	2
1	||	2
2	𝍸 |||	8
3	𝍸 ||	7
4	𝍸 |	6
5	||||	4
6	|	1

2

a i 30 ii 49.999... iii 20

b i 100 ii 199.999... iii 100

c i 22 ii 23.999... iii 2

d i 1000 ii 2499.999... iii 1500

If the data had been rounded to the nearest integer (there is no indication that it had) the answers would have been different, for example: (a) i 29.5 ii 49.5 iii still 20.

3

a

Time, t seconds	Tally	Frequency
$10 \le t < 20$	|||	3
$20 \le t < 30$	𝍸 ||	7
$30 \le t < 40$	𝍸 𝍸	10
$40 \le t < 50$	𝍸 𝍸	10
$50 \le t < 60$	|||	3
$60 \le t < 70$	𝍸	5
$70 \le t < 80$	||	2

b The highest value was 71, near to the low end of the last class interval. The possibility of very high values if further data was to be collected is very unlikely due to the regular nature of the way traffic lights work.

Note that you could have, looking at the data, made some of the classes of different size. For example, $30 \le t < 40$ and $40 \le t < 50$ could, if desired, have been split into classes of width 5 s and the final two classes could have been one of length 15 s instead.

Exercise 3.2

1

a 2
b Rain in the day followed by a cloudy but dry night.
c 5 + 3 + 3 + 4 = 15

2

a Answers in bold:

	Raisins	Apple	Banana	Total
Years 2–4	15	1	**5**	21
Years 5–6	**10**	8	**16**	**34**
Years 7–11	3	**18**	24	45
Total	28	**27**	**45**	**100**

b 45 per cent

3

Film type/ screen	Screen 1	Screen 2	Screen 3
Children's			
Comedy			
Adventure			

(Additional row / column for totals possible)

Note that the rows and columns can be either way around.

Exercise 3.3

1 23.9 per cent

2 1.64 m and 1.79 m

3 Males as 76.0 – 70.9 = 5.1 years increase whereas females as 80.5 – 76.8 = 3.7 years increase

4 Northumberland National Park

5 False, it has the 15th best in the UK

6 15

7

a i 1.89 ii 1.67 iii 1.77
b Yorkshire and the Humber
c South-East
d The axis does not start at 0; it starts at 1.5 making differences seem much greater than they are.

4 Chapter 4

Exercise 4.1

1

Monday	
Tuesday	
Wednesday	
Thursday	
Friday	

2 Example calculation of angle UK:
9 + 6 + 12 + 7 + 5 + 6 = 45 families

360 ÷ 45 = 8 degrees per family

So UK = 9 × 8 = 72 degrees

Other angles:

US = 48 degrees

Spain = 96 degrees

France = 56 degrees

Canary Islands = 40 degrees

Other = 48 degrees

Pie Chart Showing holiday destinations

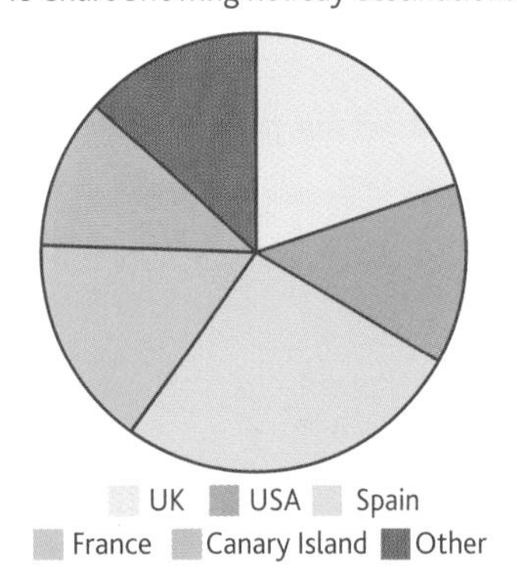

3

a 60
b 90
c Three-and-a-half symbols drawn for Thursday.
d As the total number of DVDs is 360, one degree on the pie chart will stand for one DVD hired.

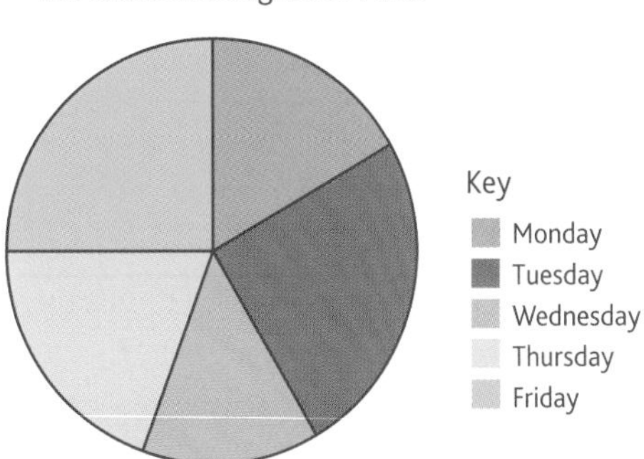

4 $\frac{\pi r^2}{6^2} = \frac{150}{100}$ (where r is the desired radius value)

so $r = \sqrt{\frac{6^2 \times 150}{100}}$

$= \sqrt{54}$

$= 7.3484692...$

$= 7.3$ cm (1 d.p.)

Spot the mistake

A key, a title and a scale are all missing, plus labels for the horizontal axis.

Exercise 4.2

1

a White coffee
b Black coffee
c 23 + 16 = 39
d i

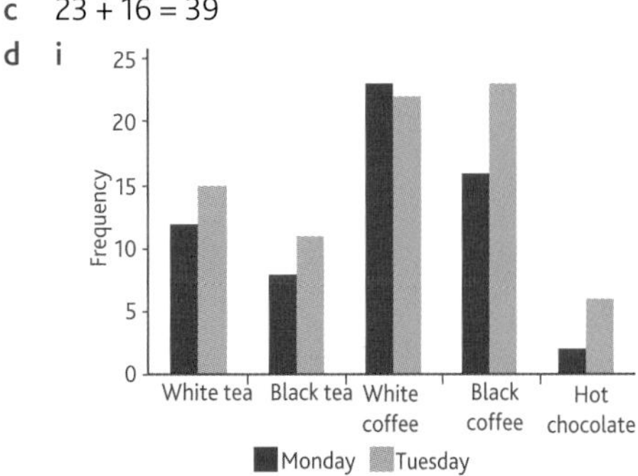

ii (There are many options for the answers, so examples are given.) Hot chocolate was the least popular of these drinks on both days. White coffee was the most popular on Monday but Black coffee was the most popular on Tuesday.

2

a

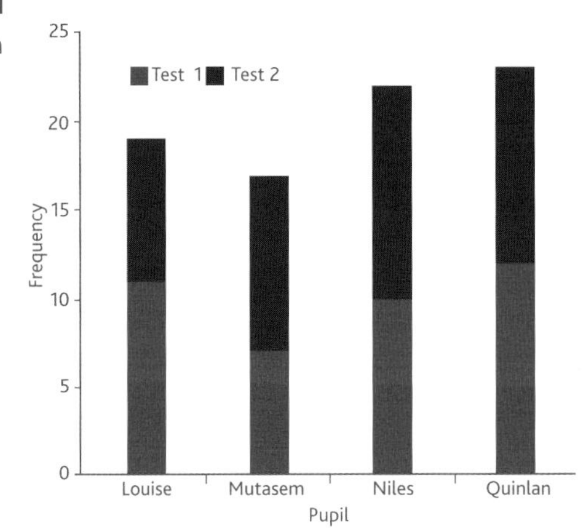

b The overall total score is clear straight away from the component bar chart.

3

a

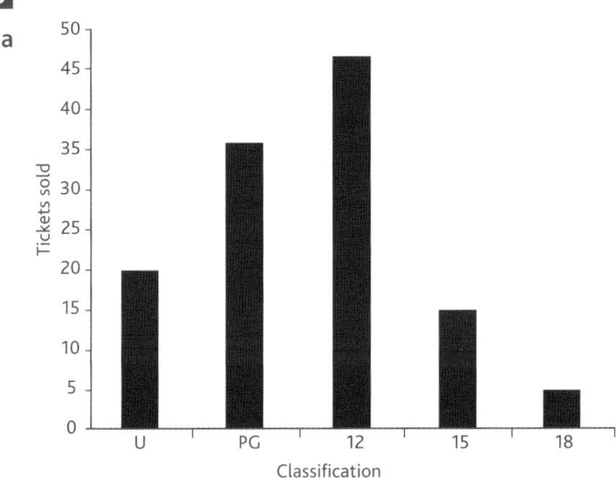

b Example calculation: $U = \frac{20}{120} \times 360 = 60$ degrees

Other angles (in degrees): 105, 135, 45, 15

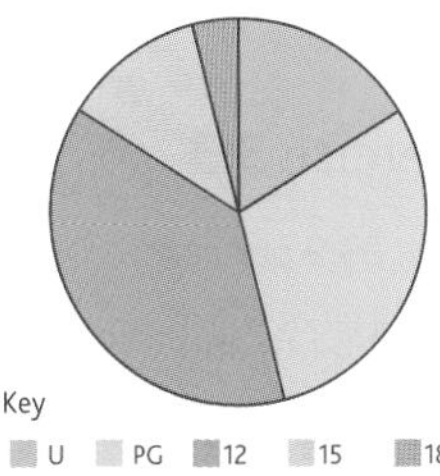

c It is difficult to justify a pie chart as it is more work and harder to compare frequencies – best answers will relate to a more visual comparison of proportions. For example, it is easier to see what proportion of tickets were sold for each type of film.
d Quicker, easier to draw and interpret and shows a clear comparison of frequencies.

4

a 10
b 4 + 4 = 8
c (5 × 2) + (2 × 3) + (10 × 4) + (4 × 5) + (4 × 6)
= 10 + 6 + 40 + 20 + 24
= 100 (this assumes that 'full' means every seat, not just every table)

d

Number of tables

12 10 8 6 4 2 0

2 3 4 5 6

Size of table

e i

2 3 4 5 6

size of group

ii Split a couple of the tables for five into twos and threes as there were not enough tables for three but none of the tables for five were needed.

iii Many of the available and desired table numbers matched up well and this was only the data for one evening; a much larger sample should be looked at before making too many changes.

Exercise 4.3

1

a

Stem	Leaves
0	7 8 9
1	0 1 2 2 2 6 6 7 9 9
2	0 1 2 3 3 4 4 7
3	0 4 4

Key 0 | 7 represents 7 minutes waiting time

b $34 - 7 = 27$ minutes

c Only one-third of these patients have been seen with waiting times below the target time (8 out of 24). However, this is only one morning's data and it is likely that waiting times are strongly linked over one morning. More data is needed.

2

a

Men	Stem	Women
9 9 8 7	0	9
8 7 7 6 3 1 0	1	5 5 6 7 9 9
2 2 1	2	0 2 6 7 9
	3	0 2

Key 1 | 2 | 0 represent 21 men and 20 women

b There were clearly more women than men in the theatre. Some rows had 30 or more women but never more than 22 men; whereas several rows had fewer than 10 men and only one row had fewer than 10 women.

c It is unlikely that the theatre was full unless the later rows were considerably smaller than the front ones, as there were fewer people in total in rows L, M and N.

3

a £184 000

b £209 000

4 Statement 1 – true – count the leaves on the RHS = 30.

Statement 2 – true – lightest is 0.4 kg, which is 400 g as there are 1000 g in a kg.

Statement 3 – false – the heaviest second-class parcel was 6.9 kg.

Statement 4 – true – the most common second-class parcel was 3.7 kg and if you count the leaves less than this on the first-class side you will find that there are 20 of them.

Statement 5 – cannot tell – the two heaviest parcels were definitely second class but there is a 6.4 kg parcel on both first and second class and, as both have been rounded to the nearest 0.1 kg, it is not possible to tell which of the two is the heavier from the stem-and-leaf diagram.

Exercise 4.4

1

a 2000

b 2000

c i Most of 15–19-year-olds will still be in education on the island, whereas 20–24-year-olds have potentially gone to work off the island, or are at university off the island.

ii Perhaps people move to the island when they retire.

2

a The lowest value in the class is 20, and the highest is 29.999 . . . as you are in the 29 group until the moment you become 30. The midpoint is then $(20 + 29.999\ldots) \div 2 = 25$

b

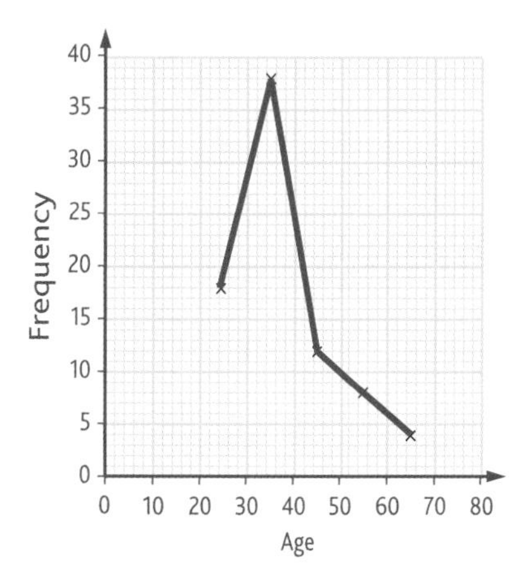

c There are 80 workers, so 25% would be 20. Over 45 years is some of the $40 \le x < 50$ class (approximately half) and all of the last two classes. This would be 4 + 8 + about 6 which is about 18 so it a reasonable statement.

3

a Frequency density = frequency ÷ class width.

For example, for first class = 6 ÷ 3 = 2 and other frequency densities are 8, 21, 20 and 1.7.

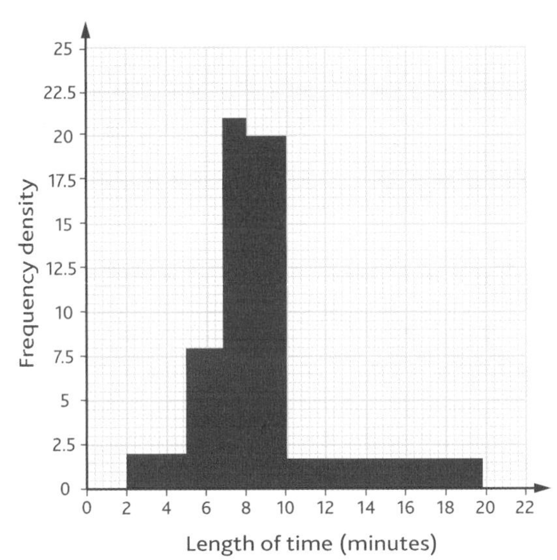

b No, 40 out of 100 is not the majority even though it was the modal group.

c 8 + 21 + 40 + 1/5 of 17 = about 73 so not quite but very close to 75 per cent and this is only an estimate due to the nature of grouped data, so it is a close call.

4

a

Weight, w kg	Frequency
$8 \le w < 16$	2.5 × 8 = 20
$16 \le w < 20$	5 × 4 = 20
$20 \le w < 22$	6.5 × 2 = 13
$22 \le w < 28$	3.5 × 6 = 21

b 21 ÷ 2 = 10.5 so 10 or 11 is the best estimate.

5

a The class runs from 89.5 s to 109.5 s so the class width is 109.5 – 89.5 = 20

b Frequency densities are 0.3, 0.7, 2.2, 2.35, 0.32 and 0.02. Bars are plotted on 89.5–109.5, 109.5–119.5 and so on.

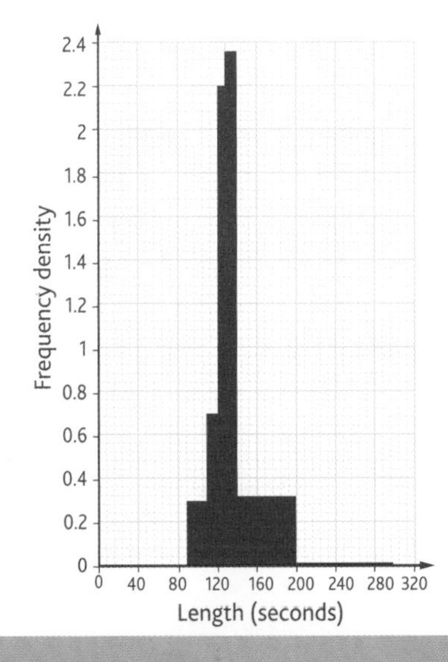

5 Chapter 5

Spot the mistake

1 The median is wrong and should be 2.

2 It's not possible to get 2.09 people! The mode shows the number which occurs the most times, it does not mean that the mode occurred more than half the times, which is what his statement means.

Exercise 5.1

1

a i 35 ÷ 5 = 7 ii 96 ÷ 8 = 12 iii 365 ÷ 12 = 30.42 iv –53 ÷ 10 = –5.3

b i 7 ii 11.5 iii 31 iv –5.5

c i no mode ii 11 iii 31 iv –4 and –6

2 71 × 4 = 284

3 71 as $\frac{(67 + 71)}{2}$ gives 69

4

a mean = 57 ÷ 10 = 5.7 mm

median = 1.5 mm

mode = 0 mm

b Mean using all the data, or median indicating many days have low rain (not mode – misleads).

5

a mean

b median or mode

c mean

d mode

6

a Any four numbers that total 20, for example 4, 5, 5, 6.

b Any five numbers where the middle number is 6 when in order, for example 4, 5, 6, 7, 8.

c Any six numbers where the middle two numbers are both 7 or are either side of, and same distance from, 7 – for example, 4, 5, 6, 8, 9, 10.

d Many possible answers, for example 4, 6, 10, 12.

7

a The mean will not have changed – disc has mean value.

b The median might have changed – depending on other discs. The mode might have changed – depending on other discs.

c None of the measures will definitely have changed.

8

a The mode will not have changed – disc value is already the mode.

b The median might have changed – depending on other discs.

c The mean will definitely have changed and will be lower than 4.

9 Subtract 1000 from each value ⇒ 8, 7, 6, 8, 3, 3, 10, 9, 11, 4. Work on this data then add 1000 back on. Mean = 1006.9, median = 1007.5.

10

a Subtract 23 000. Mean = 23 063.

b Tmes by 100 000 then subtract 100. Mean = 0.00133625.

11

a $\sqrt{1.08 \times 1.1} = 1.08995$ giving an average increase of 8.995%.

b This value is slightly less than the arithmetic mean of 1.09, which indicates a 9% increase on average not 8.995%.

12 Let x be the unknown number, then $\sqrt[3]{3 \times 8 \times x} = 6$

then $24x = 216$ so $x = 9$.

Spot the mistake

Abby cannot be correct, as she has found the mean to be 8 for data that takes values from 1 to 5 inclusive.

Barry's answer is more than halfway through the data range, yet most of the values are 1s and 2s, with fewer 3s, 4s and 5s.

Neither Abby nor Barry has used the frequencies in the correct way, as they did not multiply the frequency value by the number of bottles. For example, in the first row there are 12 instances of the result 1, so the total frequency for result 1 is 12, and so on.

The correct answer is found by dividing the total of all the values by the number of values:

$$\frac{(12 \times 1 + 14 \times 2 + 6 \times 3 + 7 \times 4 + 1 \times 5)}{(12 + 14 + 6 + 7 + 1)}$$

$$= \frac{(12 + 28 + 18 + 28 + 5)}{40}$$

$$= \frac{91}{40}$$

$= 2.275$

Exercise 5.2

1

a $42 + 20 + 8 + 2 + 3 + 6 + 1 = 82$

b $1 \times 20 + 2 \times 8 + 3 \times 2 + 4 \times 3 + 5 \times 6 + 6 \times 1 = 90$

c $90 \div 82 = 1.0976$

d 0 (the 41st = 0 and the 42nd = 0)

e 0

f Ian scores more runs per ball then Geoffrey. However, more information is required to make a genuine comparison between the two players.

2 For company B, the mean = 230 ÷ 50 = 4.6 minutes late (see table below).

Number of minutes late	Frequency	Value × frequency
0	22	$0 \times 22 = 0$
5	12	$5 \times 12 = 60$
10	14	$10 \times 14 = 140$
15	2	$15 \times 2 = 30$

For company B, the median (25.5th value) is 5 minutes late. Therefore we can say that the mean and the median both show that company B is late by fewer minutes on average than company A.

3

Time taken for some children to solve a simple puzzle (seconds)	Frequency	Midpoint	Frequency × midpoint
$0 \le t < 50$	15	25	375
$50 \le t < 100$	28	75	2100
$100 \le t < 150$	20	125	2500
$150 \le t < 200$	27	175	4725
$200 \le t < 250$	10	225	2250
	total = 100		total = 11 950

a $11\,950 \div 100 = 119.5$ s

b The median is approximately the 50th value. This is the seventh out of 20 in the $100 \le t < 150$ group.
Estimate of median is therefore $(\frac{7}{20} \times 50) + 100 = 117.5$ s

c 

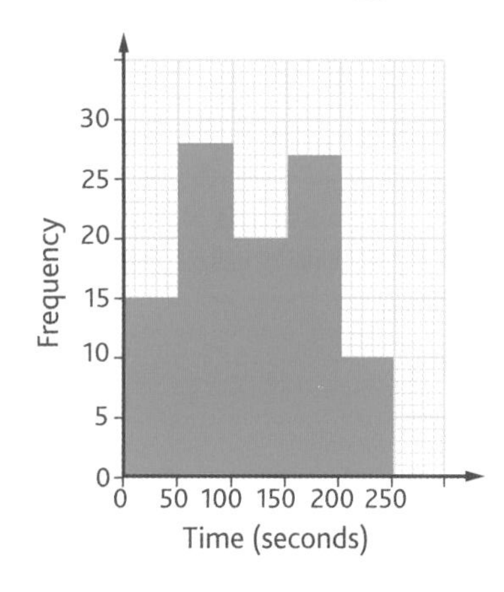

d From the diagram below estimate of mode is approximately 80 s.

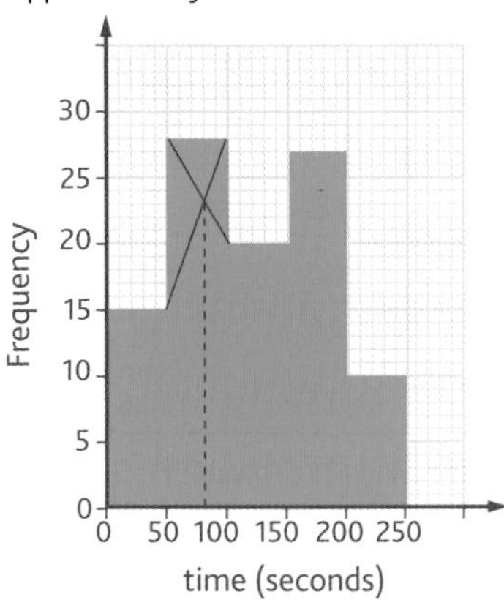

e Median line should be 35 per cent to the right between the vertical lines at 100 and 150 = 117.5. This is (and should be) exactly the same as the estimate in (b).

4

a 52

b

Length of bolt, l, (mm)	Frequency	Midpoint	Frequency × midpoint
$30 \le l < 31$	15	30.5	457.5
$31 \le l < 32$	28	31.5	882
$32 \le l < 33$	20	32.5	650
$33 \le l < 34$	27	33.5	904.5
$34 \le l < 35$	10	34.5	345
	total = 100		total = 3239

Mean = 3239 ÷ 100 = 32.39

This is within half a millimetre of the desired length, therefore the machine should not be scrapped.

5

a Toptaste mean = 315 ÷ 45 = 7
Properpreserve mean = 724 ÷ 91 = 7.96

b Robert is wrong as Properpreserve's mean is higher and as the data is based on a sample of which we have no knowledge of its fairness the use of 'definitely' is ill-advised anyway.

6 $\sum fx = 47 \times 65 = 3055$

7

a

Weight of plums, w, grams	Frequency	Midpoint	Frequency × midpoint
$8 \le w < 10$	15	9	135
$10 \le w < 12$	28	11	308
$12 \le w < 14$	20	13	260
$14 \le w < 16$	27	15	405
$16 \le w < 20$	10	18	180
	total = 100		total = 1288

1288 ÷ 100 = 12.88

b The fiftieth value is the seventh out of 20 in the $12 \le w < 14$ group

estimate of median = $\frac{7}{20} \times 2 + 12 = 12.7$

c These plums seem to be Victoria plums. The mean is only 0.12 g below the expected value and the median is only 0.2 above the expected value giving quite strong evidence in favour of these being Victoria plums.

d Mean should be 10.88 g, median should be 10.7 g. It now seems unlikely that these are Victoria plums.

Exercise 5.3

1

a 9 + 12 + 8 + ...+ 16 = 112
112 ÷ 9 = 12.4

b 12.4

c The end plant in a row is likely to be producing less than average due to being walked on or similar – the sample is not representative of all plants. The sample size used to make this estimate is quite small and therefore the estimate may well be unreliable.

d Take a random sample of 30–50 plants from different areas of the field; or a random sample of x plants from each row.

2

a $\frac{51}{85} = 0.6$

b 0.6

c A good sample size on the night chosen but only one night considered and eating habits may be different on different nights. Reliability is therefore questionable.

3

Number of stations watched	Frequency	Frequency × value
1	38	38
2	43	86
3	13	39
4	4	16
5	2	10
	total = 100	total = 189

189 ÷ 100 = 1.89 so this is the estimate of the mean number of stations watched by the population.

4

a 0.4

b Her friends are likely to be similar people with less chance of this being a random sample.

c Obtain secondary data from driving schools.

5

a $\frac{64}{80} = 0.8$

b 0.8

c

Component width, *w*, (mm)	Frequency	Midpoint	Frequency × midpoint
$33 \leq w < 34$	6	33.5	201
$34 \leq w < 35$	31	34.5	1069.5
$35 \leq w < 36$	33	35.5	1171.5
$36 \leq w < 37$	5	36.5	182.5
$37 \leq w < 38$	4	37.5	150
$38 \leq w < 39$	1	38.5	38.5
			total = 2813

2813 ÷ 80 = 35.1625 mm

6

a $\frac{8}{45} = 0.178$

b

Income (£)	Frequency	Midpoint	Frequency × midpoint
10 000–19 999	8	15 000	120 000
20 000–29 999	17	25 000	425 000
30 000–39 999	16	35 000	560 000
40 000–49 999	4	45 000	180 000
			total = 1 285 000

1 285 000 ÷ 45 = £28 556

c 45 × 4 = 180

d Use of one town in this way is highly unlikely to give a representative sample on which the whole of the UK can be based.

6 Chapter 6

Spot the mistake

$P_{50} = Q_2 = D_5$ is correct but $P_{75} = Q_3 = D_7$ is not as D_7 is 70 per cent along the data whereas the other two are 75 per cent along the data.

Exercise 6.1

1

a i 8 − 1 = 7

ii 18 − 11 = 7

iii 8 − 2 = 6

b i 7 − 3 = 4

ii 17 − 11.5 = 5.5

iii 5 − 2 = 3

2

a 54 − 0 = 54

b 22 − 1 = 21

c The interquartile range.

3 UQ = 36 + 4 = 40. UQ − LQ = 12 so LQ = 40 − 12 = 28

4 Firstly estimate LQ. ¼ along data = 25th value

This is 10 out of 28 in the $31 \leq l < 32$ group.

$= \left(\frac{10}{28} \times 1\right) + 31 = 31.357\ldots$

Next, UQ is ¾ along data = 75th value

This is 12 out of 27 in the $33 \leq l < 34$ group

$= \left(\frac{12}{27} \times 1\right) + 33 = 33.444\ldots$

So interquartile range is 33.444 . . . − 31.357 . . . = 2.087 . . .

= 2.09 mm (notice that you should not use rounded values in the middle of the calculation – only at the end).

5

a Median = $D_5 = 22$

b $D_9 - D_1 = 32 - 13 = 19$

c $27 - 16 = 11$

d Yes, the 20–80 percentile range is 11 so the IQR should be a bit less than this.

Exercise 6.2

1

a i 2.608
ii 802.4
iii 1.582
iv 62.01

b i 6.8
ii 643 830.8
iii 2.503
iv 3844.75

2 The data in part ii, as there was one value considerably larger than the others, which affects the size of the standard deviation significantly.

3 Mean = $\frac{812}{50} = 16.24$

standard deviation = $\sqrt{\frac{25\,080}{50} - 16.24^2}$

$= 15.42$

4 Mean for Richard = £23 383.33 Standard deviation for Richard = £3498.29. Alan and Richard have similar means with Richard slightly higher performing. Alan's figures are much more varied than Richard's, on a month-by-month basis.

It is difficult to judge on these figures alone. Alan's mean looks to have been rounded to the nearest £1000, so the only thing to go on is the standard deviation. Possibly choose Richard because his mean is slightly higher and based on his standard deviation his sales appear to be much more consistent (reliable).

Exercise 6.3

1

a Long left tail, peak after 100.

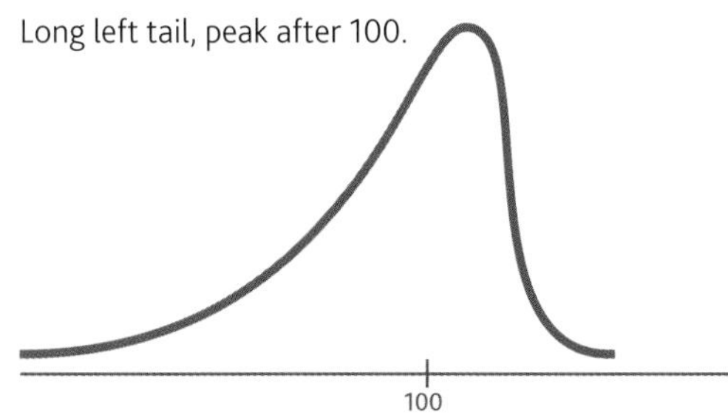

b Symmetrical about 5.

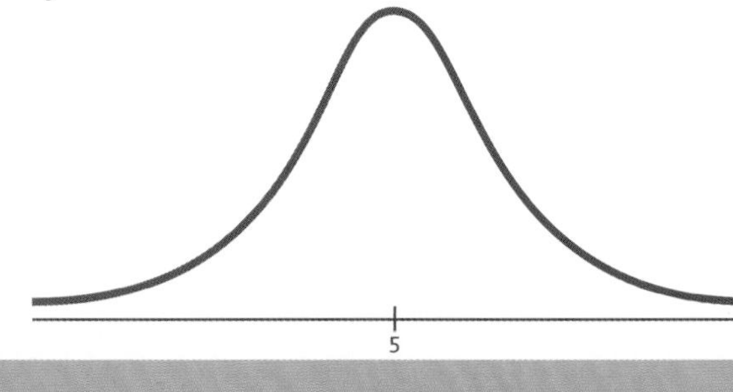

c Long right tail, peak to the left of 45.

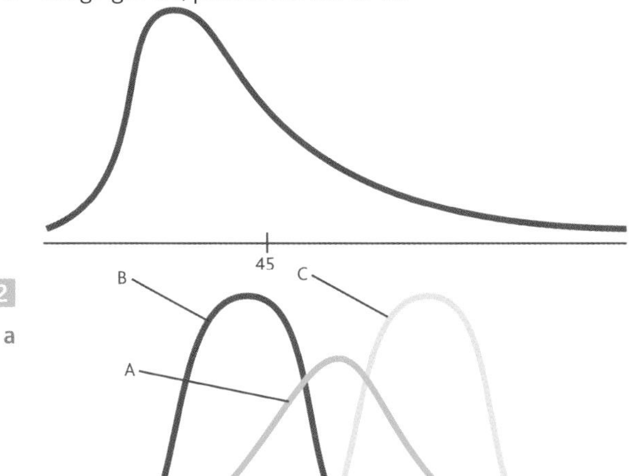

2

a Curve A is symmetrical about 80 and lower than other curves.

Curves B and C are identical in shape. Spread is symmetrical and centred over 60 and 100 respectively.

Note that all curves should approach the axis but not actually land.

b Curve A: between 40 and 120.

Curve B: between 40 and 80.

Curve C: between 80 and 120.

3 Data set 1 is negatively skewed as the mean < median < mode.

Data set 2 is Normal as mean = median = mode.

Data set 3 is positively skewed as the mode < median < mean.

4 Data set 1 = –1 Data set 2 = 0 Data set 3 = 0.5

5

a i $\frac{120 - 135}{10} = -1.5$

ii $\frac{h - 180}{18} = -1.5$ so $h = 153$ cm

b $\frac{202.5 - 180}{18} = 1.25$

$\frac{h - 135}{10} = 1.25$ so $h = 147.5$ cm

6 Standardised scores for each test are as follows.

Memory = –1, Logic = 0.25, Numeracy = –2 and Literacy = 2

So, in order of performance (best first), the tests are literacy, logic, memory and numeracy.

7 Let the mean be x and the standard deviation be y. The information given allows two equations to be constructed

$\frac{60 - x}{y} = 1.5$ so $1.5y + x = 60$ and $\frac{50 - x}{y} = -1$ so $x - y = 50$

Solve the simultaneous equations to give $y = 4$ and $x = 54$.

7 Chapter 7

Spot the mistake

Clive has incorrectly joined the points with increasing lines. As the data is discrete, the points should be joined in a series of steps.

Exercise 7.1

1

a

Time spent, t (hours)	Cumulative frequency
$t < 2$	1
$t < 3$	5
$t < 4$	10
$t < 5$	19
$t < 6$	31
$t < 7$	47
$t < 8$	50

b

Length of time, l, minutes	Cumulative frequency
$l < 5$	6
$l < 7$	22
$l < 8$	43
$l < 10$	83
$l < 20$	100

2

Volume of liquid in bottle, ml	Frequency
$490 \leq v < 500$	14
$500 \leq v < 510$	15
$510 \leq v < 520$	36
$520 \leq v < 530$	22
$530 \leq v < 540$	13

3

a

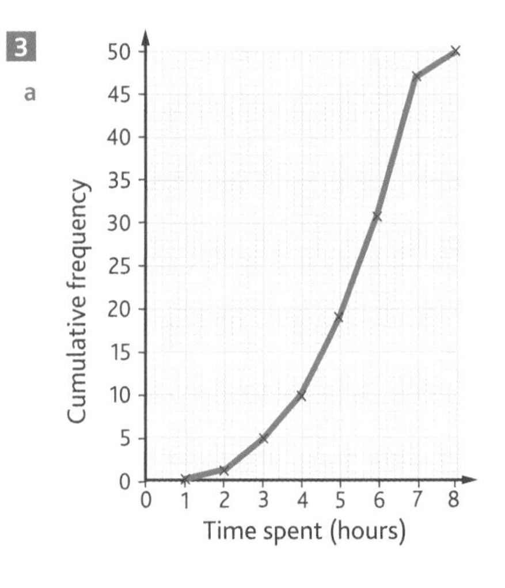

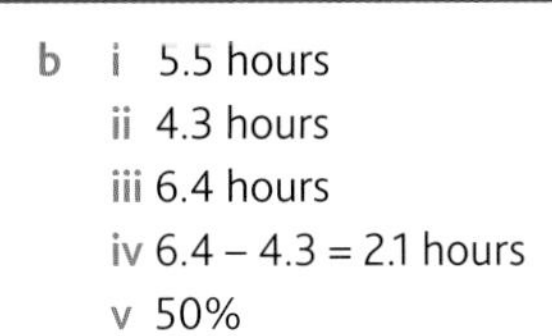
b i 5.5 hours
ii 4.3 hours
iii 6.4 hours
iv 6.4 – 4.3 = 2.1 hours
v 50%

4

a

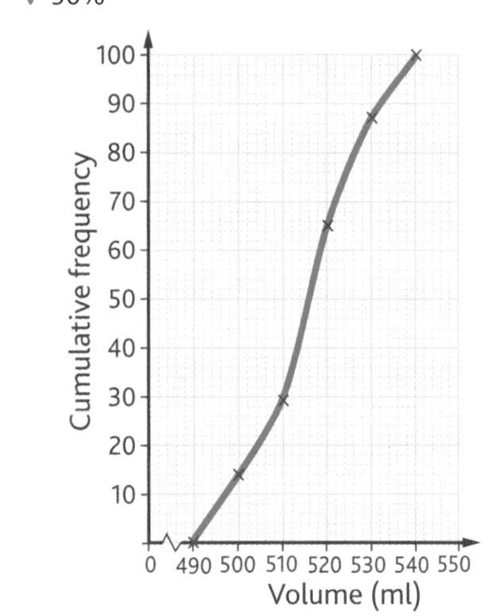

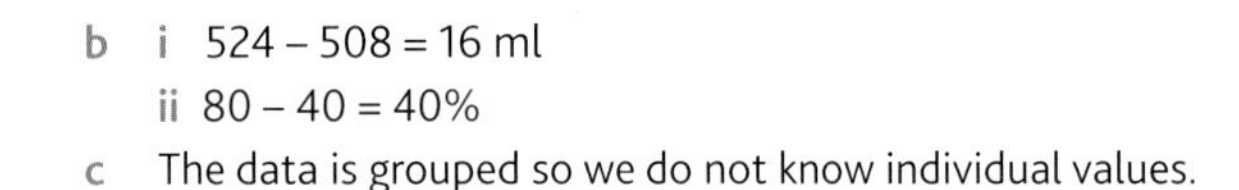
b i 524 – 508 = 16 ml
ii 80 – 40 = 40%
c The data is grouped so we do not know individual values.

5

a There is no data below the first class interval, which begins at 89.5.

b

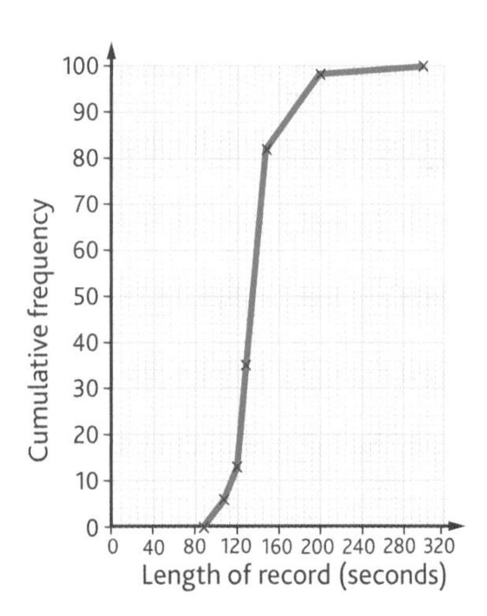

c i About 134 s
ii 144 – 124 = 20 s
iii 92%

6

a

b median = 1

IQR = 2 – 1 = 1

c D_9 = 180th value = 3

D_1 = 20th value = 1

3 – 1 = 2

d This is because there is so much data for value = 1 thus 10th percentile and 25th percentile are the same.

7

a

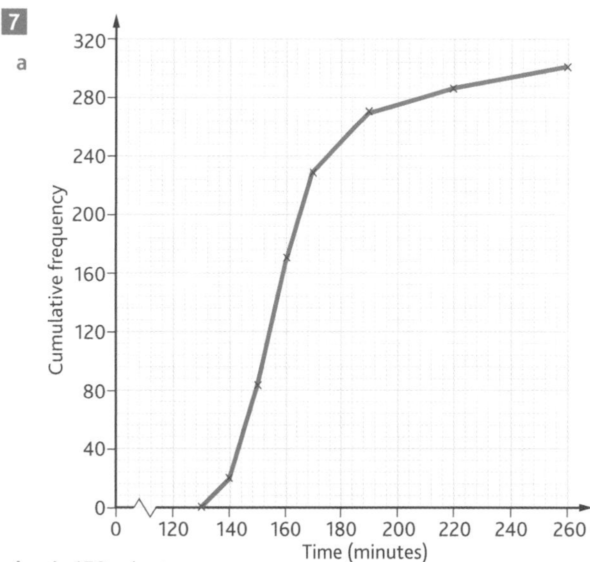

b i 158 minutes

ii 169 – 149 = 20 minutes

iii Read of from 35 per cent of 300, i.e. 105th along = 152 minutes

c 250

d Positive

Exercise 7.2

1

a

b Negative skew as the median is nearer to the upper quartile than it is to the lower quartile.

2 Some of the possible answers are given.

It takes the supporters using the bus longer on average to get home than those using the underground.

The overall range of times is very slightly shorter for bus passengers, meaning slightly less varied times for the whole data.

The interquartile range is shorter for underground passengers, meaning that the central 50 per cent of times are less widely spread.

Underground times are symmetrical, bus times are negatively skewed.

3 Here is a set of data about the time in seconds spent by 19 cars at a red traffic light:

30 22 45 67 12 35 60 55 16 44 48 37 33 28 65 49 61 30 44

a

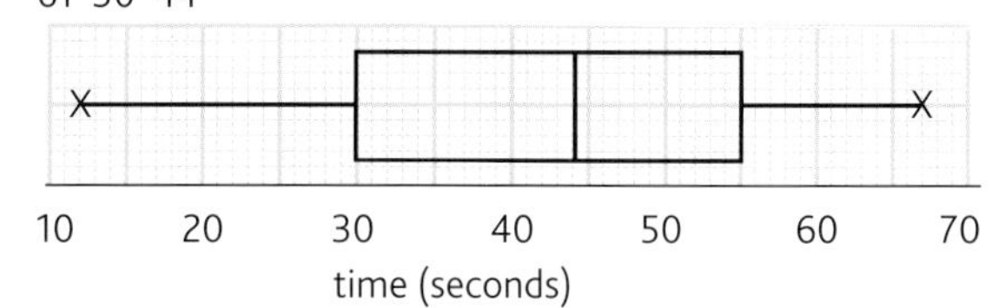

b Slightly negatively skewed.

4

a This value should probably be discarded, as there could well be an external reason for the low number of visitors. However, more information would be useful in your reasoning, such as the weather and other local events.

b Keep the value, it is only just an outlier and represents a journey that took particularly long. Journey times will be naturally unlikely to produce low outliers but occasionally will produce high outliers when there is a problem on the roads.

c Discard this value. It is such a huge outlier and bears no resemblance to usual values from this distribution. It is likely that either this value was recorded incorrectly, or the machine filling the packets malfunctioned at this point.

d Worthy of note, this value should not be discarded and is clearly a feasible part of this distribution.

5

a Topcab LQ = -2, UQ = 3, IQR = 5. Tolerances for outliers are below –9.5 and above 10.5, so there are two high outliers, namely 11 and 24.

Fastcar LQ = –1, UQ = 5, IQR = 6. Tolerances for outliers are below –10 and above 14, so there are no outliers.

b

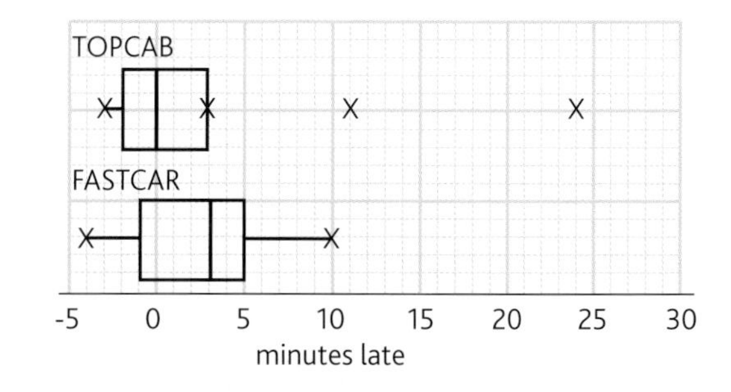

c Topcab are on average on time, with 0 minutes late. However, their data is much more widely spread than Fastcar, and Topcab has two high outliers indicating that on two occasions the taxi was very late. Fastcar are on average 3 minutes late but have no high outliers.

Given the context you could choose Fastcar, as being 3 minutes late (on average) is unlikely to cause any problems, whereas Topcab's high outliers are such that appointments / trains etc. could be missed.

Alternatively, simply based on the measures, you could decide to choose Topcab. The important issue is justifying your choice.

8 Chapter 8

Spot the mistake

The scatter graph does appear to show some negative correlation between the two variables but clearly it is ludicrous to think that the name of a month can have anything to do with the temperature, so this is an example of a spurious relationship that cannot be used to make any predictions.

Exercise 8.1

1

a Strong negative correlation

b No correlation

c Nonlinear correlation

2

a

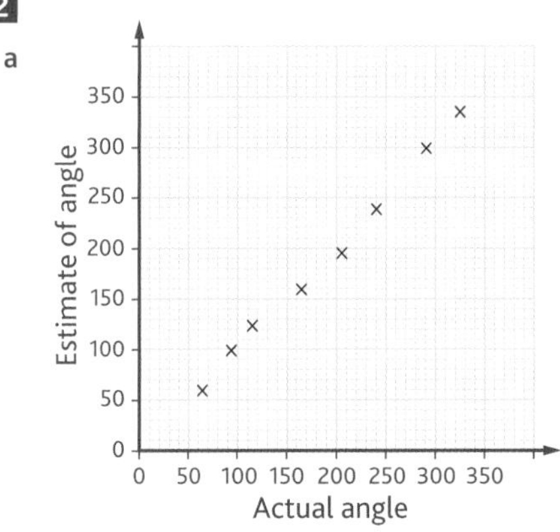

b Strong positive correlation

c

350
300
250
200
150
100
50
0
Estimate of angle
0 50 100 150 200 250 300 350
Actual angle

d i 140 degrees

ii 28 degrees

e The 140 degrees estimate is more likely to be accurate as it is interpolation; 30 degrees is extrapolation.

3 Extrapolation is the estimating of values beyond the size of the data seen so far. A line of best fit may not be valid beyond the extremes of the data as the pattern seen may simply not continue, making extrapolation unreliable.

4

a

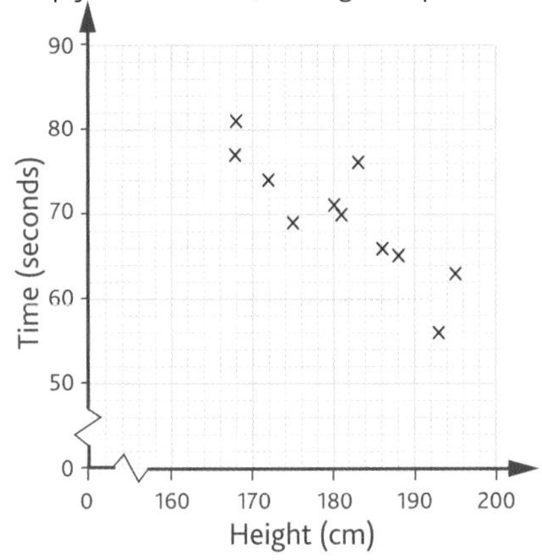

b Moderate negative correlation

c Double mean point is (180.8, 69.8). The line of best fit passes through this point, which is marked with a dot in a circle in the diagram below.

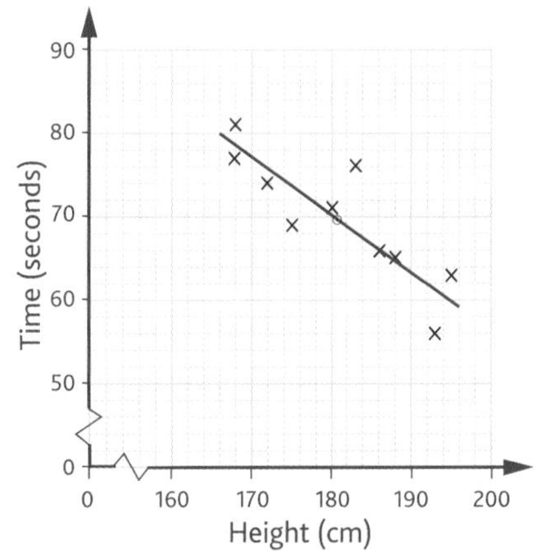

d i Around 72 seconds.

ii Around 81 seconds.

e The estimate for 177 cm is more reliable, as 165 is outside the range for which we have information, so the relationship may be different at that point.

f It is a small sample / the correlation is not that strong / there will be lots of other factors involved.

g Firstly, find the gradient of the line by choosing 2 points on the line, say (170, 77) and (180.8, 69.8). (The second point is the double mean point.)

$$\text{Gradient} = \frac{69.8 - 77}{180.8 - 170} = -\frac{2}{3}$$

The intercept cannot be read off as there is a break mark on the axes (it is *not* 91).

The equation is $y = -\frac{2}{3}x + c$

When $x = 170$, $y = 77$. Substitute into this equation to

get $c = 190\frac{1}{3}$

So equation is $y = -\frac{2}{3}x + 190\frac{1}{3}$

5

a The equation will depend on the exact position of the line of best fit. For guidance, it should be not far from $y = x$

b m is the number of degrees the estimate goes up for each degree the actual angle goes up.

Exercise 8.2

1

a There is a weak to moderate correlation between age of a child and pocket money, meaning that older children tend to get more pocket money.

b There is no correlation between score in an examination and the shoe size of the person taking the examination, meaning there is no connection between these two variables.

c There is perfect positive correlation between the length of a run and the time taken to run it, meaning that longer distance runs always take longer to run.

d There is a quite strong negative correlation between age and price of a car, meaning that the older a car is the less it tends to cost.

2

a −0.8 to −1

b +0.4 to +0.6

c −0.1 to +0.1

3 Any scatter diagram with the points in a straight line with negative gradient (strictly speaking a perfect straight line is not necessary for Spearman to be −1).

4

a

Height (m)	1.54	1.46	1.60	1.87	1.49	1.79	1.65
Weight (kg)	78	64	61	89	66	84	60
Rank (ht)	3	1	4	7	2	6	5
Rank (wt)	5	3	2	7	4	6	1
d	−2	−2	2	0	−2	0	4
d^2	4	4	4	0	4	0	16

$\sum d^2 = 32$ and $n = 7$ leading to SRCC = 0.429

b

Number of pets	4	0	1	2	9	3
Number of cars	2	2	1	1	0	3
Rank (pets)	5	1	2	3	6	4
Rank (cars)	4.5	4.5	2.5	2.5	1	6
d	0.5	−3.5	−0.5	0.5	5	−2
d^2	0.25	12.25	0.25	0.25	25	4

$\sum d^2 = 42$ and $n = 6$ leading to SRCC = −0.2

c

Age	18	21	22	25	25	34	41	55	63
100m time (seconds)	11.3	10.8	0.6	11.0	11.5	11.1	11.8	12.6	14.2
Rank (age)	1	2	3	4.5	4.5	6	7	8	9
Rank (time)	5	2	1	3	6	4	7	8	9
d	–4	0	2	1.5	–1.5	2	0	0	0
d^2	16	0	4	2.25	2.25	4	0	0	0

$\sum d^2 = 28.5$ and $n = 9$ leading to SRCC = 0.763

d

Engine size	1.6	1.6	1.6	1.4	1.1	2.1	1.8	2.8
Miles per gallon	46	59	44	54	59	31	35	28
Rank (eng)	4	4	4	2	1	7	6	8
Rank (mpg)	5	7.5	4	6	7.5	2	3	1
d	–1	–3.5	0	–4	–6.5	5	3	7
d^2	1	12.25	0	16	42.25	25	9	49

$\sum d^2 = 154.5$ and $n = 8$ leading to SRCC = –0.839

5

a

Person	A	B	C	D	E	F	G	H	I	J
Final %	78	63	68	96	63	45	80	73	61	92
Salary £	32 000	26 000	34 000	35 000	33 000	27 500	38 000	45 000	30 000	39 000
Rank (%)	7	3.5	5	10	3.5	1	8	6	2	9
Rank (salary)	4	1	6	7	5	2	8	10	3	9
d	3	2.5	–1	3	–1.5	–1	0	–4	–1	0
d^2	9	6.25	1	9	2.25	1	0	16	1	0

$\sum d^2 = 45.5$ and $n = 10$ leading to SRCC = 0.724

b There is a moderate correlation between final percentage and salary after one year, meaning that the higher the score in the examination, generally the higher the salary being earned after one year.

6 The product moment correlation coefficient is affected by outliers as it uses the actual data in its calculation, whereas Spearman's would simply record the rank, so being highest by 1 or 1000 does not change the ranking.

7 –1, 0.94, –0.67, –0.44, 0.35, 0.06

8 The greater the number of shops the greater the number of pedestrians.

9 Chapter 9

Spot the mistake

Even though the data is weekly, they only actually open for four days a week and therefore they should find four-point moving averages not seven-point moving averages.

Exercise 9.1

1 Herring had very large stocks which dropped dramatically in the 1960s, recovering temporarily around 1990, and then again more recently.

Haddock have had comparatively low stocks with a peak around 1970 and a small recovery in the early 2000s. Cod stocks have always been low in comparison, and continue to reduce to very low levels.

2

a $\frac{56 + 58 + 62 + 64}{4} = 60$ and so on giving 61, 63, 65.5, 68, 71.5, 74 and 76

b $\frac{93 + 87 + 90}{3} = 90$ and so on giving 86, 83, 78, 77, 75, 72, 69 and 66

c $\frac{143 + 165 + 173 + 153 + 148}{5} = 156.4$ and so on giving 161.8, 159.4, 158.4, 158.6, 164.4, 161.2, 164.2 and 165.2

3

a

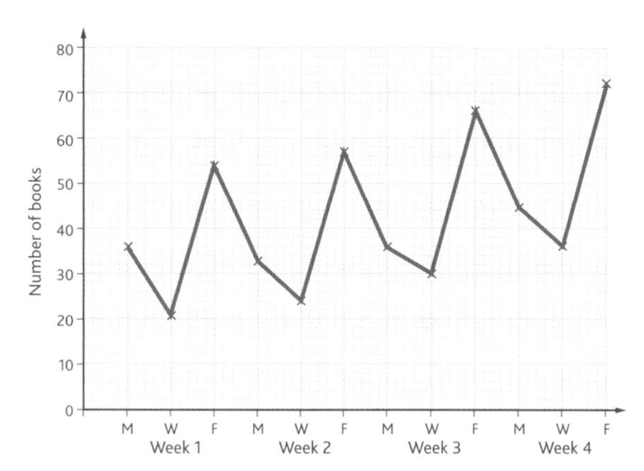

b The library is open three days a week.

c $\frac{36 + 21 + 54}{3} = 37$ and so on giving 36, 37, 38, 39, 41, 44, 47, 49 and 51

d This requires moving averages to be plotted.

e This requires a trend line to be drawn.

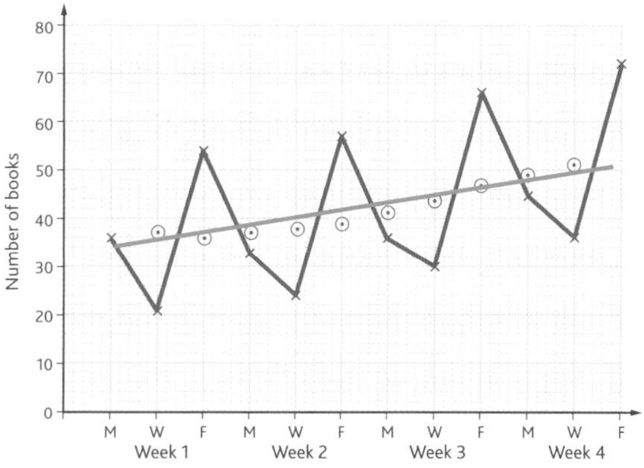

f Friday is the most popular day for books being borrowed, Wednesday is least popular. There is a general increase in the number of books being borrowed each week during this period.

4

a

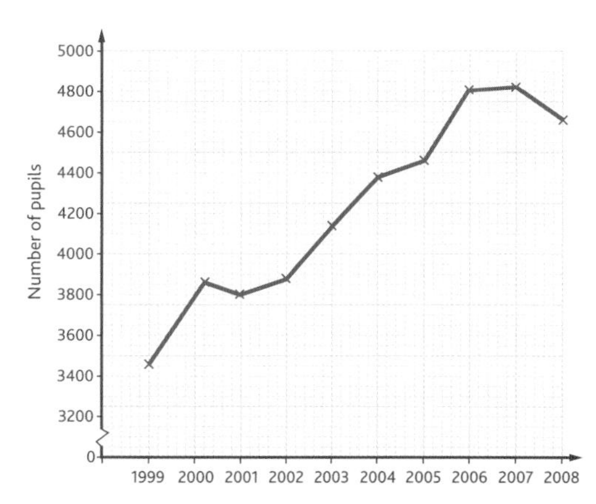

b $\frac{3455 + 3860 + 3806 + 3874}{4} = 3749$ and so on giving 3918, 4051, 4217, 4451, 4625 and 4693.

c

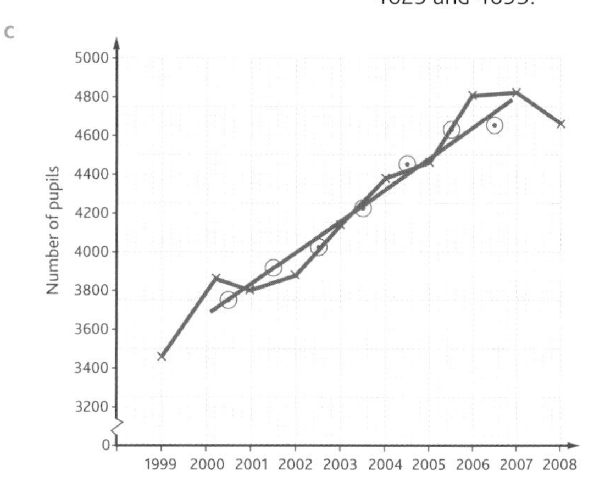

d There is a clear upward trend in the data, as shown by the moving averages. The data, however, does not simply go up every year; twice the figures have gone down year on year. This may be due to random fluctuations.

Exercise 9.2

1 The actual data is shown along the bottom of the chart. The cumulative data is shown diagonally upwards on the chart. The rolling/running total for the whole of the previous inclusive time period is shown at the top of the chart.

2 One diagram contains lots of different aspects of data instead of having to refer to two or more diagrams.

3

a

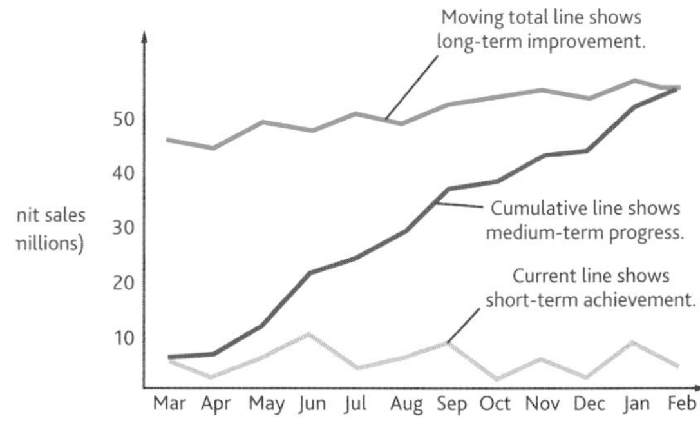

Source : http://syque.com

b A = units sold this month; B = units sold since March; C = units sold over the last 12 months.

4

a

Day	Electrical items sold	Cumulative frequency	Total for last seven days
Monday	23	23	245
Tuesday	21	44	242
Wednesday	28	72	235
Thursday	26	98	220
Friday	33	131	222
Saturday	42	173	213
Sunday	36	209	209

b

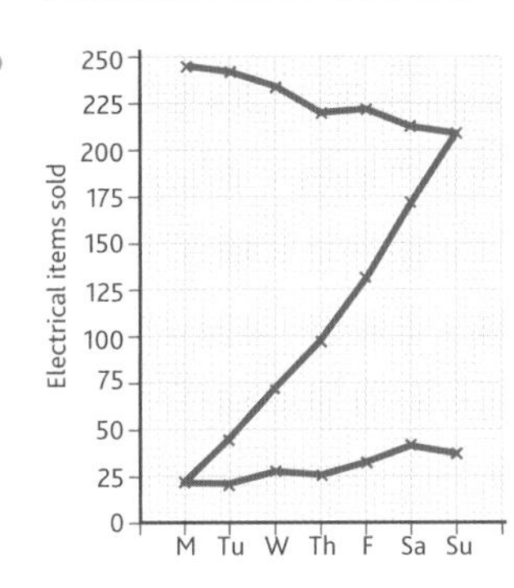

c The weekly sales of electrical items increase as the week goes on. Sales this week are, overall, down on the previous week.

5

a Three seasonal effect values are 11, 10 and 4 (approximately). Therefore average

seasonal effect is $\frac{11 + 10 + 4}{3} = 8.333\ldots$

b Extended trend line gives value about 112 when read off for Monday of week 4. 112 + 8.333 … = £120 to nearest pound.

6

a Male time series graph:

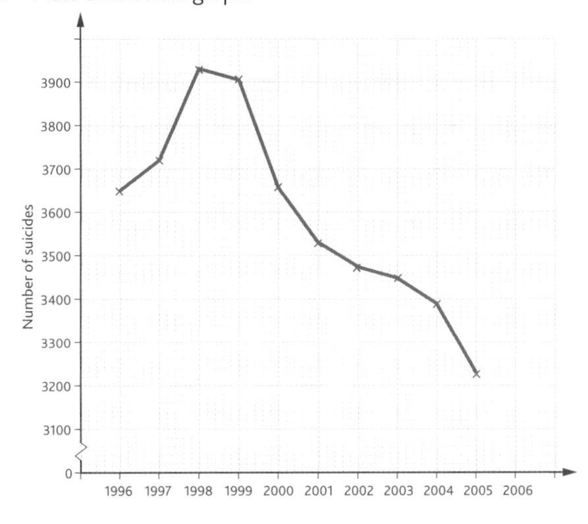

Female time series graph:

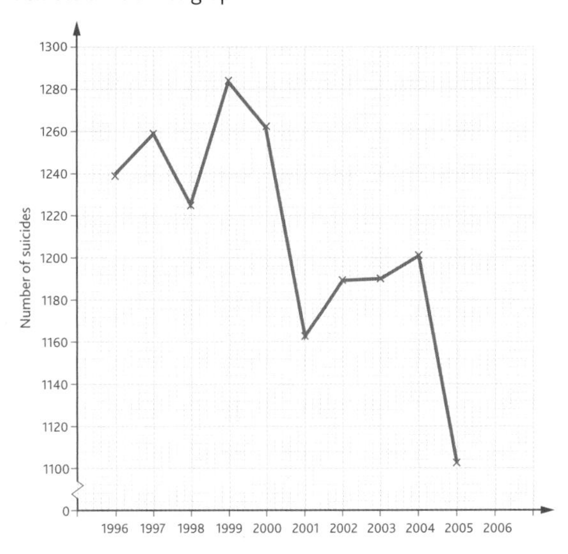

b For males, first moving average $= \frac{3654 + 3722 + 3929 + 3904}{4}$

= 3802 and so on, giving 3804, 3756, 3641, 3528, 3461 and 3384.

Giving this graph and estimated trend line:

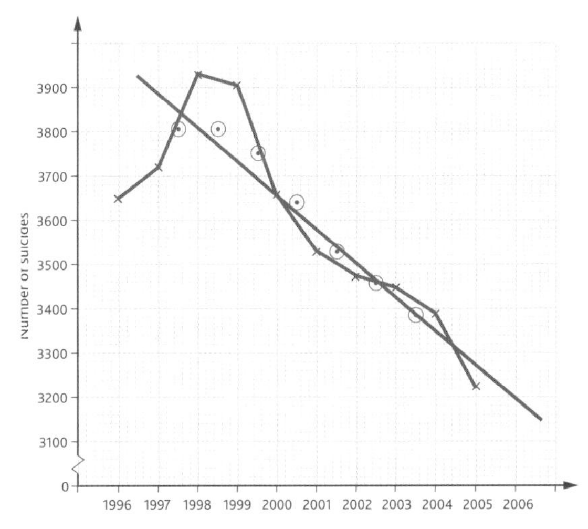

For females, the first moving

average is $\frac{1239 + 1259 + 1225 + 1284}{4}$

= 1252 and so on, giving 1258, 1234, 1226, 1204, 1190 and 1177.

Giving this graph and estimated trend line:

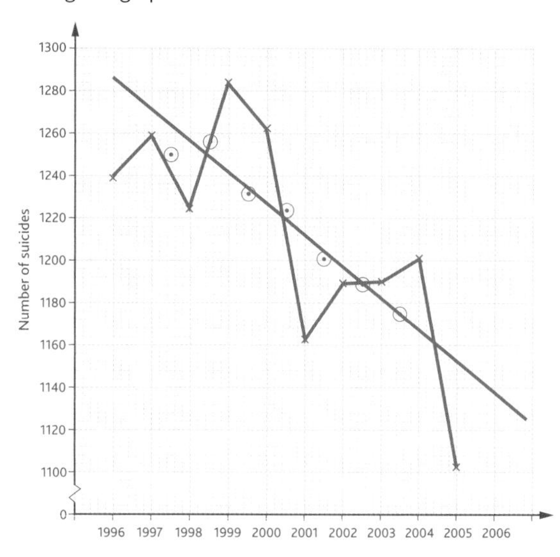

c This data is annual (yearly), so there is no seasonal effect present in either gender.

d Males = 3200. Females = 1140. (Both are approximate as they are read from trend line without any additional seasonal effects to be taken account of.)

7

a

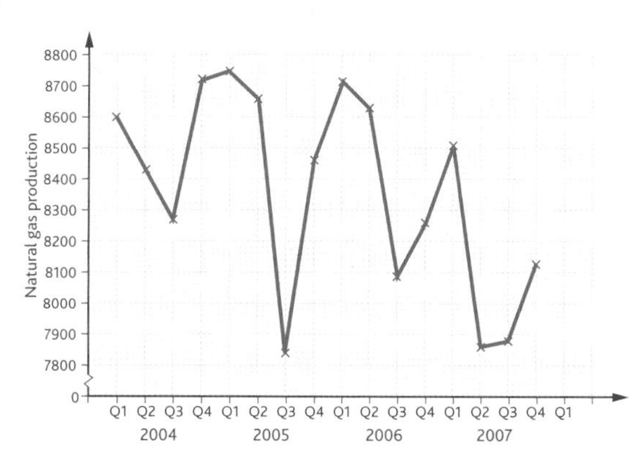

b First moving average = $\frac{8600 + 8425 + 8275 + 8714}{4} =$

8504 and so on giving 8540, 8599, 8490, 8426, 8418, 8409, 8470, 8420, 8367, 8176, 8124, 8091

c

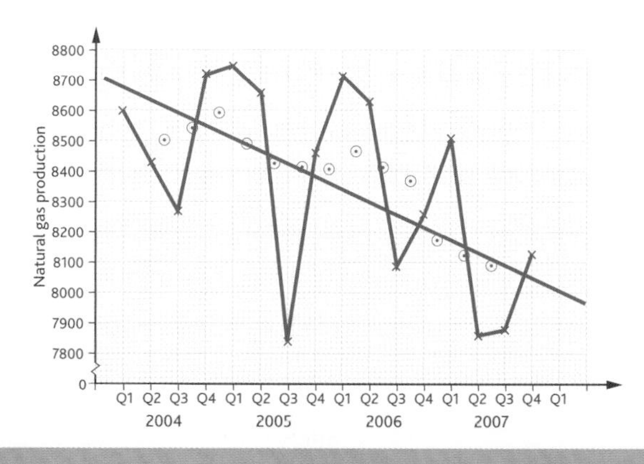

d For Q1 seasonal effects are (approximately) –80, +230, +380 and +330 giving an average seasonal effect of 215.

e 8010 + 215 = 8225 units

f The data is seasonal with troughs in Q3, and there is a general trend of a reduction in production over these years.

10 Chapter 10

Exercise 10.1

1 The dates of increase tend to come in clusters and short bursts and the rate of increase was much greater at the end of the 1970s than at most other times. The misleading graph smoothed out these sudden changes.

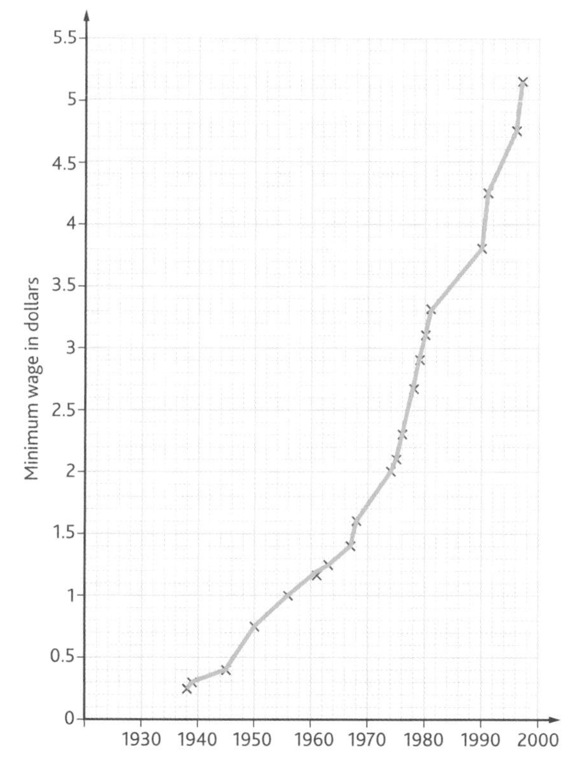

2

a

i The vertical axis begins at 106, vastly overemphasising the differences between the years.

ii 1991, 1993, 1996 and 1998 are missing. Also 2000 is not in the 1990s.

iii The size of the doctors is completely wrong, vastly exaggerating the differences. It is almost a 3D diagram but this has not been taken into account when calculating the size of each diagram. There is also too much information on one chart. Is it the ratio, the percentage, or the number of doctors that the diagram is supposed to be showing?

iv There is little right with this – only two years of data, incorrectly joined and also exaggerated by a break mark on the vertical axis.

v There is no reason why the plots on this chart should be like a scatter graph.

b

i Vertical scales must always start at zero; however a break mark is probably appropriate.

ii Insert the data for the missing years (all of which are much lower values) and remove 2000.

iii Use the fact that these are 2D diagrams to calculate appropriate values for the height and width of the doctors.

iv Different form such as bars needed. In the current form need we more years' data to assess any possible trends and so forth.

v Data would be far better ordered, and using bars or 'sticks' labelled with each country name – definitely not made to look like a scatter diagram.

a 20 TVs = 64 cm^2 so Jeff's 16 TVs = $\frac{64}{20} \times 16 = 51.2$ cm^2

$\sqrt{51.2} = 7.16$ cm

b 16 TVs = 125 cm^3 so John's 20 TVs = $\frac{125}{16} \times 20 = 156.25$

$\sqrt[3]{156.25} = 5.39$ cm

4 140 portions = 72 cm^3 so class J's 168 portions = $\frac{72}{140} \times 168$ = 86.4 cm^3

So $8 \times 3 \times h = 86.4$, $h = 3.6$ cm.

Spot the mistake

A pictogram is for qualitative data. The ages are continuous quantitative data and therefore should not be illustrated using a pictogram.

Exercise 10.2

a Some of the largest countries (such as Canada and Russia) have very low population density. Major population density centres are in Europe and Asia. The one large country with high population density is India.

b There is a large difference in the size of the groups in each shading colour, but the map is effective in showing differences. It would be interesting to see a similar map with equal-sized groups (perhaps going up in 200s).

a They show very different things – one showing the population of each area, the other showing the population per hectare, which is a measure of population density.

b A large number of people live in the south of the region and the north of the region away from the river. The large areas of population density are towards the centre of the region around the river. The areas of high population density and the areas with simply a large number of people in them do not correspond very well at all.

3

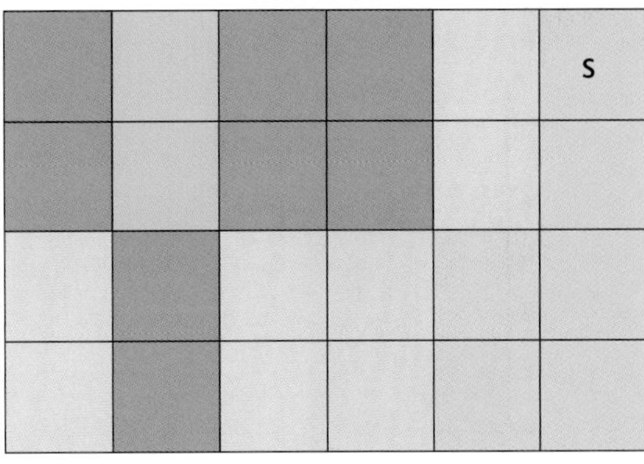

4 One possible key might be:

Key

1–5 white

6–10 light grey

11–15 darker grey

16+ black

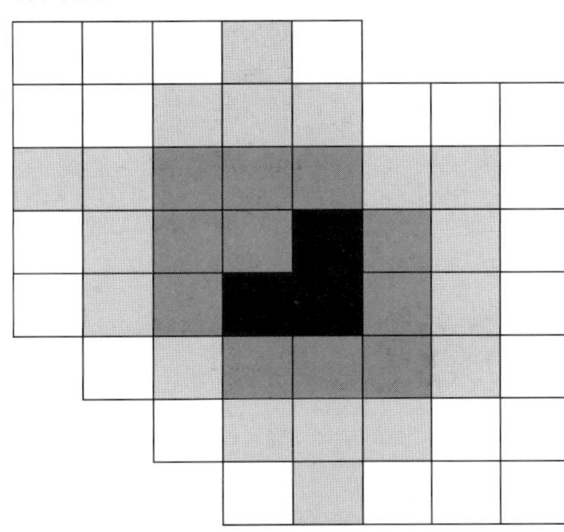

5

a Location of accidents in the home

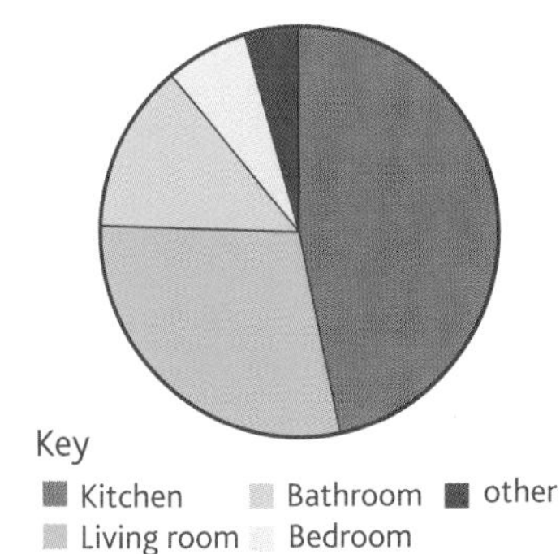

b Proportions of each amount are easier to see.

6

a There are a few options for the group widths – around 70 is likely.

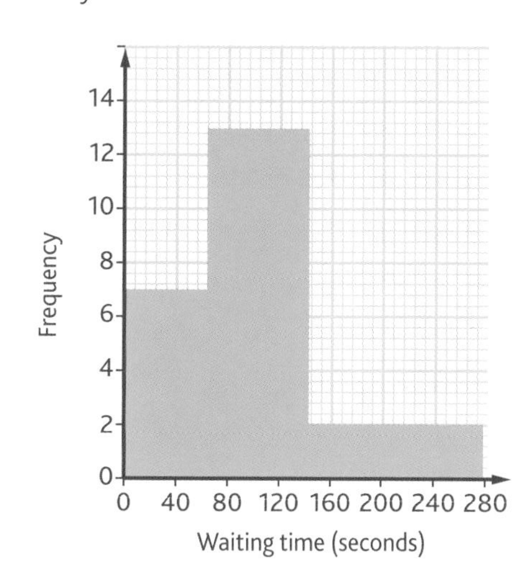

b Clearly focuses on this variable and shows the data about the waiting times only. Shows clearly that most customers only wait a short time – the scatter diagram suggests long waits for long queues.

c Still possible to get the exact values of the waiting times from the scatter diagram.

d i A simple frequency diagram for discrete data, such as a vertical line diagram or a dot plot.

ii

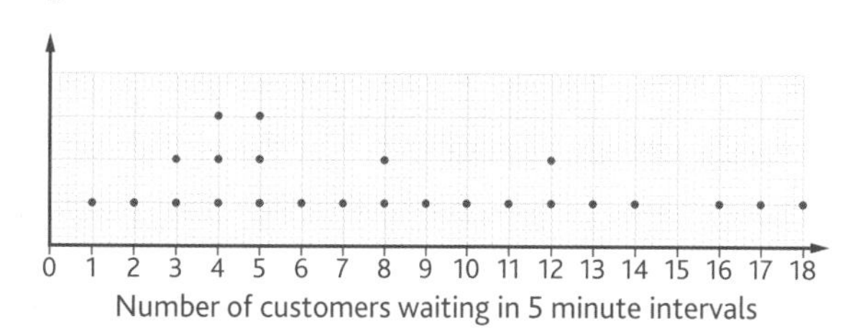

7

a 2002–2005, 2008–2009

b 1999–2001, 2006–2007

c 2001 the output gap was –3.5% of GDP.

11 Chapter 11

Exercise 11.1

1

a $\frac{35}{39} \times 100 = 89.7$

b $\frac{32}{29} \times 100 = 110.3$

c Let x be the price in 2009 then $\frac{x}{32} \times 100 = 105$ so $x = 33.6$ which is £33.60. (Always remember that, if the answer is money you must give the correct money form or you will lose a mark.)

2

a i $\frac{104}{100} \times 24\,000 = 24\,960$

ii $\frac{120}{100} \times 24\,000 = 28\,800$

b Between 2002 and 2003 there was a drop in visitors as the index number was lower for 2003 than 2002.

c $\frac{31\,200}{24\,000} \times 100 = 130$

3

a $\frac{21}{3} \times 100 = 700$

b $\frac{11}{5} \times 100 = 220$

c 1 kg of old potatoes with an index number of 6700. (Cheese 6300, margarine 2667, butter 2733, eggs 2800, tea 4050, sugar 2750, bread 5200, milk 3400 all to nearest whole number where necessary.)

d Let x be the cost of margarine in 1975 then $\frac{x}{5} \times 100 = 460$ so $x = 23$p.

e Let y be the cost of the bread in 1950 then $\frac{16}{y} \times 100 = 800$ so $y = 2$p.

4 First value is same as Q1a 89.7, second value is $\frac{29}{35} \times 100 =$ 82.9 and so on giving 100 and 110.3.

5 First value is obviously 104, second value is $\frac{110}{104} \times 100 =$ 105.8 and so on giving 96.4 and 113.2.

6 First value is $\frac{1\,227\,529}{1\,199\,910} \times 100 = 102.3$ and so on giving 101.8 and 102.5.

7 The games console soon cost more than its original price and one year after it was brought out was 25% more expensive than originally. Even 2 years later it was still 10 per cent more than originally. In years 3 and 4 its cost was now below the original price and by year 4 it was available at just 40 per cent of its original cost.

8 $\left(103 \times \frac{85}{100}\right) + \left(118 \times \frac{12}{100}\right) + \left(106 \times \frac{3}{100}\right) = 104.89$

9

a Recreation and culture, transport and restaurants and hotels.

b Education, health and communication.

c $\left(\frac{10.6}{100} \times 115\right) + \left(\frac{4.6}{100} \times 102\right) + \ldots + \left(\frac{11.1}{100} \times 110\right) = 107.55$

Spot the mistake

Edith has not taken into account the ages of the relative populations of the two places. A retirement village is likely to have a very old age profile and you would expect a much higher death rate than normal. It is essential to use standardised death rates to genuinely compare the two places.

Exercise 11.2

1 $\frac{450}{9000} \times 1000 = 50$

2

a $\frac{88}{40\,000} \times 1000 = 2.2$

b $\frac{64}{40\,000} \times 1000 = 1.6$

c The statement is probably true but cannot be definitely true as you do not know if for some reason a number of people have just migrated out of the town.

3

a Let x be the population in Babytown at the beginning of the year.

Then $\frac{375}{x} \times 1000 = 25$ so $x = 15\,000$.

b No, you need information about the death rate as well as numbers of people coming into and going out of the town.

4

a Valeport is more likely to have the higher crude birth rate as a higher proportion of the population is of the ages where producing children is more likely, 16–30 and 31–45.

b The standardised birth rate for Valeport is likely to be lower than its crude birth rate as it takes this information about the younger age profile into account.

c To be able to make a fair comparison between the two towns without the undue influence of the younger age profile of Valeport.

Exercise 11.3

1

a At first the machine (A) was fine, but there is a trend to production of bags of weight below the action limit. Production needs to be stopped for this machine with (probably minor) adjustment/repair necessary before resumption.

b Machine B is producing perfectly acceptable bags, and no action is required other than the usual monitoring to continue.

c Machine C is producing quite wide fluctuations in the weights of bags, and needs some attention. Stop production and repair the machine.

2 Machine A: the range of bounces is good, low values and no action is necessary for this machine.

Machine B: the range of bounce heights often is well above the action limit so the machine needs some adjustment to improve the consistency of the product.

Machine C: the range of bounces is good for most samples but there is a trend of increasing inconsistency within the samples. This machine could do with some minor adjustment and then a heightened level of surveillance for the next period.

12 Chapter 12

Exercise 12.1

1

a C

b B

c D

d E

e A

2

a $\frac{2}{4}$ or $\frac{1}{2}$

b $\frac{4}{52}$ or $\frac{1}{3}$

c $\frac{12}{52}$ or $\frac{3}{13}$

d $\frac{1}{52}$

e $\frac{1}{52}$

f $\frac{1}{52}$

g $\frac{51}{52}$

h 0

3 0.8

4 No, the probability that he gets up straight away is 0.1, which means that he can be expected to get up straight away once out of every 10 mornings on average.

5 0.2

Spot the mistake

The events ‘even’ and ‘prime’ are not mutually exclusive, as 2 is both even and prime.

P(even or prime) = P (2 or 3 or 4 or 5 or 6) = $\frac{5}{6}$

Exercise 12.2

1

a

	1	2	3	4	5	6
1	0	1	2	3	4	5
2	1	0	1	2	3	4
3	2	1	0	1	2	3
4	3	2	1	0	1	2
5	4	3	2	1	0	1
6	5	4	3	2	1	0

b i $\frac{6}{36} = \frac{1}{6}$

ii $\frac{10}{36} = \frac{5}{18}$

iii 0

iv $\frac{24}{36} = \frac{2}{3}$ (or any other equivalent fraction)

v $\frac{16}{36} = \frac{4}{9}$

2

a Assuming he only listens to one radio station at a time, then P(radio 1) = 1 – (0.1 + 0.3) = 0.6, if he does not listen to any other radio stations.
If he does listen to other radio stations, then P(radio 1) must be less than 0.6.

b He is unlikely to listen to more than one radio station at a time.

3

a Yes, you cannot get a head and a tail when a coin is tossed – it has to be one or the other.

b Yes, it is not possible to get a 6 and a 3 at the same time.

c No, the king of hearts is both a king and a heart.

d Yes, a card cannot be both a 3 and a 5 at the same time.

e Yes, unless there are some counters with more than one colour on!

4

a $\frac{4}{15}$

b $\frac{10}{15}$ or $\frac{2}{3}$

c $\frac{11}{15}$

Spot the mistake

Seamus has multiplied 0.3 by 0.3 incorrectly. He should work out 0.3 × 0.3 = 0.09. Also, the two events may not be independent.

Exercise 12.3

1

a $\frac{1}{2} \times \frac{1}{2} = \frac{1}{4}$

b $\frac{1}{6} \times \frac{1}{13} = \frac{1}{78}$

c $\frac{2}{6} \times \frac{1}{52} = \frac{2}{312} = \frac{1}{156}$ (Remember that you do not lose marks for not cancelling fractions down, unless a question specifically asks you to.)

2

a i 0.7 × 0.4 = 0.28

ii 0.4 × 0.3 = 0.12

b You assumed that Janet and John's choices were independent.

3 There are lots of possible answers to this question. Here are some of them. Reasons for not being independent: Ian and his sister like to keep each other company, or may go shopping together. Ian and his sister do not like to be seen together, so they will travel separately. Reasons for being independent: they go into town at different times.

4

a 0.6 × 0.6 = 0.36

b 0.4 × 0.4 = 0.16

5

a $\frac{1}{10} \times \frac{1}{10} = \frac{1}{100}$

b $\frac{1}{10} \times \frac{1}{10} \times \frac{9}{10} \times \frac{9}{10} \times \frac{9}{10} = \frac{729}{100\,000}$

(= 0.00729)

6 $\frac{1}{2} \times \frac{1}{2} \times \frac{1}{2} = \frac{1}{8}$

7

a $\frac{1}{2}$

b $\frac{1}{2} \times \frac{1}{2} = \frac{1}{4}$

c $\frac{1}{2} \times \frac{1}{2} \times \frac{1}{2} = \frac{1}{8}$

Exercise 12.4

1

a

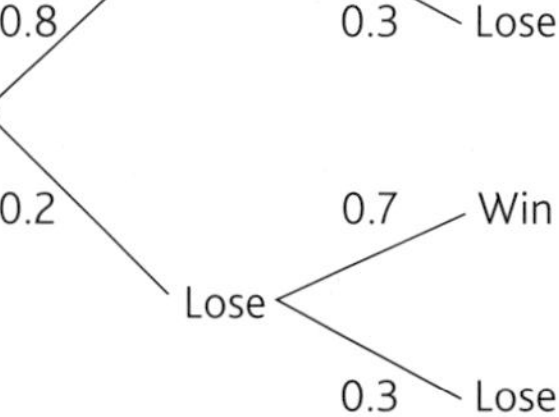

b $0.8 \times 0.7 = 0.56$

c $0.2 \times 0.3 = 0.06$

d $1 - 0.06 = 0.94$

2

a 1st attempt 2nd attempt

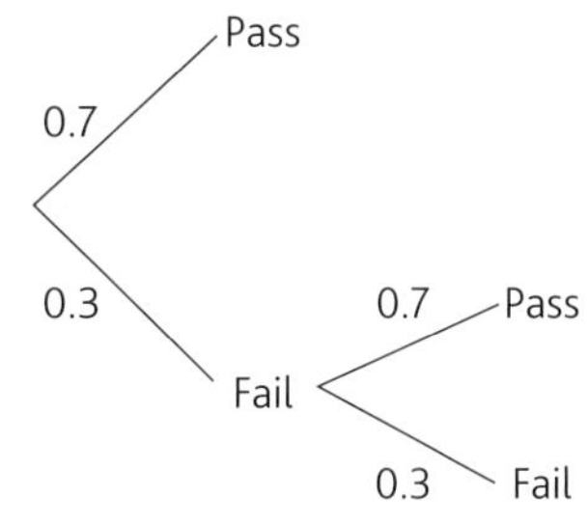

b To pass after two attempts he must have failed the first time.

$0.3 \times 0.7 = 0.21$

c $0.3 \times 0.3 = 0.09$

3

a Ist roll 2nd roll

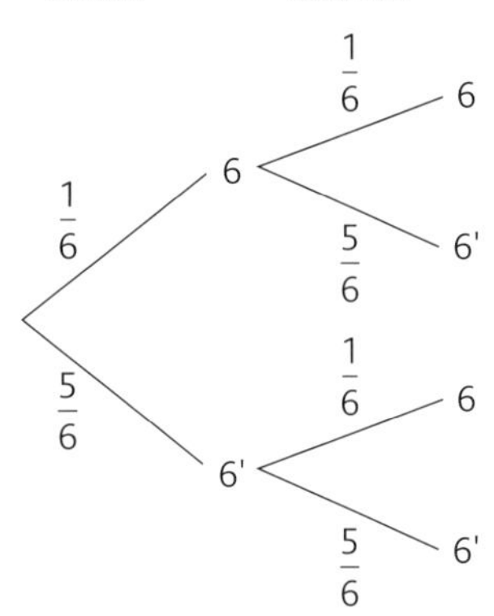

b $\frac{1}{6} \times \frac{1}{6} = \frac{1}{36}$

c $\frac{1}{6} \times \frac{5}{6} + \frac{5}{6} \times \frac{1}{6} = \frac{5}{36} + \frac{5}{36} = \frac{10}{36} \left(= \frac{5}{18}\right)$

d $1 - \frac{5}{6} \times \frac{5}{6} = 1 - \frac{25}{36} = \frac{11}{36}$

4

a

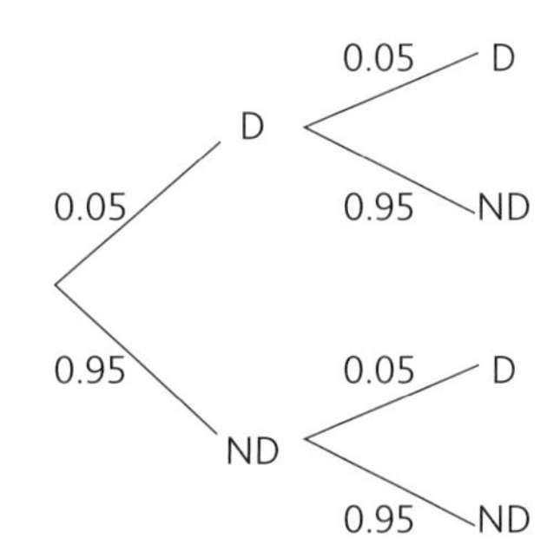

b P(D and D) $= 0.05 \times 0.05 = 0.0025$

c P(1 defective out of 2) $= 0.05 \times 0.95 + 0.95 \times 0.05 = 0.0475 + 0.0475 = 0.095$

d Given three bulbs tested, batch is rejected if P(1 defective out of 2 and next is defective) $= 0.095 \times 0.05 = 0.00475$

e P(Batch rejected) = P(rejected after 2) + P(rejected after 3) $= 0.0025 + 0.00475 = 0.00725$

5

a

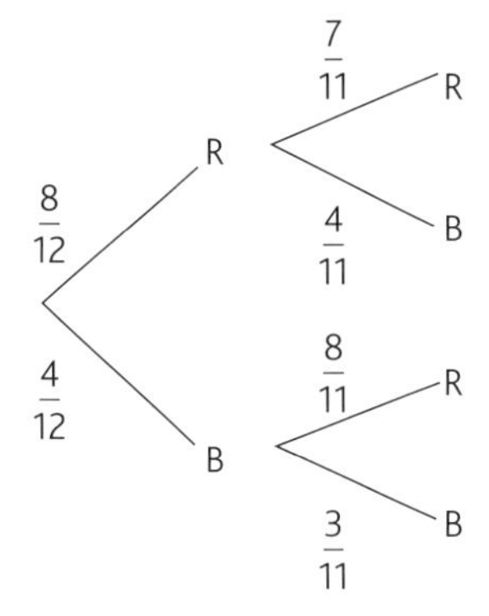

b i P (Both Blue) $= \frac{4}{12} \times \frac{3}{11} = \frac{12}{132} = \frac{1}{11}$

ii P(one of each) = P(RB) + P(BR) =

$\frac{8}{12} \times \frac{4}{11} + \frac{4}{12} \times \frac{8}{11} = \frac{32}{132} + \frac{32}{132} = \frac{64}{132} = \frac{16}{33}$

6

a

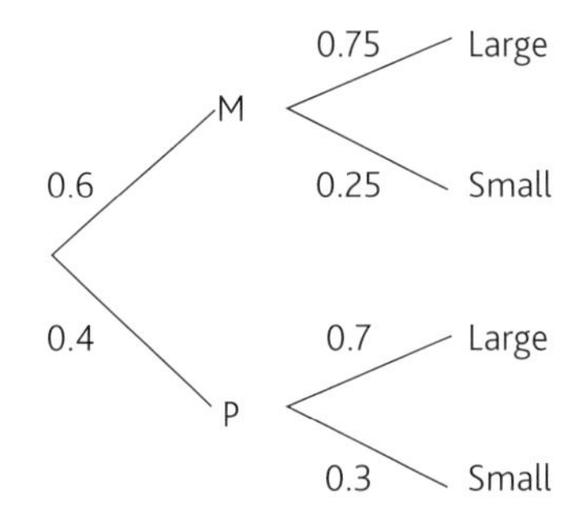

b i P(small and milk) $= 0.6 \times 0.25 = 0.15$

ii P(small) = P(milk and small) + P(plain and small) $= 0.6 \times 0.25 + 0.4 \times 0.3 = 0.15 + 0.12 = 0.27$

7

a Ist selection 2nd selection

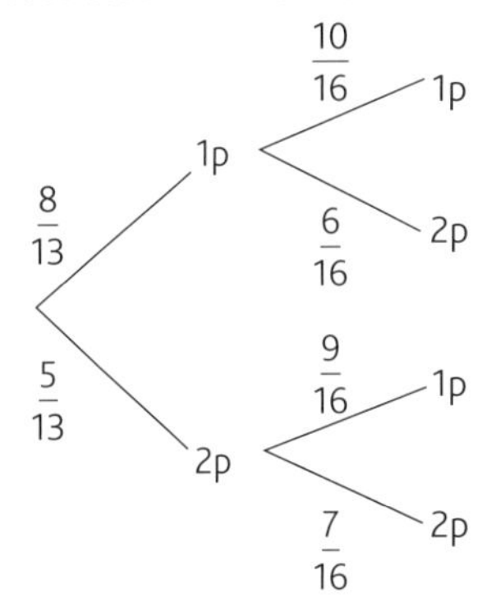

b For the money to be unchanged, the same amount has to be taken on the first and second selection. P(unchanged) = P(1p and 1p) + P(2p and 2p)

$= \frac{8}{13} \times \frac{10}{16} + \frac{5}{13} \times \frac{7}{16}$

$= \frac{80}{208} + \frac{35}{208}$

$= \frac{115}{208}$

c For the money to be unchanged after three transfers, then 2p in total must go each way.

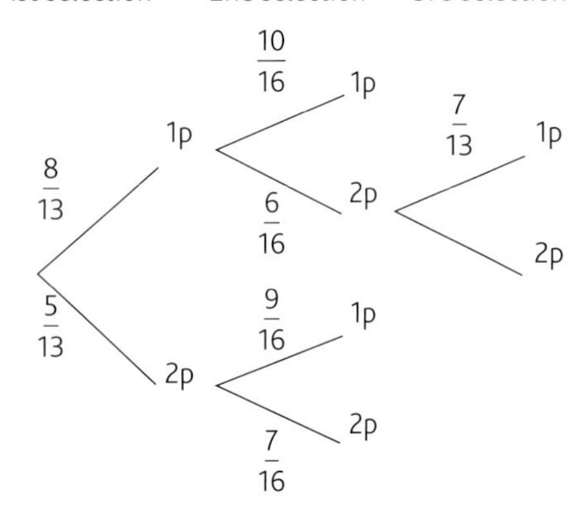

Notice, it is only necessary to extend one part of the tree.

P(Unchanged after 3) $= \frac{8}{13} \times \frac{6}{16} \times \frac{7}{13}$

$= \frac{336}{2704} = \frac{21}{169}$

Exercise 12.5

1 Table 1 A completed version of Table 6 from Chapter 12. Found in 12.5, Exercise 12.5. Set in same style as before, but can be reproduced smaller.

Row	Number of cars in the row	Number of cars more than 4 years old	Relative frequency of cars more than four years old
A	10	4	0.4
B	20	6	0.3
C	15	3	0.2
D	10	2	0.2
E	10	0	0
F	15	3	0.2
G	20	4	0.2
H	10	5	0.5
I	20	7	0.35
J	10	4	0.4

b The best estimate uses all the data.

Total of 'Number of cars in the row' column = 140

Total of 'Number of cars more than 4 years old' column = 38

Relative frequency = $\frac{38}{140}$. This is the best estimate of the probability.

2

a Malcolm is unlikely to be correct. There are two considerations here.

i Malcolm's new die may be a fair die, in which case the probability of getting a 6 is $\frac{1}{6}$, in which case Malcolm is not correct.

ii Malcolm's new die may not be a fair die, in which case relative frequency would provide an estimate of the probability of getting a 6. However, a total of six rolls is not sufficient to obtain a reliable estimate. So, Malcolm is not likely to be correct.

b Malcolm needs more evidence, so he would need to roll the die more times.

3

a

	Male	Female	Totals
Love PE	12	11	23
Not so keen on PE	3	4	7
Totals	15	15	30

b i $\frac{15}{30}$

ii $\frac{15}{30}$

iii $\frac{23}{30}$

iv $\frac{11}{30}$

v $\frac{12}{15}$

4 Expected frequency = $0.9 \times 180 = 162$

5

a i 35

ii 14

iii 9

iv 10

b i $\frac{25}{66}$

ii $\frac{43}{66}$

iii $\frac{3}{66}$

c i $\frac{14}{35}$

ii $\frac{7}{35}$

6

a Table B

b Most of the frequencies are close to 50 ($\frac{1}{5} \times 250 = 50$, the expected frequency if the spinner is fair).

c $\frac{1}{5} \times 200 = 40$

Exercise 12.6

Due to the nature of the exercise, every time you set up and run these simulations you will obtain different results. It is suggested that you check the work of other students and that they check yours.

Spot the mistake

Simon has the best estimate of the two. The mistake they have made is that Simon should have replaced his papers back into the box and mixed them up! Without doing this there are fewer pieces of paper and fewer marked ones. In fact, David should have worked out: $\frac{12}{20} = \frac{48}{n-50}$ if Simon were to keep his not quite as easy and not as reliable as the sample is smaller!

Ibrahim is correct; a better estimate could be obtained by combining the two sets of data. In which case the calculation would be: $\frac{27}{70} = \frac{60}{n}$ so number of pieces of paper $= 60 \times \frac{70}{27}$

Exercise 12.7

1 $\frac{100}{n} \approx \frac{5}{100}$ so, $n \approx \frac{100 \times 100}{5}$, and $n \approx 2000$

2

a $\frac{50}{n} \approx \frac{6}{80}$, so $n \approx \frac{50 \times 80}{6} \approx \frac{4000}{6} \approx 666.\bar{6}$

So there are about 670 fish in the lake.

b They could catch more fish.

3

a $\frac{10}{n} \approx \frac{2}{32}$, so $n \approx \frac{10 \times 32}{2} = 160$

b His first sample is small. He only stands outside his house for the second sample. Buses usually follow one route, so he is likely to see the same buses repeatedly in his second sample. The company may own many buses that do not follow routes in his town. He assumes that buses are randomly allocated to bus routes at the start of the day; this may not be the case as minibuses may be used for routes on narrow streets or routes that are less popular.

4

a $\frac{27}{n} \approx \frac{12}{40}$, so $n \approx \frac{27 \times 40}{12} = 90$

b $\frac{x}{375} \approx \frac{4}{50}$, so $x \approx \frac{377 \times 4}{50} = 30$

Glossary

A

Arithmetic mean: the correct name for the mean, as it is usually worked out.

Average: the general name given to the measures of central location – an idea of a typical value from the data.

Average seasonal effect: the mean of the calculated seasonal effects from the graph.

B

Back-to-back stem-and-leaf diagram: a stem-and-leaf diagram where the stem is down the centre and the leaves from two distributions are either side for comparison.

Bar chart: a frequency diagram where the height of a bar represents the frequency of an item.

Base year: the year that is taken as the time from which all other years are compared, with the base year having value of 100.

Bias: the name given to sample data that does not fairly represent the population it comes from.

Bimodal: a set of data that has two modes is said to be bimodal.

Bivariate data: data that has two variables.

Box-and-whisker diagram: a diagram that shows the position of the minimum, the lower quartile, the median, the upper quartile and the maximum values from a set of data.

Box plot: the short name for a box-and-whisker diagram.

C

Categorical data: data that is grouped into categories.

Causality: causality means that the changes in one variable are as a direct result of changes in the other variable and not some other factor.

Census: the data is obtained from every member of the population.

Chain base numbers: a set of index numbers where the base year changes each time to be taken as the year before the year for which the chain base number is found.

Changing the origin: another name for scaling.

Choropleth map: a diagram with different colours, or shades of colours, to represent different amounts of items over a given area, usually an area of land.

Class intervals: a better statistical wording for groups.

Class width: the upper class bound minus the lower class bound.

Clusters: groups of items that have a distribution within them similar to the populations from which they come.

Component bar chart: a frequency diagram for two or more sets of data with each set of data wholly contained within one bar.

Composite bar chart: another name for a component bar chart.

Confounding variable: a variable that is unexpected and interferes with the variables under consideration.

Continuous data: data that is measured on a continuous scale.

Control charts: graphs showing the measures found during quality assurance so that they can be monitored for appropriateness.

Correlation: the statistical word for a connection between two variables.

CPI (Consumer Price Index): the CPI is an increasingly used measure of inflation similar to but containing some different items from the RPI.

Crude rate: the number of times an event such as a birth or death occurs per thousand of the population.

Cumulative frequency: the total frequency up to and including a particular value.

Cumulative frequency curve: a cumulative frequency diagram for continuous data with the points joined by a continuous curve.

Cumulative frequency diagram: the name given to any diagram that shows the cumulative frequencies for a distribution.

Cumulative frequency polygon: a cumulative frequency diagram for continuous data with the points joined by straight lines.

Cumulative frequency step polygon: a cumulative frequency diagram for discrete data with the points joined in steps.

D

Deciles: deciles split ordered data into 10 equal groups.

Discrete: discrete values are exact with no rounding necessary to record them.

Discrete data: data that takes separate values.

Dispersion: another word for spread.

Double mean point: a point plotted at the values of the means of the two variables – the point through which the line of best fit must pass.

Dual bar chart: a multiple bar chart with specifically two sets of data.

E

Empirical probability: another term used for experimental probability.

Equally likely: outcomes that are equally likely; all have the same probability.

Event: something that takes place, such as rolling a die, tossing a coin, taking a card from a pack.

Experimental probability: the name given to a relative frequency when it is used as an estimate for a probability.

Explanatory variable: the variable you control, whose effect you wish to investigate.

Extraneous variable: a variable that may affect the outcome of an experiment.

Extrapolation: an estimate made often using a line of best fit from outside the data values known.

F

Frequency: the frequency of an item is the number of times it has occurred.

Frequency density: the frequency density is used to complete a histogram and is calculated using actual frequency divided by class interval width.

Frequency diagrams: all charts or diagrams that compare the frequencies of objects.

Frequency polygon: a frequency diagram for continuous data with a line joining the midpoints of the class intervals using the appropriate frequencies.

Frequency table: a table showing the number of times values or items have occurred in a set of data.

G

GDP (gross domestic product): the GDP is a measure of the income and output of a country's economy.

Geometric mean: an average where the n values in a data set are multiplied and the nth root is taken.

Gradient: the steepness of a line.

Groups: a range of values within which data lies.

H

Histogram: a diagram for continuous data with bars as rectangles whose areas represent the frequency.

Hypotheses: plural form of 'hypothesis'.

Hypothesis: a statement you make that you think may be true.

I

Independent: two events are independent if the outcome of the second is not affected by the outcome of the first.

Index number:
$$\frac{\text{current value of item}}{\text{value of item in base year}} \times 100$$

Intercept: the value at which a line passes through the y-axis.

Interdecile range: the ninth decile minus the first decile (D_9–D_1). This gives the range of the central 80 per cent of the data.

Interpercentile range: the difference between two percentiles.

Interpolation: an estimate often made using a line of best fit from within known data values.

Interquartile range: the upper quartile minus the lower quartile.

L

Leaves: the figures to the right of a stem-and-leaf diagram indicating the final digit of the data.

Line of best fit: a straight line drawn through a double mean point to represent the apparent connection between two variables.

Linear transformation: where data is divided or multiplied by a number, as well as possibly being scaled, in order to make it easier to calculate measures.

Lower class bound: the lowest value that could be present in a particular class interval.

Lower quartile: the lower quartile (Q_1) is the value ¼ along a set of data.

M

Mean: the total of all the data values divided by the number of data values.

Median: the middle number in a list of ordered data.

Modal class: another name for the modal group.

Modal group: the class interval with the most values within it.

Modal value: another name for the mode.

Mode: the number that occurs in a list most often.

Moving average: the mean of one whole cycle of data in a time series that moves through the data one value at a time.

Multiple bar chart: a frequency diagram where two or more sets of data are presented in groups of bars for comparison.

Mutually exclusive: events that are mutually exclusive cannot occur at the same time; if two events A and B are mutually exclusive, then P (A or B) = P (A) + P (B)

N

Negative correlation: negative correlation exists when as one variable increases the other variable decreases.

No correlation: there is no correlation between two variables when the value of one seems to have no connection to the value of the other.

Normal distribution: a symmetrical continuous set of data that has special features, often naturally occurring such as heights and weights.

O

Open-ended class: a class interval where the upper class bound is not defined.

Opinion polls: surveys that gather the opinions of the respondents.

Ordered: put from lowest to highest, or the other way round.

Ordered stem-and-leaf diagram: ordered means the leaves are put in order of size to complete the construction of a stem-and-leaf diagram.

Outcome: the result of an event such as getting a 6 when a die is rolled, or a Head when a coin is tossed.

Outlier: any value that does not fit in well with the rest of the data: the value is either less than the lower quartile – 1.5 × the interquartile range; or more than the upper quartile + 1.5 × the interquartile range.

Output gap chart: a chart showing how the real output of a country is compared to the potential GDP for that country, usually over a period of time.

P

Pearson's measure of skew: $\dfrac{3(\text{mean} - \text{median})}{\text{standard deviation}}$

Percentiles: percentiles split ordered data into 100 equal groups.

Pictogram: a pictogram uses symbols to represent items of data.

Pie chart: a circular diagram where the angle of a sector represents the frequency of an item.

Pilot survey: a small-scale survey carried out before the main survey.

Population: every possible item that could occur in a given situation.

Positive correlation: positive correlation exists when as one variable increases the other variable also increases.

Primary data: data that you collect yourself.

Probability: a measure of how likely an event is to occur.

Product Moment Correlation Coefficient (PMCC): the PMCC is a way of calculating a value for correlation that uses the actual size of the data for the variables.

Proportional pie chart: the areas of proportional pie charts are used to represent and compare the frequency of items from two different samples.

Q

Qualitative: data that measures a quality such as taste or colour.

Qualitative data: non-numerical data that describes a situation.

Quality assurance: quality assurance is where quality is checked by obtaining samples of the products and seeing whether the means, medians or ranges of these samples are within appropriate levels.

Quantitative: data that is a quantity and is therefore numerical.

Questionnaire: a set of questions that may be asked verbally or in writing.

Quota: a specific number of people or items with a particular feature.

R

Random sample: another name that implies the use of simple random sampling.

Range: the highest value minus the lowest value.

Rank: the rank of a piece of data is its position in the ordered list of that data.

Raw data: data that is unsorted in any way.

Relative frequency: the frequency with which one outcome of an experiment occurs relative to the total number of times the experiment is carried out.

Representative: a representative sample will have the same characteristics as the population it is taken from and will be free from bias.

Respondent: the person who replies to a questionnaire.

Response variable: the variable that depends on changing the explanatory variable; the outcome that you measure.

RPI (Retail Price Index): the RPI is a measure of inflation used to determine changes in costs and prices.

S

Sample: part of a population from which information is taken.

Sample frame: the list of all the members of a population available to appear in the sample.

Sample space diagram: a diagram that shows all the possible outcomes in a table.

Scaling: adding or subtracting the same value to or from each number in a set of data, to make it easier to calculate measures.

Scatter diagram: a visual way of showing bivariate data.

Seasonal effects: the fluctuations in data that occur in any kind of regular time pattern, such as particular times of a week or year.

Secondary data: data that someone else has collected.

Simple random sample: every member of the population has an equal chance of being in the sample, and every possible sample has an equal chance of occurring.

Size of seasonal effect: actual value less the value on the drawn trend line.

Skew data: data that is unequally spread either side of the data.

Spearman's rank correlation coefficient (SRCC): the SRCC is a way of calculating a value for correlation that uses the ranks of the two variables in a formula.

Spread: a general term indicating the amount by which different values in a data set are close to each other or not.

Standard deviation: the square root of the variance.

Standardised rates: the number of times an event occurs per thousand of the population, taking into account the age profile of the population.

Standardised score:
$\frac{\text{actual score} - \text{mean}}{\text{standard deviation}}$

Stem: the left part of a stem and leaf diagram indicating the main digit or digits of the data.

Stem-and-leaf diagram: a frequency diagram that uses the actual values of the data split into a stem and leaves, with a key.

Strata: different sections of a population such as gender or age groups.

Stratified random sample: a sample with consideration of the strata within the population – individuals from within each stratum are chosen using simple random sampling.

Symmetrical data: data that is equally spread either side of the centre.

T

Time series: any set of data that has values recorded at different times.

Trend line: a line of best fit drawn by eye through the plotted moving average points.

Two-way table: a table that shows information about two features of data at the same time.

U

Unordered stem-and-leaf diagram: unordered means the leaves are not put in order of size.

Upper class bound: the highest value that could be present in a particular class interval.

Upper quartile: the upper quartile (Q_3) is the value ¾ along a set of ordered data.

V

Variance: the average of the squared deviations from the mean.

Vertical line diagram: a diagram that is like a bar chart but with pencil thin bars and is more often used for discrete quantitative data.

W

Weighted index number: if an index number is calculated for items made up of many sub items with differing importance (weighting), a weighted index number will take into account the importance of each sub item in the calculation.

Z

Z-chart: a time series graph that also shows cumulative data and the running total for the data over a complete cycle.

Index

Table of random numbers

51	38	42	50	40	27	59	68	04	27	38	22	91	04	73	82	85	40	82	84
09	56	76	51	04	73	94	30	16	74	69	59	04	38	83	98	30	20	87	85
55	99	98	60	01	33	06	93	85	13	23	17	25	51	92	04	52	31	38	70
72	82	45	44	09	53	04	83	03	83	98	41	67	41	01	38	66	83	11	99
04	21	28	72	73	25	02	74	35	81	78	49	52	67	71	40	60	50	47	50
87	01	80	59	89	36	41	59	60	27	64	89	47	45	18	21	69	84	76	06
31	62	46	53	84	40	56	31	74	76	52	53	72	95	96	06	56	83	85	22
29	81	57	94	35	91	90	70	94	24	19	35	50	22	23	72	87	34	83	15
39	98	74	22	77	19	12	81	29	42	04	50	62	34	36	81	43	07	97	92
56	14	80	10	76	52	38	54	84	13	99	90	22	55	41	04	72	37	89	33
29	56	62	74	12	67	09	35	89	33	04	28	44	75	01	57	87	45	52	21
93	32	57	38	39	36	87	42	72	55	73	97	98	36	57	41	76	09	11	68
95	69	51	54	43	19	20	49	57	25	90	55	26	20	70	98	43	73	56	45
65	71	32	43	64	67	22	55	65	65	48	86	10	88	20	12	40	18	49	25
90	27	33	43	97	84	20	57	49	91	41	20	17	64	29	60	66	87	55	97
90	29	42	45	61	34	30	13	30	39	21	52	59	28	64	98	08	76	09	27
99	74	06	29	20	55	72	70	11	43	95	82	75	37	90	24	77	43	63	21
87	87	66	91	16	97	51	50	61	36	96	47	76	68	49	11	50	56	51	06
46	24	17	74	97	37	39	03	54	83	34	00	74	61	77	51	43	63	15	67
66	79	81	43	40	92	84	72	88	32	83	24	67	01	41	34	70	19	26	93
36	42	94	58	83	30	92	39	18	40	03	00	12	90	32	37	91	65	48	15
07	66	25	08	99	27	69	48	85	32	16	46	19	31	85	02	86	36	22	96
93	10	05	72	18	26	36	67	68	48	31	69	68	58	93	49	45	86	99	29
49	50	63	99	26	71	47	94	32	71	72	91	34	18	74	06	32	14	40	80
20	75	58	89	39	04	42	73	37	93	11	07	28	77	91	36	60	47	82	62
02	40	62	09	00	71	09	37	80	44	50	37	32	70	20	38	71	86	75	34
59	87	21	38	29	78	72	67	42	83	65	21	54	79	66	42	47	86	31	15
48	08	99	66	43	38	28	13	50	25	47	93	11	15	07	84	28	30	19	07
54	26	86	75	44	15	20	39	20	03	58	54	80	29	62	53	06	97	71	51
35	35	58	45	23	58	63	66	09	62	80	92	14	55	81	41	21	48	87	34
73	84	90	49	01	21	90	29	57	06	68	73	51	10	51	95	63	08	57	99
34	64	78	00	92	59	67	74	58	48	92	09	42	20	40	37	63	80	58	93
68	56	87	47	63	06	24	71	41	98	79	06	07	18	58	29	16	49	67	37
72	47	05	42	88	07	27	55	58	74	82	08	42	28	26	48	25	32	00	31
44	44	96	75	89	57	12	60	42	38	77	36	45	69	21	68	32	70	04	96
28	11	57	47	61	57	89	88	62	18	93	67	57	32	96	72	21	17	13	54
87	22	38	88	91	99	16	08	17	76	27	47	52	14	98	86	35	68	23	85
44	93	14	59	67	40	24	10	11	63	40	47	07	56	14	22	62	74	93	39
81	84	37	25	90	43	56	62	94	58	49	03	84	22	57	22	47	98	86	37
09	75	35	21	04	47	54	08	98	44	08	16	44	86	69	71	20	52	64	94
77	65	05	04	22	18	20	10	81	87	05	69	43	70	96	76	42	05	21	10
19	06	51	61	34	03	61	55	98	58	83	50	01	48	99	85	08	67	15	91
52	91	87	07	19	62	32	28	04	91	42	48	65	24	86	09	87	68	55	51
52	47	25	14	93	91	75	51	49	26	49	41	20	83	30	30	43	22	69	08
52	67	87	40	63	41	91	86	10	47	80	70	56	87	25	86	89	94	21	42
66	25	71	73	78	60	50	62	91	04	95	97	64	16	71	31	32	80	19	61
29	97	56	42	56	90	16	75	74	95	99	26	01	63	25	16	54	18	54	46
15	25	03	68	92	45	53	00	06	29	46	43	46	66	27	12	85	05	22	44
82	08	65	67	64	13	51	14	38	28	24	30	39	62	20	35	23	90	57	36
81	35	03	25	87	24	83	59	04	67	51	52	26	21	69	75	87	28	06	50